# FOOD SYSTEMS
# MODELLING

———

# FOOD SYSTEMS MODELLING

CHRISTIAN J. PETERS
USDA, Agricultural Research Service, Food Systems Research Unit,
Burlington, VT, United States

DAWN D. THILMANY
Department of Agricultural and Resource Economics and Regional Economic
Development Institute, Colorado State University, Fort Collins CO,
United States

ACADEMIC PRESS

An imprint of Elsevier

elsevier.com/books-and-journals

Academic Press is an imprint of Elsevier

125 London Wall, London EC2Y 5AS, United Kingdom
525 B Street, Suite 1650, San Diego, CA 92101, United States
50 Hampshire Street, 5th Floor, Cambridge, MA 02139, United States
The Boulevard, Langford Lane, Kidlington, Oxford OX5 1GB, United Kingdom

**British Library Cataloguing-in-Publication Data**
A catalogue record for this book is available from the British Library

**Library of Congress Cataloging-in-Publication Data**
A catalog record for this book is available from the Library of Congress

ISBN: 978-0-12-822112-9

For Information on all Academic Press publications visit our
website at https://www.elsevier.com/books-and-journals

*Publisher:* Charlotte Cockle
*Acquisitions Editor:* Megan R. Ball
*Editorial Project Manager:* Maria Elaine D. Desamero
*Production Project Manager:* Joy Christel Neumarin Honest Thangiah
*Cover Designer:* Matthew Limbert

Typeset by Aptara, New Delhi, India

# Contents

## 8 Using input-output analysis to estimate the economic impacts of food system initiatives   157

Becca Jablonski, Jeffrey K. O'Hara, Allison Bauman, Todd M. Schmit and Dawn D. Thilmany

## 9 Environmental Input-Output (EIO) Models for Food Systems Research: Application and Extensions   179

Patrick Canning, Sarah Rehkamp and Jing Yi

## 10 Modeling biophysical and socioeconomic interactions in food systems with the International Model for Policy Analysis of Agricultural Commodities and Trade (IMPACT)   213

Keith Wiebe, Timothy B. Sulser, Shahnila Dunston, Richard Robertson, Mark Rosegrant and Dirk Willenbockel

A companion website for this book is available at:
https://www.elsevier.com/books-and-journals/book-companion/9780128221129

# Contributors

**Allison Bauman**  Colorado State University, Department of Agricultural and Resource Economics, Fort Collins, CO, United States

**Nicole Tichenor Blackstone**  Friedman School of Nutrition Science and Policy, Tufts University, Boston, MA, USA

**Larissa Calancie**  Friedman School of Nutrition Science and Policy, Tufts University, Boston, MA, United States; ChildObesity180

**Patrick Canning**  USDA, Economic Research Service, Food Economics Division, Washington, DC, United States

**Kate Clancy**  Food Systems Consultant and Visiting Scholar at the Center for a Livable Future, Bloomberg School of Public Health, Johns Hopkins University

**Jill K. Clark**  The Ohio State University

**Jennifer E Cross**  Colorado State University, Sociology

**Shahnila Dunston**  International Food Policy Research Institute, Washington, DC 20005, USA

**Christina Economos**  Friedman School of Nutrition Science and Policy, Tufts University, Boston, MA, United States; ChildObesity180

**Darcy A. Freedman**  Case Western Reserve University

**A. Frehner**  Farming Systems Ecology group, Wageningen University and Research, Wageningen, Netherlands; Department of Socioeconomics, Research Institute of Organic Agriculture FiBL, Frick, Switzerland

**Alannah R. Glickman**  The Ohio State University

**Miguel I. Gómez**  Charles H. Dyson School of Applied Economics and Management, Cornell University, Ithaca, NY, USA

**Erin Hennessy**  Friedman School of Nutrition Science and Policy, Tufts University, Boston, MA, United States; ChildObesity180

**Clare Hinrichs**  The Pennsylvania State University

**Rick J. Hogeboom**  Multidisciplinary Water Management Group, Faculty of Engineering Technology, University of Twente, the Netherlands; Water Footprint Network, the Netherlands

**Becca Jablonski**  Colorado State University, Department of Agricultural and Resource Economics, Fort Collins, CO, United States

**Olivier Jolliet**  Environmental Health Sciences, School of Public Health, University of Michigan, Ann Arbor, MI, USA

**Maarten S. Krol**  Multidisciplinary Water Management Group, Faculty of Engineering Technology, University of Twente, the Netherlands

**Jess Kropczynski**  University of Cincinnati

**A. Muller**  Department of Socioeconomics, Research Institute of Organic Agriculture FiBL, Frick, Switzerland; Institute of Environmental Decisions, Federal Institutes of Technology Zurich ETHZ, Zurich, Switzerland

**Thomas Nemecek**  LCA Research Group, Agroscope, Switzerland

**Charles F. Nicholson**  School of Integrative Plant Science, Cornell University, Ithaca, NY, USA; Charles H. Dyson School of Applied Economics and Management, Cornell University, Ithaca, NY, USA

**Jeffrey K. O'Hara**  U.S. Department of Agriculture, Agricultural Marketing Service, Washington, DC, United States

**Christian J. Peters**  USDA, Agricultural Research Service, Food Systems Research Unit, Burlington, VT, United States

**Sarah Rehkamp** University of Hawai'i at Mānoa, Honolulu, Hawaii, United States

**Richard Robertson** International Food Policy Research Institute, Washington, DC 20005, USA

**Sarah Rocker** The Pennsylvania State University

**Mark Rosegrant** International Food Policy Research Institute, Washington, DC 20005, USA

**Meagan Schipanski** Colorado State University, Soil and Crop Sciences

**Todd M. Schmit** Cornell University, Dyson School of Applied Economics and Management, Ithaca, NY, United States

**Joep F. Schyns** Multidisciplinary Water Management Group, Faculty of Engineering Technology, University of Twente, the Netherlands

**Christina Skonberg** Simple Mills, Sustainability & Strategic Sourcing, Chicago, IL, United States

**Timothy B. Sulser** International Food Policy Research Institute, Washington, DC 20005, USA

**Dawn D. Thilmany** Department of Agricultural and Resource Economics and Regional Economic Development Institute, Colorado State University, Fort Collins, CO, United States

**Greg Thoma** Ralph E. Martin Department of Chemical Engineering, University of Arkansas, Fayetteville, AR, USA

**Mariko Thorbecke** Anthesis Group, Analytics, Boulder, CO, United States

**H.H.E. Van Zanten** Farming Systems Ecology group, Wageningen University and Research, Wageningen, Netherlands

**Keith Wiebe** International Food Policy Research Institute, Washington, DC 20005, USA

**Dirk Willenbockel** Institute of Development Studies, University of Sussex, Brighton, BN1 9RE, UK

**Jing Yi** Dyson School of Applied Economics and Management, Cornell University, Ithaca, NY, United States

# Acknowledgements

We are grateful for the support we received throughout the course of this project, and we have many people and several organizations to thank.

First, we thank the 39 contributors to *Food Systems Modeling* for their time, effort, expertise, esprit de Coeur, and belief in the purpose of this book. Beyond their commitment to writing a chapter, the contributing authors and coauthors strove to make this a cohesive volume. Three virtual workshops were held over the course of a year to identify synergies and overlaps between chapters, fill gaps in the book's content, discuss challenges in writing for a broad audience, determine the best chapter order, and brainstorm the essential material to include in the introductory and concluding chapters. Contributing authors provided peer reviews of each other's work, helping to ensure continuity in how related topics were covered, integrating cross-references where appropriate, and assuring accessibility to an interdisciplinary audience. Each chapter went through multiple rounds of review, and contributing authors incorporated feedback from graduate students, fellow contributors, and their demanding editors.

Second, we thank the 11 Tufts University graduate students who participated in the student review of chapters: Nayla Bezares, Dana Bourne, Anna Bury, Jason "Forest" Kashdan, Cora Kerber, Andrew May, Ashley McCarthy, Rajratna Sardar, Kate R. Schneider, Natalie Volin, and Meredith J. Young. We appreciate the time, effort, and care they invested in reading the manuscripts and considering the accessibility of the material to an interdisciplinary, graduate student-level audience. Additional thanks are due to Cora, Andrew, and Ashley, who served as student leaders in the review process, convening discussions of each chapter manuscript with the other reviewers and summarizing the key points for the editors.

Third, we thank the staff at Elsevier for their guidance and support throughout the process. In particular, we thank Megan Ball for encouraging us to submit a book proposal and Elaine Desamero for guiding us during the writing and submission of manuscripts. Their understanding and willingness to adjust timelines as we navigated the tricky business of writing a multiauthor volume during a global pandemic was invaluable.

Fourth, we thank our home institutions (Colorado State University for Dr. Thilmany, USDA Agricultural Research Service and, formerly, Tufts University, for Dr. Peters) for providing the time to work on this endeavor. We appreciate the opportunity to synthesize how different disciplines approach the task of modeling food systems and to communicate this knowledge to the field.

Fifth, we recognize the important grounding that the community-based initiatives mentioned across the book contributed to individual chapters. In addition, we thank Jen Faigel with Boston's CommonWealth Kitchen and Dawn Plummer with the Greater Pittsburgh Food Policy Council for sharing how their plans and programs provide examples of bringing models to communities in action, which is discussed in the concluding chapter.

Finally, we thank the USDA National Institute of Food and Agriculture (Award No.

2020-67023-30692) for a conference grant to help us write an accessible and cohesive book. While the pandemic prevented an in-person gathering of authors, these resources enhanced our virtual author workshops and supported the student review of chapters. This grant helped us to write a better book.

# 1

# Using models to study food systems

*Christian J. Peters[a] and Dawn D. Thilmany[b]*

[a]USDA, Agricultural Research Service, Food Systems Research Unit, Burlington, VT, United States. [b]Department of Agricultural and Resource Economics and Regional Economic Development Institute, Colorado State University, Fort Collins CO, United States

## 1.1 Introduction

Food systems have existed for millennia. Since the agricultural revolution enabled the first cities to form, people have grown crops, raised livestock, caught fish, milled grains, pressed oils, preserved perishables, baked breads, butchered meats, and turned these raw and processed ingredients into food people eat. Goods often changed hands and forms many times before they were transformed into meals, and these exchanges involved people and institutions, like merchants and governments, who were not directly, or at least not solely, involved in the production, processing, transport, or preparation of food. Agriculture and fisheries depended, and still do, on natural resources, like soil, and natural processes, like the water cycle, for their fundamental productivity. These complex networks of people, institutions, plants, animals, and natural resources evolved to supply food to civilizations. They have existed physically for a very long time, but only recently have we begun to conceive of them as food systems.

Why the change of perspective? As you will see in this book, the field of food systems has emerged to weave together insights gained from many separate disciplines into a cohesive understanding of the whole. For most of the 20th century, tremendous depth of knowledge was gained through reductionist approaches, and this knowledge led to some impressive, practical outcomes like increasing crop yields. However, persistent and vexing societal problems that are commonly aligned with food system activities, like climate change and obesity, have proven resistant to reductionist interventions. Food systems, as a field, takes a more holistic view in the hope of better understanding the problems and thereby improving outcomes and avoiding unintended consequences.

Of course, this is easier said than done. Food systems are so large and multi-faceted, they complicate experimentation. Hence, both mental and mathematical models provide a means of trying to organize what is known about food systems. Mental models allow us to frame how we believe food systems work, while mathematical models convert these concepts into virtual food systems, allowing us to run simulations before intervening in real ones. Models, however, are no panacea. They can seem impenetrable if not communicated effectively. Different models can (and do) at times yield seemingly contradictory results, so there may be an on-going need for evaluation and refinement. In part, this is because different disciplines see the world from different perspectives and design models well-suited to answer certain questions but not others. We do not understand models from outside our own fields, making us skeptical. Without being able to see from other perspectives, we fail to grasp the connections and linkages that make food systems so hard to predict yet so interesting to consider.

This book is meant to help you better understand why different modeling approaches are used to study food systems and what you can, and cannot, learn from each of them. We try to capture the full scope of the field of food systems, sometimes digging into the details of the methods, but in other cases simply introducing new frontiers where disciplines are better aligning and coordinating to address transdisciplinary food system issues. Our goal is to expand your understanding of the promise and perils of using models to study and evaluate food systems so that you can read critically and apply your modeling knowledge carefully, whether you are a researcher or a practitioner.

## 1.2 Models and why we use them

A model is a simplification of reality. We use models all the time to understand the world and to guide our decisions. In fact, models are so ubiquitous, that we often use them unconsciously, particularly **mental models** – our notions of how the world works. Many types of models are used in the sciences. Physical models help to envision things too small to see with the naked eye, like the plastic models used in organic chemistry, or too large to see up close, like a globe is to the Earth. Statistical models estimate the mathematical relationship between an observed phenomenon and factors that we think influence that phenomenon. Lab and field experiments model real phenomena under controlled conditions. Models are everywhere.

This book concerns itself primarily with computational models. Such models simulate biophysical, economic, and social processes in food systems in digital form, sometimes called "in-silico". Crop models, for example, mathematically simulate the growth of a plant over a growing season as it 'reacts' to simulated temperature, precipitation, daylength, and other environmental conditions. These models rely on wide range of data and incorporate theories of how the world works, often integrating ideas from multiple disciplines. The modeling simplifications that we make matter, and the assumptions we make, often through the lens of a specific discipline, matter more than we have realized previously. As a result, one cannot put blind trust in a model, but models should not be dismissed either. Rather, to use models responsibly, one must understand a model conceptually, including its strengths and limitations, and know how to interpret the results.

To illustrate, consider a budget as an example of a simple model. Budgets are used to manage all kinds of resources, from household finances to nutrients in a watershed to national greenhouse gas emissions. A budget estimates future inflows and outflows of a resource that you wish to manage for some purpose, such as saving money for a major purchase or keeping emissions of a pollutant below a given threshold. You do not know what the actual inflows and outflows will be, so you must estimate them from past observations, such as past bank statements. You may even adjust these estimates by making assumptions about conditions you expect to change in the future, such as getting a raise in salary or increased prices due to inflation. You know your estimates will almost certainly be wrong; they will not equal your exact income and expenses to the penny. However, the budget will probably be close enough to keep you on track and showing whether or not you are moving in the right direction. In other words, it will be close enough to increase your savings for that big purchase.

Of course, there will be times when the budget is thrown off by something unexpected. Perhaps your household incurred a major expense, like an expensive car repair or a medical bill. Perhaps the watershed experienced a hundred-year storm that increased runoff relative to historical levels, causing much greater losses of nutrients from farms. Whatever the case, a carefully constructed budget can be thrown off by phenomenon that were outside the scope of your consideration, but the baseline estimates are still a useful metric from which to compare what was expected to what happened.

The same is true with all models. When carefully constructed with enough accuracy and precision for the goals we are trying to pursue, models can help us better understand the world and to make informed decisions. However, we must always bear in mind that models are simplifications, and good modelers will present results in a way to assure they are not mistaken for reality. Acknowledging assumptions, keeping models as simple as possible, using models for their intended purpose, explaining the uncertainty in the results, and clarifying what remains unknown can help non-modelers avoid putting undue faith in model outputs while still gaining useful perspective (Saltelli, Bammer, Bruno, Charters, and Di Fiore, 2020).

## 1.3  Models in food systems

A system is an interconnected set of elements that is coherently organized and achieves some purpose (Meadows, 2008, p.11). Like models, systems can be found everywhere you look, and the more you practice looking at the world through a systems lens, the more often you will see systems at play in all realms of society. Modelers especially see the world in terms of systems, perhaps because to create a model, one must explicitly identify the elements the model contains, how the elements are organized, and the phenomena to be simulated.

Food systems are "supply chains operating within broader economic, biophysical and sociopolitical contexts" (National Research Council, 2015, p.6). A supply chain consists of the tangible entities involved in food production, processing, and distribution, entities like farms, grain elevators, mills, food manufacturers, trucking companies, wholesalers, and grocery stores. Given this complexity, you might mistake a supply chain to represent food systems. However, systems are governed by rules, and some of the 'rules' that govern food systems exist beyond the bounds of a supply chain. Biophysical factors, like annual weather patterns, and sociopolitical factors, like national trade policies, also influence supply chain behavior.

In many ways, food systems are analogous to ecosystems. Like ecosystems, food systems are complex networks of communities, composed of living organisms, interacting with one another and their environment. While the analogy is imperfect since food systems are shaped by human desires and capacities, the comparison may be useful for thinking about what models have to offer as a way of understanding the world.

Within ecosystem science, Oreskes (2003) distinguishes three purposes for models – "organizing data, synthesizing information, and making predictions." The situation is similar within the field of food systems. People, especially non-modelers, often focus on the predictive value of models. However, creating reliably accurate and precise computational models, like we have for weather, require substantial investments of time and money. Such investment is only warranted if precise estimates are needed by large numbers of people. Yet, even modest investments can still create useful models that may be sufficient for the task at hand. For example, simply framing a system allows one to reveal gaps in knowledge and elevate understanding of how systems function to provide insight to guide judgment.

Indeed, modeling can be a valuable tool precisely when systems are complex and difficult to predict. Harrison, Lin, Carroll, and Carley (2007) argues, in the context of management and organizational research, that models can be used to better understand human systems. In addition to prediction, they offer up six other reasons to model: proof of concept, discovery, explanation of behavior, critique of existing theory, prescription of a better management approach, and empirically guiding future research. Like Oreskes' organizing data and synthesizing information, these six reasons all describe a learning process. Models are tools for learning.

## 1.4 Types of models used to study food systems

As the contents of this book might suggest, there is an evolving taxonomy of food systems models that can be framed and categorized in a variety of ways, from the scope of the model (both in terms of space and time) to consideration of the sector of interests, field of study or outcome it is intended to assess. For this same reason, there are a variety of schematics and Figs. now available to describe food systems, but the example presented below (Fig. 1.1) is particularly relevant for how this book and its chapters focus their contributions.

When first envisioning the modeling contributions we wanted to highlight in this book, we identified a diverse set of studies of food systems behavior, relationships and assessments, varied in the scale and orientation with which they chose to study food systems. We arrived at four orientations that are used throughout this book to model and study food systems (Fig. 1.1). Ovals (blue) indicate the three major stages of food supply chains: production, distribution, and consumption. Clouds (gray) indicate sources and sinks that are often considered beyond the scope of the food supply chains, but which are clearly part of food systems. The rectangles (brown) represent topics that are commonly at the heart of public policy goals (natural resources and ecosystem services, economies and infrastructure, communities, public health, and well-being), and systems that exist beyond the bounds of the food system. Linkages and interactions with these issues are commonly integrated into models to understand the performance and outcomes related to private and policy levers used in food systems. Dashed lines indicate the orientation and different scales at which to examine these stages, resources, actors and linkages within food systems: sector-specific (orange), supply chain (red), linkages (dark blue) and the full food system (green).

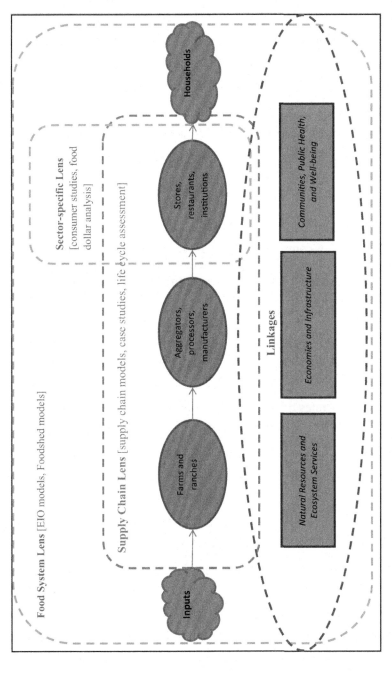

**FIGURE 1.1  Four orientations for studying food systems.** Ovals (blue) indicate the three major stages of food supply chains: production, distribution, and consumption. Clouds (gray) indicate sources and sinks that are often considered beyond the scope of supply chains, but which are part of food systems. The rectangles (brown) represent the broader ecological and societal systems that are strongly aligned with food systems. Dashed lines indicate different foci and scales at which to examine food systems: sector-specific (orange), supply chain (red), linkages (blue) and the full food system (green).

The purpose of this Fig. is to convey that there is no *single* correct scale at which to examine the food system and that one will inevitably choose to focus on certain outcomes or links to related systems *depending on the purposes of the model.* It is acceptable for a model to be incomplete, after all, models simplify reality. However, anyone trying to understand the food system holistically must bear in mind the limits of any single approach.

## 1.5 Stage of food production

As demonstrated by the array of studies presented in this compilation, models exist to study food systems at multiple scales and stages, and the choice of how to frame the study may be driven by earlier research. Yet, in the case of a topic as essential to societal well-being as food, researchers are continually being challenged by increasingly complex and timely priorities and agendas from local, national, and global interests.

### 1.5.1 Single stages of the food system

Given the expertise and rigor commonly expected among researchers, there is a tendency to focus on a single stage or sector of the food system, where there is a standardized set of data, methods and narrowly defined science to assess outcomes for that stage of production, distribution or consumption. And, before systems of models can begin to be assembled and integrated, it is necessary for there to be some level of understanding of how each stage of the food system does change, perform, and respond under varying environmental, market and policy conditions. The early years of development within the food systems community of practice were spent perfecting these single stage models, with occasional linkages to a closely aligned stage or issue that allowed for initial exploration of how to broaden the scope of analysis.

Consequently, Fig. 1.1 illustrates that modeling any single activity in the food system will only provide useful information on a component piece of a relatively complex system, so there has been increasing attention and efforts to develop and refine models that better integrate a wider array of food systems activities to allow for more informative findings.

### 1.5.2 Supply chains

As characterized in Fig. 1.1, food supply chains are commonly depicted as moving from the farm- or ranch-gate to the consumer through supply chains, with some attention to the volumes, prices and "markups" that occur at each stage of the chain. However, as several of the chapters in this book will illustrate, these supply chains could be more realistically defined as multiple interacting sectors (producers, wholesalers, retailers and consumers), and each composed of multiple enterprises interacting at each of those stages.

While supply chains are most commonly aligned with business and economic analysis, and research focuses on market performance metrics, such as costs, prices, profitability and efficiency, there are efforts to integrate linkages that would allow for evaluation of tradeoffs between economic and other social outcomes Still, economic assessment allows for a "price" or "marginal cost" to be assigned to a choice available in systems, so that tradeoffs can be evaluated in a standardized way. For example, a food shopper may want to understand the additional cost of buying a gallon of milk that is produced using production practices that capture 25 percent more carbon. These outcomes are important to policy makers as well as the consumer households and communities that seek food systems that align with their visions for a quality of life through the potential for spillover health, economic and environmental benefits.

### 1.5.3 Broader food systems

One can quickly arrive at the conclusion that, given all the interesting linkages and trade-offs, food systems research requires a fully systematic approach (Fig. 1.1). Public or private interventions in food systems to improve outcomes often operate under the assumption that there are well defined, transparent and clear signals sent to all actors in the food system, but the complexity of such systems does not always facilitate clear communication and differing actor motivations (incentives) can lead to unintended outcomes.

## 1.6 Three major types of models

### 1.6.1 Biophysical models

Early food systems work commonly focused on the agricultural production occurring at the farm and ranch level. With their focus on the biophysical flows or impacts or processes within food systems, they represent some of the most tangible metrics with which models can measure tradeoffs. For example, several of the chapters contained in this volume focus on the land and water needs and carrying capacity within food systems, with interesting linkages and tradeoffs explored between these natural resources and economic (price and policy incentive), social (environmental impact) and health (dietary) outcomes.

### 1.6.2 Socio-Economic models

Perhaps the most common "denominator" that is included in any food systems model to allow for a clear way to measure tradeoffs is the estimated economic values, costs or benefits underlying tradeoffs in the system. Although a focus on the exchange of goods and economic impacts may be too narrow of a lens with which to evaluate various outcomes, it is a common starting point with which to understand the magnitude and direction of such tradeoffs. More broadly, an assessment of community outcomes, such as equitable returns and food security often are added to socioeconomic models. For example, the design and implementation of food system interventions that target food insecure, limited resource farmers, and communities that have a history of environmental injustices may frame research questions and models that distinctly measure improvement for those targeted populations.

### 1.6.3 Participatory modeling

Despite research questions that may integrate a socioeconomic lens, to make any model that aspires to impact social outcomes may benefit from blending participatory research into the modelling exercise. With a focus on the learning, capacity building and self-governance that may need to occur within communities, participatory approaches have gained favor in broad realms of research, but the field of food systems offers up some rich examples. Beyond the chapters in our book that highlight such approaches, the notable growth in food policy councils across North America, many of which conduct assessments, food plans and action agendas that include on-the-ground research and evaluation is evidence of the popularity a participatory research culture in this realm.

When choosing the portfolio of work presented in this book, we were particularly motivated to showcase the value of integrating methods from the rich food systems toolkit to illustrate how one might arrive at the "right fit" approach for a variety of research questions and issues. But, in addition to the types of sciences and models to be considered, it is important to highlight common issues that challenge all researchers.

## 1.7 Common issues with models

With any food systems topic, there are several issues to consider when choosing the model that is best suited to address the question at hand. Although the jargon within a field may frame these issues in a slightly different way, Table 1.1 shows why good research reports should point out how modelers address key issues. From deciding on the scope and complexity that your research project requires (or resources allow), to clearly delineating linkages and tradeoffs that the model will allow you to assess, good research protocols require an upfront discussion of the conceptual basis, including explicit acknowledgement of simplifying assumptions, of the model. Moreover, a good study must identify the unknowns, and consequently, there is increasing use of scenarios to report several potential alternative outcomes by modelers and of sensitivity analysis to delineate relevant ranges of uncertainty.

Table 1.1 briefly outlines the context of why these issues are pertinent to good research, what challenges may arise, and what chapters in this volume offer good examples and justification for addressing such issues in food systems modelling. For empirical research, the quality and availability of data is also an important to address and may help to define the spatial and temporal scope of one's modelling efforts. In addition to data, the inclusion of a participatory process, community engagement, and integrating place-based context will assure relevancy for those who may be called to make decisions based on such research.

## 1.8 Organization of this book

We arranged the chapters of this book to build on one another, presenting material in what we believe is the best sequence for the novice to the field of food systems modeling. This is probably the most natural order in which to read the book in its entirety. However, we recognized that anyone familiar with the field might want to jump around. Thus, each chapter stands alone, and can be read individually. To avoid redundancy, we encouraged authors to coordinate and divide up the work of presenting concepts or modeling approaches in detail. As a result, you will encounter enough information to understand an unfamiliar term, but

**TABLE 1.1** Common issues underlying food systems models.

| Issue | Overview | Area of Concern | Chapters Addressing Issue |
|---|---|---|---|
| Assumptions | The premises upon which the model is built, in other words, the things treated as given. | Although necessary to create and solve models, if these are not realistic, they may call the validity of model into question. Agent based models and participatory processes can help address concerns. | Chapters 5, 6, 7, 8, 10 & 15 |
| Level of complexity | For increased relevancy and context, additional sectors, causal links or issues are integrated into models, raising the level of complexity. | This is a Goldilocks' problem – the level of complexity should be 'just right' – too complex, and the model can be unwieldy or too difficult to interpret. Too simple, and the model will fail to capture important behavior that occurs under certain conditions. | Chapters 3, 13, 14 & 15 |
| Scope | To be solvable, models leave some issues outside their bounds. Know the realm in which the model was meant to be used. | Together with complexity, the scope of the study must be aligned with the outcomes to be evaluated and conclusions to be drawn. A few specific elements of defining scope are delineated below. | Chapters 3, 4, 10, 14 & 15 |
| Spatial and geographic scale | The spatial level of scale must be relevant to the research topic at hand. | The modeler should explain why the scale chosen is relevant to the key research question, whether it be the domain of control for the decision maker, or the jurisdiction of the policy or the spatial dimensions of outcomes and impacts. | Chapters 3, 4, 6, 7, 8 &14 |
| Temporal scale | Dynamics may or may not be important to the research question, but if change in outcomes is a focal point, it is likely a time element is needed. | The research question should motivate whether one examines current conditions, looks back at historical conditions, or forecasts the future.Data may not be available or consistently reported across time. | Chapters 3, 4, 5, 6, 7, 9, 11, 12, 13, 14 & 15 |
| Uncertainty | Scientists are used to handling uncertainty, but integrating it into models adds an element of complexity. Integrating uncertainty should be aligned with outcomes to be evaluated. | In statistics, uncertainty is estimated using tools such as confidence intervals, providing a range of outcomes. Other approaches, such as scenarios or sensitivity analysis, may be more appropriate for policy analysis. | Chapters 3, 5, 6, 8, 9, 10 & 14 |

(*continued on next page*)

**TABLE 1.1**    Common issues underlying food systems models—cont'd

| Issue | Overview | Area of Concern | Chapters Addressing Issue |
|---|---|---|---|
| Data quality and parameter validation | The availability of data, its consistency and methods used to estimate unavailable parameters are key to model validity. | Modelers should be clear about the source of the data used to run the model, the methods used to estimate parameters, and be transparent about their strengths or limitations. | Chapters 3, 6, 7, 8, 9, 10, 11, 12, 13 & 14 |
| Realm of decision-making | Modelers should clarify if the purpose of the model is to guide decision making, and if so, if it is in a private or public realm. | The scope, spatial, temporal and scenarios to address uncertainty should align with the context of decision-making. | Chapters 3, 4, 7, 10, 12, 13 & 14 |
| Relevancy and Place-based Context | Models should consider the common interest in the place being studied. | Beyond spatial considerations, place-based context might also come from participatory feedback from place-based stakeholders. | Chapters 4, 6, 8, 11, 12, 13, 14 & 15 |
| Common Linkages | Systems are made of linkages and relationships between actors, issues and levels of the food system, focusing commonly on tradeoffs. | The diversity of sciences, metrics, data and intended outcomes may present a challenge to framing and measuring linkages. | Chapters 3, 5, 7, 8, 9, 10, 13 & 14 |

authors may refer you to other chapters in the book for more detail on a topic. In addition, we included a glossary of terminology to ensure readers can understand key concepts without needing to consult every chapter.

Building on the introduction provided in this chapter, Chapter 2 explores fundamental principles and concepts to the field of food systems modeling. The chapter explains eight key ideas: sustainability, sustainable development, sustainable agriculture, resilience, food systems, sustainable food systems, systems thinking, and multi-, inter-, and transdisciplinary research. These concepts are used elsewhere in the book and suffuse the literature on food systems. Hence, the book addresses these terms at the start.

Chapters 3–6 present an array of biophysical models. Chapter 3 presents Life Cycle Assessment (LCA), the most common approach to estimating environmental impacts of food products in a holistic way. Standardized methods exist for conducting LCA, and arguably, they are the most fully developed tool available for assessing environmental impacts in food systems. However, as you will see, LCA is an evolving methodology, which benefits from research done on individual environmental impacts and natural resources. To this end, Chapters 4 presents Water Footprint Assessment, a framework for estimating the water requirements of food products. Influenced heavily by crop water use, estimating the water requirements of foods relies on a deep understanding and simulation of the cycling of water from the atmosphere to the soil through plants and back to the atmosphere. Chapter 5 distinguishes between three contrasting approaches to estimating the land requirements of foods, showing the different insights gained by considering the quality of available land for food production and the capacity to convert non-arable grazing lands and byproduct feeds into livestock products. Chapter 6 concludes the set by giving examples of biophysical models designed to estimate the capacity of agricultural resources to meet human food needs. Carrying capacity modeling estimates the number of people that could be fed from a given area of land, and foodshed analysis explores the potential for a community, state, region, or country to rely on more localized food production.

Throughout most of the world, food reaches people through markets. Accordingly, Chapters 7–10 explore economic models for understanding food systems through a market lens. Chapter 7 introduces how economic theory is embedded in three different market and supply chain models to answer questions about where food comes from, the impact of supply and demand focused policies on food consumption, and the dynamics of markets in real time. Chapter 8 presents input-output (I-O) modeling, a standardized approach to estimating economic flows, and how to apply I-O models to understand the potential economic impacts of investments or programs to support local and regional food systems. Chapter 9 builds on this foundation, showing how I-O models can also be used to explore the material flows through food systems, and thereby, to estimate the environmental impacts of food systems regarding water, land, energy, and greenhouse gas emissions. Chapter 10 describes the IMPACT model, an economic model that integrates inputs from biophysical models to answer questions about how food systems respond to changes in both economic and biophysical drivers. Collectively, these chapters cover a wide range of spatial and temporal scales, from individual states to the global scale and from a single point in time to looking decades into the future.

Chapters 11–14 show applications of models in social and community contexts. Chapter 11 presents the use of social network analysis to understand how different actors within food systems are related and how these relationships grow and change over time. Chapter 12 introduces participatory modeling and the concept of community-based system dynamics, in which researchers build models collaboratively with stakeholders to understand food systems problems at a community scale. Chapter 13 expands on these ideas, presenting the theory of stakeholder-driven community diffusion and using models to learn how to create successful food systems interventions. Chapter 14 describes the use of models, principally LCA, in the business context to either evaluate the sustainability of products or to improve sustainability performance, including the limits of focusing on individual supply chains and a need for broader strategy across the food industry.

The research discussed throughout this book often takes place in teams. Yet creating productive and collaborative teams does not happen by accident. Accordingly, Chapter 15 reflects on the qualities and dispositions that food systems scientists and teams need to succeed, based on the evaluation of interdisciplinary and transdisciplinary research projects at Colorado State University. The findings offer guidance on how to embark on a team science project and to understand the roles and relationships between its members to cultivate a team that grows in effectiveness over time.

Food systems modeling is an emerging and evolving field. The material presented herein presents a cross-section of major methods used to model food systems, but we have not captured everything, and the boundaries of the field are constantly changing. To this end, Chapter 16 serves two purposes. It takes stock of the general lessons conveyed in this book about the use of models to understand food systems and to use this knowledge to manage food systems for better social and environmental outcomes. In addition, the closing chapter looks forward, providing a quick review of a few emerging areas within food systems modeling and directions in which we would like to see the field progress. We hope this leaves the reader with some lessons on how to read the literature and to guide their future work in the field.

## Acknowledgements

This work was supported in part by the U.S. Department of Agriculture, Agricultural Research Service.

## References

Harrison, J.R., Lin, Z., Carroll, G.R., Carley, K.M., 2007. Simulation modeling in organizational and management research. Academy of Management Review 32 (4), 1229–1245.

Meadows, D.H., 2008. Thinking In Systems: A Primer. Chelsea Green Publishing, White River Junction, Vermont.

National Research Council, 2015. A Framework for Assessing Effects of the Food System. The National Academies Press, Washington, DC https://doi.org/10.17226/18846.

Oreskes, N., 2003. The role of quantitative models in science. In: Canham, C.D., Cole, J.J., Lauenroth, W.K. (Eds.), Models in ecosystem science. Princeton University Press, Princeton, NJ, pp. 13–31 2003.

Saltelli, A., Bammer, G., Bruno, I., Charters, E., Di Fiore, M., Vineis, P., et al., 2020. Five ways to ensure that models serve society: a manifesto. Nature 582, 482–484.

# 2

# The origins, definitions and differences among concepts that underlie food systems modeling

*Kate Clancy*

Food Systems Consultant and Visiting Scholar at the Center for a Livable Future, Bloomberg School of Public Health, Johns Hopkins University, United States

## 2.1 Introduction

"Definitions form the foundation of reasoning" (Faber, Jorna, and Van Engelen, 2005, p. 13). Every discipline creates terminology to describe important concepts within its field, but these terms are often unknown outside the discipline. Yet, the success of inter- and transdisciplinary (ITD) research hinges on the ability of researchers to communicate clearly and gain an understanding of each other's disciplines and definitions over time. For decades, researchers have shown that differences in definitions often lead to miscommunication and inaccurate problem identification (Kelly et al., 2019; Stock and Burton, 2011; Szostak, 2007; Thompson, 2009). Furthermore, "differences in terminology are often not just language-based but a sign that people from different disciplines assign qualitatively different meanings to the term" (Thompson, 2009, p. 286; see also Kelly et al., 2019). As language is seen as the key to effective communication in ITD research (Thompson, 2009), projects must allocate time to develop shared vocabularies (Bracken and Oughton, 2006).

The descriptions in this chapter illustrate the challenges that language issues pose to academics and other stakeholders engaged in participatory research, given each discipline's own peculiar jargon. These descriptions also present the complication that many of the fundamental concepts described here have no agreed-upon definitions even after (and perhaps because of) decades of use. Reaching agreement on the meanings of terms increases clarity for both authors and readers. Resolving language and conceptual issues requires discussion in a trusting environment, the development of a glossary if that is appropriate and useful and agreement among a team about offering a definition of a term if it differs from that in the glossary. Inter-disciplinary researchers and practitioners should develop a common understanding of key concepts, not to master another discipline but to understand the different insights that disciplines generate regarding the research question. It is also important to practice reflexive thinking, especially concerning one's own biases and assumptions.

As mentioned, there is no universal agreement or acceptance for most terms. A common or most apt definition has been supplied in the following pages, but in many cases there is no theoretically operationalized meaning, so the researcher/user must clearly define what a term means in a given application. Terms should be defined and specified with regard to a particular project's temporal and spatial scales, its operational principles and the frameworks that clearly relate to the issue being addressed (Johnston, Everard, Santillo, and Robert, 2007).

These concepts and terms undergird the goals, objectives and processes by which researchers and practitioners understand, problem-solve and intervene in the food and agricultural arena. One set of definitions revolves around "sustainable"—an ambiguous term that has been applied to many elements (for example development, agriculture), often with no clear explanation of its meanings. The second set of definitions addresses food systems and systems thinking. These are the structure and processes utilized to explain and study complex systems.

Choosing the appropriate and relevant meaning of a term to frame a modeling exercise can be aided by knowing its etymology. Such knowledge provides perspective about a term's most effective use, lets the user see subtle differences with similar words and allows the user to choose the correct definition for the exercise. Importantly, knowing a term's origin and history and its often ambiguous contemporary meanings, should encourage readers of the modeling literature to pay attention to if and when the developers of a model use certain terms and make clear the definitions for that particular model.

Readers should also be aware that although these concepts often underpin the objectives of a project, the terms themselves are not always acknowledged explicitly. This is often the case for publications reporting on models. Researchers who utilize systems thinking use frameworks that explicitly address these concepts more often.

## 2.2 Origins and definitions of terms

### 2.2.1 Sustainability and related concepts

The concept of sustainability—the ability to be maintained at a certain rate or level (Lexico.com, 2021)—was acknowledged in early world history. The term was first used in the beginning of the 18th century regarding the decline of forest resources. In the 19th century, Malthus brought attention to the possible problems of population growth and in 1866, William Stanley Jevons wrote about the "coal question" (du Pisani, 2006). Early in the 20th century, the same alarm was sounded about oil. The term appeared in *A Blueprint for Survival* (Goldsmith and Allen, 1972) and *The Limits to Growth* (Meadows, Meadows, Randers, and Behrens III, 1972) and later was linked to development (du Pisani, 2006, discussed below).

Despite extensive studies over the last 50 years, there is no agreed-upon definition of sustainability. In fact, as of 2007, there were already 300 different definitions of sustainability and sustainable development (Johnston et al., 2007), in large part because the sustainability discussion arose from broadly different schools of thought historically (Purvis, Mao, and Robinson, 2019). It is defined differently within and between cultures and has changed over time, particularly with the gradual addition of economic and social pillars/perspectives to the original environmental resource focus (Purvis et al., 2019). The three pillars have been accepted conceptually but they have had little theoretical development, which makes it difficult to operationalize the concepts (Purvis et al., 2019). Hasna (2010) argues that "primarily, the divergence of definitions evolved from the confluence of many themes including the ecosphere, local and global vision, the consequences of humanity's influence on a planetary scale, human social welfare, human values, knowledge of ethics over different periods of time, management of resources (limitations) from cradle to cradle and social and cultural contexts through promotion of social justice and environmental awareness" (2010, p. 266). Sustainability has been described as a paradigm, a shared ethical belief, a system property and a management methodology. It is both positivist and normative—leaning to the latter. Put simply, it has been defined as "sustaining a healthy economic, ecological and social system for human development" (Mensah, 2019, p. 5).

### 2.2.2 Sustainable development

The coinage of the term sustainable development (SD) is in dispute. In 1972, Barbara Ward and Renée Dubos urged in their book *Only One Earth* that developers must recognize "what should be done to maintain the earth as a place suitable for human life not only now, but also for future generations" (in Satterthwaite, 2006, p. 10). This definition became popularized later. In 1981, the United Nations General Assembly Resolution, which set up the World Commission on Environment and Development (WCED, the Brundtland Commission), used the phrase sustainable development (du Pisani, 2006) and in 1987 the Commission defined it as "development that meets the needs of the present without compromising the ability of future generations to meet their own needs" (WCED, 1987, p. 36). It a systems approach to growth and development to manage natural, produced and social capital for the welfare of their own and future generations. A contemporary definition defines it as "an effort to guarantee a balance among economic growth, environmental integrity and social well-being (Mensah, 2019, p. 6).

Over the years, the term has collected many ambiguous or distorted definitions (Johnston et al., 2007). Furthermore, sustainability and SD are frequently conflated, leading to competing and conflicting notions of both and "allowing the fuzzy concept to be utilized by any actor for their own means" (Purvis et al., 2019, p. 11). Although many people see them as synonyms, they can be distinguished: Sustainability is the goal or endpoint of a process called sustainable development: 'sustainability' refers to a state, sustainable development refers to the process of achieving the state (Diesendorf, 2000; Gray, 2010). The ambiguity of SD definitions causes it to be equated too frequently only with economic growth, or with natural resources. When only one of the pillars is the focus of a program or research project, the idea of their necessary integration is undermined.

### 2.2.3 Sustainable agriculture

In 1988, Lockeretz pointed out that "sustainable agriculture is not so much a new idea as a synthesis of ideas originating from a variety of sources out of various motivations" (p. 180). Discussion about the concept began in the early 20th century: the phrase was coined in the 1960s by McClymont and in 1977 the first International Federation of Organic Agriculture Movements (IFOAM) conference was titled "Towards a Sustainable Agriculture."

---

**BOX 2.1**

### 1990 Farm Bill Definition of Sustainable Agriculture

An integrated system of plant and animal production practices having a site-specific application that will over the long term:
- satisfy human food and fiber needs
- enhance environmental quality and the natural resource base upon which the agricultural economy depends

- make the most efficient use of nonrenewable resources and on-farm resources and integrate, where appropriate, natural biological cycles and controls
- sustain the economic viability of farm operations enhance the quality of life for farmers and society as a whole (NSACnd)

---

It became clear in the 1960s that there were many different views and definitions of the concept. Not surprisingly, supporters of sustainable agriculture come from diverse backgrounds, academic disciplines and farming practices and their definitions often have been in conflict (Francis 1990, in Gold, 2007). In 1984, Douglas laid out the three different contexts for agricultural sustainability: food sufficiency, ecological integrity, or social sustainability (in Thompson, 2010, p. 218). More recent works describe sustainable agriculture in multiple ways, including a farming system that offers an alternative to conventional agriculture's ideology, an objective property of an individual agricultural system and simply the ability for a farm to continue over time (Siebrecht, 2020). The concept has many components (e.g. the 1990 Farm Bill definition (NSAC) shown in Box 2.1.)

Because the sustainable agriculture arena was and remains filled with people who are interested in the same goals for different reasons, Lockeretz argued that there was "a need for labels for distinct terms, but imprecise usages have resulted in these labels becoming more or less interchangeable" (1988, p. 175). This also meant that substantially distinct and independent goals such as low-input, organic, different from conventional, ecologically sound and many other types of agricultural production must be achieved in their own right (Lockeretz, 1988). It was observed that at any point in time a concept like sustainable agriculture is a compromise and will remain fluid, driven by changes in politics, ideology, science, community values, etc. (Gold, 2007). The Natural Resources Conservation Service (NRCS) general manual states that sustainable agriculture is a way of practicing agriculture that seeks to optimize skills and technology to achieve long-term stability of the agricultural enterprise, environmental protection and consumer safety. It is achieved through a number of different management strategies to conserve resources while providing a sustained level of production and profit (Gold, 2007). Interestingly, several experts have observed that the conundrum of different disciplinary definitions becomes a benefit when considering the development of models as a way to assess agricultural sustainability using multiproduct criteria, data sources and interactions in a complex system (Craheix et al., 2015; Dougherty, 2017).

Because sustainability consists of a set of concepts that are fundamentally different in nature, there will not be one universal definition (Siebrecht, 2020), so the meaning of sustainable agriculture must be specified carefully in any type of project or research undertaking.

This characterization will depend on the kind of practical problem that the research team is trying to understand (Thompson, 2010), such as challenges to the global fish supply, how to increase the demand for fruits and vegetables, or nitrogen dynamics in agroecosystems (IOM, 2015).

## 2.3 Systems concepts

A system has multiple interacting components that can be interrelated (SERC, 2018), as well as a boundary that distinguishes it from the environment in which it operates. Feedback loops between the internal components (in the food and agriculture sphere, subsystems such as land use, processing and transportation) and between the system and its wider environment underpin how the system behaves and evolves. A system can also be described as a set of parts that behave in a way that an observer views as coordinated to accomplish one or more goals (Wilson and Morren, 1990). A systems model is a simplified tool to help understand and visualize complex sets of relationships and its design is dictated by the questions it aims to help address. Such a model is a human construct to aid understanding and is not a model of reality (Woodhill, 2019).

### 2.3.1 Food systems

Though the term food systems was probably used earlier, one early use was in the set of reports promulgated by the National Commission on Food Marketing in 1966, after which it was amplified in two volumes titled *The Food Manufacturing Industries* (Connor, Rogers, Marion, and Mueller, 1985) and *The Organization and Performance of the U.S. Food System* (Marion, 1986). The global food system is composed of large-scale interconnected systems of systems that include ecosystems (both natural and agricultural); climate; food processing, distribution and retail networks; data and information systems; social-economic systems (including education, financing and cultural subsystems); and other elements. They operate within a wide context of human systems and natural systems that create a set of external drivers, including population, technological developments, markets, politics and many others (Woodhill, 2019). Food systems are multiscale, ranging in spatial extent from local to global; in temporal terms from slow to fast and short to long; in jurisdiction through various levels of organized political institutions; and in constitution through laws and regulations (Cash et al., 2003; Ericksen et al., 2010). They are complex, dynamic and multifunctional—and inescapably interconnected—so they create complex analytical and policy challenges (Fig. 2.1).

All food systems involve the interplay of human and natural (socio-ecological) systems and as such are complex adaptive systems. This means that they have high degrees of complexity, uncertainty and continuous adaptiveness to ever-changing environments and may behave in ways that cannot be entirely predicted and controlled through human endeavor (Woodhill, 2019). Food systems are too often defined as simply supply chains but are more accurately understood as systems when "the aggregate of food-related activities and the environments (political, social, economic and natural) within which these activities occur" is included (Pinstrup-Andersen and Watson, 2011).

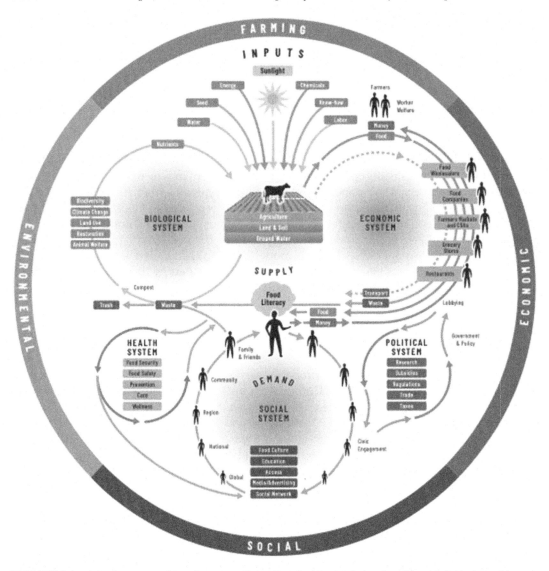

**FIGURE 2.1**    A food system and its subsystems (Reproduced with permission from Nourish Initiative n.d.).

Food systems are not well defined (their misidentification as only supply chains being a case in point) and many of the connections among multiple actors and stakeholders have not been identified. Very little is known about how they work at different levels and only partial knowledge is available to help decision-makers advance programs and policies that can lead to sustainable and resilient systems (Béné et al., 2019). There are multitudes of food systems with similar components, but they differ in scale, geography, politics, culture and many other attributes. Importantly, any given food system offers multiple potentially competing and complementary points for intervention (Foran et al., 2014).

## 2.3.2 Sustainable food systems

The first usage of sustainable food systems (SFS) was probably in the late 1980s, sometime after the phrase sustainable diets was coined (Clancy, 1983; Gussow and Clancy, 1986). Like other terms described here, SFS has multiple definitions, frameworks and narratives developed by scholars and practitioners from different disciplines, such as agriculture, nutrition, social ecology and agroecology, which emphasize different dimensions of sustainability (Béné et al., 2019). The most widely cited definition is "a food system that delivers food security and nutrition for all in such a way that the economic, social and environmental bases to generate food security and nutrition for future generations are not compromised" (High Level Panel of Experts, 2014), but this has been criticized as too abstract and hard to operationalize. The FAO clarifies the three dimensions of SFS as (1) economically sustainable if the activities conducted by each food system actor or support service provider are commercially or fiscally viable; (2) socially sustainable when there is equity in the distribution of the economic value added, taking into account vulnerable groups; and (3) environmentally sustainable when the impacts of food system activities on the surrounding natural environment are neutral or positive (FAO, 2018).

Because the concept of sustainability is poorly defined and applied in research and practice, based on fairly narrow interpretations, it follows that sustainable food systems are probably being defined in ambiguous ways. To overcome this problem, researchers and practitioners need to recognize the multi-dimensional nature of the concept of sustainability, which includes the environment, economics, public health, nutrition, governance and other aspects (Mason and Lang, 2017). Another critical element is the recognition in research and practice of the trade-offs that occur between competing or conflicting dimensions (Béné et al., 2019). Those studying food systems or their subsystems must be very clear about the particular elements under consideration as they utilize a variety of methods, including modeling.

## 2.3.3 Systems thinking and modeling

The evolution of systems science has brought a panoply of discoveries that aptly start with Charles Darwin's thoughts in 1859 on "how plants and animals are bound together by a web of complex relations" (in Wilson and Morren, 1990, p. 67). In the 1940s, Bertalanffy developed General Systems Theory, one of three components of which was systems science, "a discipline consisting of the scientific explorations and theory of systems in the various sciences such as biology, sociology, economics, etc." (Strijbos 2010 in Hieronymi, 2013, p. 581). In 1956, Forrester created systems dynamics, an approach to understanding the nonlinear behavior of complex systems using stocks, flows, feedback loops and time delays (Rogers, 2021). The term systems thinking was coined in 1987 by Richmond and defined succinctly in 2015 as a "holistic (integrative) versus analytic (dissective) thinking; recognizing that repeated events or patterns derive from systemic structures which, in turn, derive from mental models; recognizing that behaviors derive from structure; a focus on relationships versus components; and an appreciation of self-organization and emergence" (Monat and Gannon, 2015, p. 11). It operationalizes the fact that understanding a system means examining the linkages and interactions between the elements that compose the whole system: the process of systems thinking explores interrelationships, perspectives and boundaries (Williams, 2018).

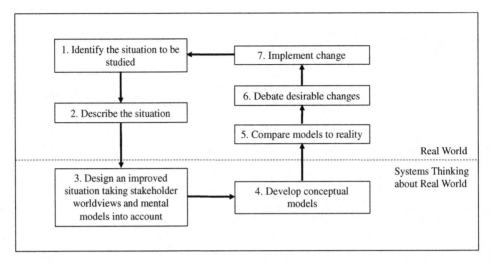

**FIGURE 2.2**  The seven stages of soft systems methodology (Wilson and Morren, 1990).

In systems-based inquiry, there are two major approaches, hard and soft systems, with myriad frameworks, elements, perspectives and analytical tools. A hard-system approach arose in the context of machine-based or hardware-dominated systems (Naughton 1984 in Wilson and Morren, 1990, p. 73). This approach assumes the objective reality of systems, a well-defined problem that can be quantified, technical factors in the forefront, a scientific approach to problem solving, including simulations and an ideal solution (Cairns, 2020). Bawden (1991) gave the example of a hard approach as perceiving a dairy farm as a system and posing the question "What is the ecology of the farm?" (1991, p. 2367).

In the 1980s, Peter Checkland developed soft systems thinking to address complex problems with a large social component that do not lend themselves to being quantified and involve actions by people with different worldviews, all factors that make solutions not obvious (Wilson and Morren, 1990). Unlike hard systems, soft systems approaches assume that organizational problems are poorly defined, that stakeholders interpret problems differently, that human factors are important and that the outcomes are learning and better understanding rather a solution (Cairns, 2020). In Bawden's example, a soft approach would ask the question about a dairy farm as "Is this messy and complex situation associated with changing what seem to be the issues perceived as problematic?" (p. 2367). Fig. 2.2 shows the seven stages of soft systems methodology.

One often-used soft systems model is called the iceberg model. It consists of a hierarchy of levels of understanding a system, with observable events at the top, followed by patterns and trends, then underlying structures (what is causing an observed pattern) and finally the mental models (deeply held beliefs and assumptions) of stakeholders (Christiaens, 2018; Monat and Gannon, 2015) (Fig. 2.3).

Nicholson et al. (2020) make a strong case for utilizing systems thinking and modeling tools in developing conceptual frameworks that assess the connections between food systems subsystems and other elements. Frameworks represent the relationships among food systems components using theories and concepts from multiple academic fields, in their case linking

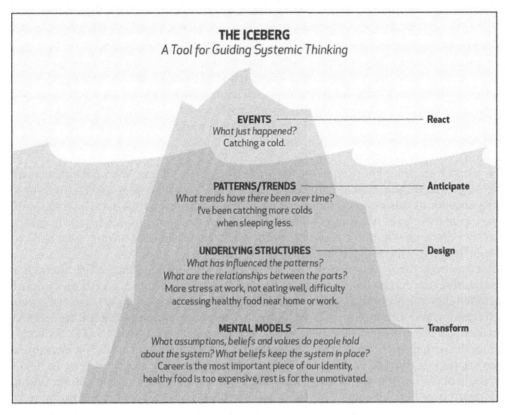

**FIGURE 2.3** The iceberg model to guide soft systems thinking (Reproduced with permission from Ecochallenge.org. Available at https://ecochallenge.org/iceberg-model/.)

agriculture with nutritional outcomes. Systems thinking can improve the understanding of the causal factors linking subsystems to each other and address dynamics and nonlinearities, where the magnitude of a response is not proportional to the magnitude of the input, in these interactions (SERC, 2018) (see Systems Properties and Their Use in Modeling).

### 2.3.4 Multi-, inter- and transdisciplinary research

Collectively, different forms of research involving more than one discipline are called integrated research (Stock and Burton, 2011), not to be confused with the USDA terminology that means a combination of research, education and extension (USDA-NIFA, 2015). All of the forms of research have been defined in different ways. The ones that follow seem to be the most apt for an understanding of food systems modeling.

Multidisciplinary studies are defined as several different academic disciplines researching one theme or problem with multiple disciplinary goals. Participants exchange knowledge but do not cross subject boundaries to create new knowledge and theory (Cronin, 2008).

The term interdisciplinary was probably first used a century ago by the Social Science Research Council (Frank, 1988). In the last 30 years, there has been an interest in interdisciplinary work across most natural and social sciences, as well as the humanities, in response to the prevalence of complex problems in all sectors (Szostak, 2015a). Interdisciplinary research was defined by a National Academy of Sciences panel as research that integrates information tools, perspectives, and/or theories from two or more disciplines to solve problems whose solutions are beyond the scope of a single discipline (NAS 2004 in Szostak, 2015b). In addition, it was defined by a set of experts as several unrelated academic disciplines crossing subject boundaries to create new knowledge and theory and solve a common research goal (Kelly et al., 2019).

The word transdisciplinary was coined in 1970 by Jean Piaget, who described it as "a higher stage succeeding interdisciplinary relationships" (in Bernstein, 2015, p. 1). As with all the concepts in this chapter, there are multiple definitions of the term. Cronin defines transdisciplinary studies as "projects that both integrate academic researchers from different and related disciplines and nonacademic participants to research a common goal and create new knowledge and theory" (2008, p. 4).

What has been made abundantly clear by myriad authors working in this arena is that integrated research conducted by teams of natural and social scientists is critically important to sustainability science, but it offers multiple challenges, including (1) a lack of coherent shared framing and terminology, (2) difficulty in integrating different disciplinary methods, (3) problems in the implementation of the research process and co-creation of knowledge, (4) practitioners' engagement on a scale of information to collaborators and (5) generating and measuring impact (Brandt et al., 2013). Very extensive literature exists on how to navigate these challenges. Experts put a strong emphasis on developing excellent communication among team members by practicing trust, talking through issues and maintaining a sense of humor (see, for example, Thompson, 2009). Szostak (2007) recommends that projects start with knowing that the issue in question requires an interdisciplinary approach. Then the team should gather relevant disciplinary insights, perform a critical examination of the insights to distinguish assumptions from arguments and evidence from assertions and integrate the inputs from the various disciplines, which "calls for creativity, intuition and inspiration" (p. 12).

## 2.4 Differences between sustainability and resilience and food systems and systems thinking

During the past few decades in which the terms just discussed have been in use, widespread confusion has arisen around two sets of them in particular. One is the difference between sustainability and resilience; the other is the difference between food systems and systems thinking. This section offers a deeper examination of these two sets of terms whose conflation can invite multiple issues in problem identification, interpretation and methodology choices.

### 2.4.1 The difference between sustainability and resilience

Resilience is the capacity to recover quickly from difficulties (Lexico.com, 2021). It is characterized as a loosely organized cluster of concepts and a way of enabling exchanges across disciplines (Tendall et al., 2015). The first scientific use of the concept was in the 19th century. In the 1940s and 1950s, the concept emerged in psychology; the engineering/physics interpretation arose in the 1960s and 1970s at the same time that ecologists started to use resilience to describe some aspects of ecosystem dynamics around equilibrium. Holling's, 1973 definition has been widely adopted: "a measure of the ability of ecosystems to absorb changes of state variables, driving variables and parameters and still persist." Since then, the concept of resilience has become increasingly popular in other disciplines and subdisciplines, has been applied in social contexts and has been a key component of the idea of social-ecological resilience that emerged in the late 1990s (Béné et al., 2019).

In this century, the focus has moved to resilience thinking as "an approach for understanding complex adaptive systems and a platform for inter- and transdisciplinary research with an emphasis on social-ecological systems" (Folke, 2016, p. 44). Resilience thinking adds two other dimensions to the definition: the adaptive capacity to learn, combine experience and knowledge, adjust responses to changing external drivers and internal processes and continue operating (Folke et al., 2010); and a transformative capacity to create a fundamentally new system when ecological, economic, or social structures make the existing system untenable (Walker, Holling, Carpenter, and Kinzig, 2004). The Resilience Alliance (2015) offers an amalgamated version- "the capacity of a social-ecological system to absorb or withstand perturbations and other stressors such that the system remains within the same regime, essentially maintaining its structure and functions. It describes the degree to which the system is capable of self- organization, learning and adaptation" (Holling, 1973 and Walker et al., 2004).

It functions through a combination of different capacities: absorptive capacity, leading to persistence; adaptive capacity, leading to incremental adjustments/changes and adaptation; and transformative capacity, leading to transformational responses (Béné et al., 2019). Meuwissen et al. (2019) adapt these to sustainable agriculture by understanding the three capacities as robustness, the ability to withstand stresses (e.g., by employing genetic and species diversity; see Peterson, Eviner, and Gaudin, 2018); adaptability, the ability to change the composition of various elements such as inputs to respond to stress without changing structures (e.g., irrigation); and transformability, the ability to change the internal structure or feedback mechanisms (e.g., integrating animals and crops on a farm; see Lengnick, 2015).

Resilience is not the same as sustainability, but its similarity can lead to confusion between the terms. Consensus on the definitions of both concepts is lacking, as shown by the description by Marchese et al. (2018) of three frameworks being used to look at sustainability and resilience. The first is resilience as a component of sustainability—describing resilience as an integral part of the larger concept of sustainability, with sustainability the primary objective. The second is sustainability as a component of resilience—resilience as the ultimate objective, with sustainability a contributing factor. The third framework describes resilience and sustainability as concepts with separate objectives that can complement or compete with each other.

Whatever framework underpins research or interventions, it is critical that the differences are not neglected (Ungar et al., 2020). As Grubinger (2012) puts it, "There is no bright line between the two concepts but they aren't the same". Unfamiliarity with the distinctions leads to problems in implementation, such as failures by decision-makers or researchers to capitalize on synergies or account for competing objectives, which can result in incomplete understandings and project outcomes and may cause future conflict, as has been seen in

**TABLE 2.1**    Similarities between Sustainability and Resilience.

| Assumptions | Harmony between human society and the natural environment is possible |
|---|---|
| Research Foci | Social and ecological systems; climate change impacts; globalization; community development |
| Methods | Climate change policies and actions, especially governance; education and learning as implementation tools |
| Goals | System survivability, security and well-being (social and biodiversity); sense of place and belonging (heritage) |

Reproduced with permission from Lew et al., 2016, p. 22.

the development of climate change strategies (Marchese et al., 2018). Allen and Prosperi point out the necessity of distinguishing between sustainability and resilience when defining problems in systems dynamics modeling for food systems (2016). Their particular concern is that crucial sustainability and resilience variables be identified accurately in models examining food security.

The confusion between the concepts has two basic sources. The first is "the weak and sometimes sloppy conceptualization and definitions that researchers often use for the two terms" (Marchese et al., 2018, p. 20). Sometimes sustainability is used in the narrowest way to mean maintaining the status quo. At the opposite extreme is the tendency to define sustainability as including every possible good condition available to human societies—a conclusion that the researchers believe is the message from the Brundtland Report. In general, "sustainability is focused on the core goals of maintaining and protecting natural and cultural resources for the future and mitigating change" (Lew, Ng, Ni, and Wu, 2016, p. 20). Resilience, on the other hand, focuses on the capacity of systems to respond to both extreme disturbances and persistent stress (Folke, 2016).

The second reason for the confusion between sustainability and resilience is that the two approaches share some important goals, assumptions and methods. One mutual goal is system survivability and a mutual assumption is that human societies function in a state of harmony within the larger context of the natural world. Both approaches refer to the state of a system or feature over time, focusing on the capacity of the system under normal operating conditions to respond to disturbances: they are also linked to global political trends or global frameworks (Fiksel et al., 2006). These concepts share common objectives, such as a focus on climate change and seeking a balance between humans and nature (see Table 2.1). Research content for both may include social and ecological systems, climate change impacts, globalization, and/or community development. These commonalities make it seem like sustainability and resilience are the same thing (Lew et al., 2016).

But there are deeply fundamental differences in the way the concepts frame, study and resolve research questions. The most significant is in their basic ontological assumptions about the nature of the world: whether it is normal to be in a state of stability and balance, or in a state of change and even chaos (Lew et al., 2016). For sustainability, the goals are normative ideals, culture, environment and economic conservation, intergenerational equity and fairness. Some research foci of sustainability are the environmental and social impacts of economic development, overuse of resources and carbon footprints. For resilience, the

**TABLE 2.2**   Differences between Sustainability and Resilience.

|  | Sustainability | Resilience |
|---|---|---|
| Assumptions | Stability and balance are the norm (or are at least possible) | Nonlinear and unpredictable change and chaos are the norm |
| Goals | Normative ideals (culture, environment and economic conservation, intergenerational equity, fairness) | Strategic, dynamic and self-organizing systems; learning institutions and innovative cultures |
| Research Foci | Environmental and social impacts of economic development and growth, overuse of resources, carbon footprints | Natural and human disaster management, climate change impacts, social capital and networks |
| Methods | "Wise Use" resource management, mitigation or preservation against change, recycling and greening, education for behavior change | Reducing vulnerability and increasing physical and social capacity for change (flexibility, redundancy), system feedback and performance, education for innovation |
| Criticism | Poorly defined and highly politicized | Does not address the causes of social and environmental change |

Reproduced with permission from Lew et al., 2016, p. 22.

goals are strategic dynamic and self-organizing systems, learning institutions and innovative cultures. Research foci are natural and human disaster management, climate change impacts and social capital and networks.

There are also different research questions, methodologies and metrics (see Table 2.2), as well as intergenerational and temporal dimensions (Allen and Prosperi, 2016). For example, sustainability efforts are often focused on larger spatial and longer temporal scales than resilience. Solutions to problems are actions like prudent resource management, mitigation or preservation against change, recycling and greening and education for behavior change (Lew et al., 2016).

Resilience initiatives tend to focus on adapting to new conditions, creating innovative uses of traditional knowledge, creating new environmental knowledge and improving living conditions (Lew et al., 2016). Solutions include reducing vulnerability and increasing physical and social capacity for change (such as flexibility and redundancy), finding ways to monitor system feedback and performance and providing education for innovation. A resilience approach to the management of resources "would emphasize the need to keep options open, to view events in a regional rather than a local context and to emphasize heterogeneity" (Holling, 1973, p. 21).

## 2.4.2  The difference between food systems and systems thinking

The conflation of food systems and systems thinking (due in part to the multiple meanings of the term "systems") contributes to the problem that most food systems work does not utilize the properties of systems thinking. As described above, food systems are constructs that aid people in understanding the sets of activities and players within very complex arrangements

(Woodhill, 2019). Researchers have used different approaches to explore and understand food systems, for example identifying different research traditions in agriculture: cropping systems and integrated crop/livestock systems research, farming systems research, agroecosystems research and knowledge systems research (Southern Region Sustainable Agriculture Research and Education, 2009). Another example comes from Foran and colleagues, who identified four different conceptual frameworks for food security (which also apply to food systems): agroecology, agricultural innovation systems, social ecological systems and political ecology (2014). In addition, Nicholson et al. evaluated 36 frameworks through the lens of systems modeling to understand how the variables used in studying the linkages between agriculture and food security meet the criteria for systems modeling (2020).

Given these complexities and while food systems can be conceptualized in myriad ways, what has been widely agreed upon is that a food system at any scale is the epitome of a complex adaptive system and therefore must be assessed using systems science and thinking to be understood fully (IOM, 2015).

In contrast with food systems, systems thinking is not a construct. It has been variously defined as a set of tools, a methodology, a philosophy and an approach to integration (Allen, 2020); a perspective and a language (Monat and Gannon, 2015); a discipline, a growing field and an important complement to analytic thinking (Kim, 1999). Systems thinking is critical to an understanding of sustainability (Brazdauskas, 2019; Lanzi et al., 2020). It focuses on the identification of interrelationships between components such as climate, farming, transportation, financing and consumption and is a way to understand how agroecological systems function within natural and social systems while considering cultural and other narratives (van Berkum, Dengerink, and Ruben, 2018). Food systems consist of multiple subsystems that require more analysis, but The Economics of Ecosystems and Biodiversity project offers the critique that most scientific research focuses on components or subsystems of food systems, to the detriment of connecting the pieces of the puzzle to achieve a comprehensive understanding of reality (Müller and Sukhdev, 2018; Zhang et al., 2018).

Systems thinking does not supplant reductionist thinking and research—it complements it. But it is more complicated. Richmond described seven critical thinking skills that should be applied to model development (Richmond, 2018):

- Dynamic thinking frames a problem as a pattern of behavior over time.
- System-as-cause thinking is needed to construct a model, including its boundaries, to explore how a behavior arises.
- Forest thinking allows modelers to not focus on details but to group them to see the average picture of a system.
- Operational thinking is used to think about how behavior is generated and processed—not just what influences it.
- Closed-loop thinking allows seeing causality as an ongoing process
- Quantitative thinking is knowing how to quantify "soft" phenomena that cannot always be measured, such as motivation.
- Scientific thinking is used to integrate models to see where they make sense—to define testable hypotheses.

A chapter on "a systems approach" in a report on food and agriculture research for the future addresses the challenges, opportunities and barriers to success of integrated system models (National Academies of Sciences, Engineering and Medicine, 2019). A careful reading of the chapter (8) offers insights, implicitly, on the usages of different types of systems thinking. For example, assumptions made about the differences in crop productivity may focus on only a few drivers when systems-as-cause thinking calls for multiple drivers. High degrees of uncertainty across multiple system elements requires operational thinking and scaling up models from field to landscape will be helped by forest thinking.

It is important to note that not all agricultural and food research or modeling efforts are candidates for systems approaches. They should be used when the issue is important, the problem is chronic and familiar and people have unsuccessfully tried to solve the problem (Goodman, 2018) and when there are multiple actors and multiple causes of the problem, there are competing or conflicting interests and there is no single explanation for what is causing the problem (Australian Prevention Partnership Centre, 2018).

There are two necessary components of the application of systems theory and systems thinking to food systems modeling and other related efforts. The first is clearly identifying the specific food system that is being assessed or proposed for intervention, such as the distribution system of fluid milk in a multistate region, or the actors in a city who engage with multiple subsystems of obesity prevention. The second is explicitly recognizing and utilizing the properties of systems (described in the next section) in the planning and execution of the models. The IOM committee's ideal framework for assessing food systems suggests that when choosing properties for a study, researchers should: (1) allow consideration of effects across the full food system under study, (2) account for several types of potential effects across the domains (social, economic, environmental and health) and dimensions (quantity, quality, distribution and resilience), (3) capture systems dynamics such as feedback interactions and heterogeneity, (4) capture processes and outcomes at the scale suited to the problem and (5) address the critical concerns of stakeholders and policymakers.

The committee was fully aware that a complete systems assessment will be rare. In cases of partial assessment, the modelers should at least acknowledge the effects of drivers outside the scope of the assessment and acknowledge what is beyond the scope of the study. They should also include as many domains and dimensions as feasible to capture as much knowledge about the system as possible. Of course, small-scale studies also make important contributions to an issue and a field. But researchers should not presume that such efforts are "systems research" and should provide the context of the bigger issue of which their inquiry is a part.

As mentioned, most work on food systems does not utilize systems thinking—researchers and practitioners often claim that they are working on food systems but mean that they are examining more than one link in a supply chain. Clarifying the distinction between the two has been muddled somewhat by the use of the term "food systems approach." The terminology appears to be an attempt to merge the two concepts, but much of the research and practice of the approach reveals a spotty use of systems properties—some studies fully embrace systems properties and some barely employ them. In the former category is the synthesis of food system studies and research projects in Europe (Achterbosch, Getz-Escudero, Dengerink, and van Berkum, 2019). These authors argue that a food system approach has great value in looking at complex sustainability challenges across multiple scales and domains. Many of the projects they review have modeling components, some of which are mentioned in the next section.

### 2.4.3 Systems properties of food systems and their use in modeling

Complex adaptive systems such as food systems possess a set of specific properties that have arisen from the social sciences, public health, biology, business, land and ecosystem management and other fields and disciplines (IOM, 2015). These properties become part of the framework that guides and undergirds the development of food systems models. Understanding and utilizing these properties yields important insights into systems behavior and can capture the current configuration of a food system, as well as the potential for future alternatives for that system. Use of the systems properties described below is critical to understand the complexity and dynamics of food systems and the behaviors within it.

Over the decades of systems thinking practice, the language describing and parsing it has proliferated. There are principles (openness, purposefulness, multi-dimensionality, counter-intuitiveness and emergent properties) (Gharajedhi in Monat and Gannon, 2015), elements (interconnections, dynamic behavior and different scales) (Arnold and Wade, 2015), key concepts (interconnections, self-organization, syntheses, emergence, feedback loops, causality and systems mapping) (Acaroglu, 2017), components (boundaries, interacting parts, reservoirs, feedback, emergent properties, nonlinear change, tipping points and resilience) (SERC, 2018) and premises that differentiate systems thinking from the normal scientific method (holism, transformation, control, communications, hierarchies/nested systems and emergent properties) (Wilson and Morren, 1990).

Another way that systems thinking has been categorized is by its properties, which are described in the IOM report A Framework for Assessing the Effects of the Food System (2015). Because the report is wholly focused on complex adaptive food systems, those properties described in IOMs Chapter 6 are summarized here. There are many properties and this is not an exhaustive list. Some of these properties have been mentioned in preceding sections; here they are more fully defined and illustrated by examples from other chapters in this book and from previously published research papers and reports.

Individuality. Complex systems contain many different actors that vary in context, motivation, scale and exposure to information or the environment. The interaction of these actors is often a key driver of system behavior. Actors at a lower scale in food systems include pathogenic bacteria and agricultural pests; individual human actors in supply chains include farmers, processors, food service providers and many others; and more aggregated actors are entities like multinational firms, universities and governments. The daily decisions made by all these actors shape the structure and the dynamics of a food system and drive the outcomes within its health, environmental, social and economic dimensions. This is well illustrated in Chapter 12 (Glickman, Clark and Freedman) as the authors explore the roles of the public and academics in participatory modeling. The community-based systems dynamics model they utilize engages the public in describing their own experiences inside their food systems, which academics then help to explain in model form

Heterogeneity. Actors in food systems may have different goals, information and environmental exposures and constraints. These in turn affect adaptation or constraints on the actors' reactions to change. Heterogeneity occurs across supply chains (Nicholson and Gomez Chapter 7) and in the model used to study Life Cycle Analyses (Thoma et al. Chapter 3). Although Thoma et al. do not use the term in the chapter, the Monte Carlo simulations they conduct to quantify uncertainty in the LCA results are needed because so much heterogeneity exists in the yogurt system under study. Another example is the many efforts to increase the public consumption of fruits and vegetables. This work involves actors along supply chains from farmers to distributors to household food shoppers, all of whose decisions will affect the likely impacts of the intervention (IOM, 2015).

Adaptation. All actors adapt to the actions of other actors or changes in the state of a system. The processes of adaptation are varied and may be illustrated by examples as varied as changing farming practices, the evolution of drug resistance by bacteria and evolving consumer preferences. Nicholson and Gomez (Chapter 7) utilize a systems dynamics model to describe the adaptations across supply chains under various conditions, including an increase in yield that leads to multiple changes across vegetable supply chains. Adaptations have impacts across subsystems and food supply chains and across space that can trigger responses from actors near or far from them. Thinking about a full set of adaptive responses that is triggered by a change or shock to a system is an important component of understanding system effects and reflecting it in model building. For example, the IMPACT model assumes implicitly that adaptation occurs over time in commodity markets and producer and consumer behavior (Wiebe et al. Chapter 10).

Feedback and interdependence. Complex systems contain distinct and often linked pathways. The pathways may cross multiple levels of a system and create feedback, which occurs "when outputs of a system are routed back as inputs as part of a chain of cause-and-effect that forms a loop" (Achterbosch et al., 2019, p. 8). The effects often lag and can be either positive (reinforcing) or negative (balancing) between components of a system.

European scientists as a whole have been ahead of their American counterparts in utilizing systems thinking and modeling applied to research, on-the-ground programs and policy. The authors of a large European-based review of food systems research insisted that the study of feedback loops be a key criterion for inclusion in their review because of its importance to understanding a whole system and its trade-off effects (Achterbosch et al., 2019). This report, sponsored by the European Commission, looked at 52 relevant European research and innovation reports in order to better formulate the knowledge needs of the Commission's research programs. Many of the cases they reference utilize models of different types (Achterbosch et al., 2019).

Models and other research methods have identified myriad cases of feedback in food systems, such as (1) grazing practices that erode vegetative cover, which decreases water and nutrient accumulation in the soil; (2) market or commodity price changes that affect consumer purchases or farmers' production decisions; (3) fertilizer run-off into waterways that affects seafood growth and supply, but because of a lag in distance does not lead to more appropriate production practices; and (4) concentration in large supermarket chains that reinforce food security problems in low-income and rural areas (IOM, 2015). Wiebe et al. (Chapter 10) describe feedback loops where the modeling framework shows the interdependence between income growth, food demand, demand for commodities and other factors. Rocker, Kropczynski and Hinrichs (Chapter 11) examine how social network analyses (SNA) characterize social and economic connections among food system entities (e.g. individuals, organizations, businesses). SNAs can be mapped to understand patterns of cooperation and coordination and the flows of information and resources between and among food systems actors.

Stocks and flows. These properties help in analyzing systems in a quantitative way. A stock is an entity that accumulates or depletes over time, such as water, a commodity, or an inventory of baked goods. A flow is the rate of change in a stock (Kim, 1999). Nicholson and Gomez (Chapter 7) note that supply chain models often use a "stock management structure" to track information. Causal loop diagrams, a tool of systems thinking, capture how variables in a system are interrelated and produce feedback to each other (Kim, 1999). In community-based systems dynamics projects, causal loop diagrams are translated by the team into stock and flow models to allow simulations to be constructed (Glickman, Clark and Freedman, Chapter 12).

Spatial complexity. Systems contain spatial organizations, such as physical geography and networks that govern actors' interactions, the speed of feedback and heterogeneity across the system. The scale at which a food system operates must be carefully chosen because it drives the identification of the external variables likely to affect the system (Allen and Prosperi, 2016). Spatial examples within food and agricultural systems include supply chains, market segmentation, ecosystems and food webs. Spatial structure directly shapes the local context and experience by actors but often affects impacts at a distance (for example, aquatic dead zones) and creates unintended effects. The maps of potential food sheds in Chapter 6 (Peters) show variations in their size that are influenced by the size of the population center and competition with other cities in the vicinity.

Dynamic complexity. Feedback, interdependence and adaptation produce dynamics in systems with specific properties. One of these is nonlinearity (in which there is not a straight-line relationship between one variable and another) and another is "leverage points" that yield large effects from relatively small changes in a systems' configuration. One example of a nonlinear relationship is a transshipment model that examines the costs of localizing the fluid milk supply (Nicholson, Gómez, and Gao, 2011), in which the relationship between increased localization and costs appears to be nonlinear. The interaction of social, economic and ecological systems, each of which demonstrate nonlinear processes, can lead to even stronger nonlinearities in response to overall systems change: for example, the diminished quality and quantity of water in rural and urban areas due to crop and animal production and food manufacturing.

Another dynamic complexity example comes from the food transportation sector in a study of regional food freight (Miller et al., 2016). This multidisciplinary team, aided by practitioner advisors, utilized systems dynamics methods, including stock and flow diagrams, system archetypes, feedback loops and other tools to seek the root causes of the systems failures in food transportation in the upper Midwest. They were able to identify potential solutions to these challenges, exploring how to optimize food systems resilience and identifying opportunities for efficiency and diversity at a regional scale. They found several leverage points: optimizing efficiency and diversity, cropping systems diversity, distance to market, truck size, contracts, terminals and trip segments, settlement patterns and engineering innovations.

Emergent behaviors. These behaviors differ from what might be expected from the sum of individual components of the system. They emerge as the result of the placement of the frameworks or models of the system into their appropriate scales so they can be observed. Easterling and Kok (2002) have argued that the vulnerability of the U.S. agricultural production system to climate change shows the importance of recognizing the role of multiple scales simultaneously in order to uncover protective emergent properties through an entire system, rather than aggregating, for example, multiple regions across a country to measure and predict climate change effects. An excellent example of an emergent property is the one uncovered by Vadrevu et al. (2008). They designed an analytical hybrid process model to quantify an agroecosystems health (AEH) index across many spatial scales (field to landscape). They combined a key set of biophysical variables (soil health, biodiversity and topography) and social variables (farm economics, land economics and social organization) that determine the growth of functional properties. Each variable was described by combination of measurable variables such as soil organic matter, plant diversity, an elevation grid, the integration of farm and residential homeowners into social networks and the current agricultural-use value of farms in the study area. Utilizing multiple datasets in the analytical hierarchy process, they conducted pair comparisons to arrive at a simple index, then plotted the landscape patterns of estimated AEH. Overall, the socioeconomic variables and biodiversity contributed most to the AEH index; soil quality and topography contributed less. The study did uncover an emergent property: a landscape pattern in the index that was not apparent in any of the underlying data sets (2008).

## 2.5 Conclusions

Knowing the origins and definitions of key terms and concepts that suffuse the discussion about food systems allows one to be a more critical reader of modeling research. Some of the concepts described originated centuries ago, but most contemporary versions arose in the 1960s and 1970s, when the weaknesses of linear thinking were becoming obvious in many disciplines and institutions. Almost all of the terms have multiple definitions, as the early meanings were adopted and adapted by different disciplines, which defined them to fit within disciplinary epistemologies and objectives. In ITD work, it is important to recognize that disciplinary definitions can differ in significant ways and even be contradictory and to offer guidance on how to signal colleagues and readers about the specific definitions being used in particular contexts.

Training in systems thinking and properties can also enhance participation in food and agriculture modeling exercises. Not all food systems research or interventions need to employ systems thinking, but there are many benefits to doing so within interdisciplinary teams. Because a food system at any scale is the epitome of a complex adaptive system, it should be assessed through multiple properties, concepts and elements to be understood at a useful and relevant depth.

Systems thinking can improve the understanding of the causal factors linking subsystems to each other and address the dynamics and nonlinearities in these interactions. Most food systems are not well defined and many of the connections among multiple actors and stakeholders have not been identified. Little is known about how they work at different levels and only partial knowledge is available to help decision-makers advance programs and policies that lead to sustainable resilient systems. The field is wide open as the effects of climate change, food insecurity, loss of biodiversity and many other interacting issues worsen. Modeling is one of the methods needed to capture these interdependencies and their consequences.

Integrated research conducted by teams of natural and social scientists is critically important to sustainability science, but it does offer multiple challenges that are described throughout the chapter. Fortunately, guidance on how to navigate these challenges is found in a rich literature that spans 50 years. In the same vein, food system modelers can provide a real service to the field by thinking of their research reports as educational tools for systems thinking. For example, both research and applications could be better understood if the relevant terms and properties in a model were identified explicitly. Making systems thinking more evident can raise the status of system sciences in the agri-food arena, help readers more fully grasp a model's outcomes and inspire more researchers and practitioners to apply a systems thinking lens to their explorations.

# References

Acaroglu, L., 2017. Tools for systems thinkers: the 6 fundamental concepts of systems thinking. Available at: https://medium.com/disruptive-design/tools-for-systems-thinkers-the-6-fundamental-concepts-of-systems-thinking-379cdac3dc6a

Achterbosch, T.J., Getz-Escudero, A., Dengerink, J.D., van Berkum, S., 2019. Synthesis of existing food systems studies and research projects in Europe. European Commission. Directorate-General for Research and Innovation, Bioeconomy, Brussels.

Allen, T., Prosperi, P., 2016. Modeling sustainable food systems. Environmental Management 57, 956–975.

Allen, W., 2020. Systems thinking. Learning for Sustainability Services, 5 p. Available at: https://Learningforsustainability.net/systems-thinking. Accessed on 2021..

Arnold, R.D., Wade, J.P., 2015. A definition of systems thinking: a systems approach. Procedia Comput Sci 44, 669–678.

Australian Prevention Partnership Centre, 2018. Systems Thinking. Available at: https://preventioncentre.org.au/resources/learn-about-systems/

Bawden, R.J., 1991. Systems thinking and practice in agriculture. Journal of Dairy Science 74 (7), 2362–2373.

Béné, C., Oosterveer, P., Lamotte, L., Brouwer, I.D., de Haan, S., Prager, S.D., Khoury, C., 2019. When food systems meet sustainability – current narratives and implications for actions. World Development 113, 116–130.

Bernstein, J.H., 2015. Transdisciplinarity: a review of its origins, development and current issues. Journal of Research Practice 11 (1), 1–20.

Bracken, L.J., Oughton, E.A., 2006. What do you mean?' The importance of language in developing interdisciplinary research. Transactions of the Institute of British Geographers New Series 31 (3), 371–382.

Brandt, P., Ernst, A., Gralla, F., Luederitz, C., Lang, D.J., Newig, J., ... von Wehrden, H., 2013. A review of transdisciplinary research in sustainability science. Ecological Economics 92, 1–15.

Brazdauskas, M., 2019. Assessing the importance of systems thinking for sustainability-oriented education. Journal of Creativity and Business Innovation 5, 168–176.

Cairns, D., 2016. Soft systems approach. International Center for Developing Science and Technology. CSC9T4. Available at http://dl.icdst.org/pdfs/files3/8ecbfa944823956ddcfd14d71de0470d.pdf .

Cash, D.W., Clark, W.C., Alcock, F., Dickson, N.M., Eckley, N., Guston, D.H., ... Mitchell, R., 2003. Knowledge systems for sustainable development. Proceedings of the National Academy of Sciences of the United States of America 100 (14), 8086–8091.

Christiaens, W., 2018. 5.2.1 The iceberg model. KCE Process Book, Brussels. Available at http://processbook.kce.fgov.be/node/536.

Clancy, K. (1983). Moving Towards a Sustainable Diet. Presentation to Cayuga Dietetic Association and Cornell Cooperative Extension of Broome County. October.

Connor, J.M., Rogers, R.T., Marion, B.W., Mueller, W.F., 1985. The food manufacturing industries: Structure, strategies, performance and policies. Heath & Co, D.C..

Craheix, D., Bergez, J.-E., Angevin, F., Bockstaller, C., Bohanee, M., Colomb, B., ... Sadok, W., 2015. Guidelines to design models assessing agricultural sustainability, based upon feedbacks from the DEXi decision support system. Agronomy for Sustainable Development 35, 1431–1447.

Cronin, K., 2008. Transdisciplinary Research (TDR) and Sustainability. Overview report prepared for the Ministry of Research, Science, and Technology (MoRST). New Zealand. pp. 1–36.

Dellink, R., Lanzi, E., Agrawala, S., Lutz, W., Sherbov, E., Zimm, C., ... Rao, N., 2019. Chapter 4: Developing pathways to sustainability: fulfilling human needs and aspirations while maintaining human life support systems. In: Hynes, W., Lees, M., Miller, J. (Eds.), Systemic Thinking for Policy Making: The Potential of Systems Analysis for Addressing Global Policy Challenges in the 21st Century, New Approaches to Economic Challenges. OECD Publishing, Paris, pp. 21–32.

Diesendorf, M., 2000. Sustainability and sustainable development. Chapter 2. In: Dunphy, D., Benveniste, J., Griffiths, A., Sutton, P. (Eds.), Sustainability: The corporate challenge of the 21st century Sydney. Allen and Unwin, Sydney, Australia, pp. 19–37.

Dougherty, H., 2017. Opinion: How agricultural modeling affects sustainability. James Beard Foundation, New York City.

du Pisani, J.A., 2006. Sustainable development—Historical roots of the concept. Environ Sci (Ruse) 3 (2), 83–96.

Easterling, W.E., Kok, K., 2002. Emergent properties of scale in global environmental modeling—Are there any? Integrated Assessment 3 (2–3), 233–246.

Ericksen, P., Stewart, B., Dixon, J., Barling, D., Loring, P., Anderson, M., … Ingram, J., 2010. The Value of a Food System Approach. In: Ingram, J., Ericksen, P., Livermore, D. (Eds.), Security and Global Environmental Change. Earthscan, London, pp. 25–45.

Faber, N., Jorna, R., Van Engelen, J., 2005. The sustainability of "sustainability"—A study into the conceptual foundations of the notion of "sustainability. Journal of Environmental Assessment Policy and Management 7, 1–33.

Fiksel, J., 2006. Sustainability and resilience: toward a systems approach. Sustainability: Science, Practice and Policy 2 (2), 14–21.

Folke, C., 2016. Resilience (Republished). Ecology and Society 21 (4), 30p.

Folke, C., Carpenter, S.R., Walker, B., Scheffer, M., Chapin, T., Rockström, J., 2010. Resilience thinking: integrating resilience, adaptability and transformability. Ecology and Society 15 (4), 1–9.

Food and Agriculture Organization of the United Nations (FAO), 2018. Sustainable food systems: Concept and framework. FAO Technical Brief, Rome.

Foran, T., Butler, R.A., Williams, L.J., Wanjura, W.J., Hall, A., Carter, L., … Carberry, P., 2014. Taking complexity in food systems seriously: an interdisciplinary analysis. World Development 61, 85–101.

Frank, R., 1988. Interdisciplinary": the first half century. Issues in Integrative Studies 6, 139–151.

Gold, M.V., 2007. Sustainable agriculture: definitions and terms. Special Reference Briefs Series. no. SRB 99-02. Updates SRB 94-05, 18pp. (September 1999; revised August 2007).

Goldsmith, E., Allen, R., Davull, J., Allaby, M., Lawrence, S., 1972. A Blueprint for Survival. Ecosystems Ltd, London.

Gray, R., 2010. Is accounting for sustainability actually accounting for sustainability…and how would we know? An exploration of narratives of organizations and the planet. Accounting. Organizations and Society 35, 47–62.

Goodman, M., 2018. Systems thinking: what, why, when, where and how? Systems Thinker, 7pp. Available at https://thesystemsthinker.com/systems-thinking-what-why-when-where-and-how/.

Grubinger, V., 2012. Resilience and sustainability in the food system. Resilient Design Institute, 3pp. Available at https://www.resilientdesign.org/resilience-and-sustainability-in-the-food-system/.

Gussow, J.D., Clancy, K.L., 1986. Dietary guidelines for sustainability. Journal of Nutrition Education 18 (1), 1–5.

Hasna, A.M., 2010. Sustainability classifications in engineering: discipline and approach. International Journal of Sustainable Engineering 3 (4), 258–276.

Hieronymi, A., 2013. Understanding systems science: a visual and integrative approach. Syst Res Behav Sci 30, 580–595.

High Level Panel of Experts, 2014. Food losses and waste in the context of sustainable food systems. A report by the High Level Panel of Experts on Food Security and Nutrition of the Committee on World Food Security, Rome.

Holling, C.S., 1973. Resilience and stability of ecological systems. Annu Rev Ecol Syst 4, 1–23 graygold.

Institute of Medicine (IOM) and National Research Council of the National Academies, 2015. A framework for assessing effects of the food system. The National Academies Press, Washington, DC.

Johnston, P., Everard, M., Santillo, D., Robert, K.-.H., 2007. Reclaiming the definition of sustainability. Environmental Science and Pollution Research 14 (1), 60–66.

Kelly, R., Mackay, M., Nash, K.L., Cvitanovic, C., Allison, E.H., Armitage, D., … Werner, F., 2019. Ten tips for developing interdisciplinary socio-ecological researchers. Socio-Ecological Practice Research 1, 149–161.

Kim, D.H., 1999. Introduction to systems thinking. Pegasus Communications, Inc, Waltham, MA.

Lengnick, L., 2015. Resilient Agriculture. New Society Publishers, Gabriola Island, BC, Canada.

Lew, A.A., Ng, P.T., Ni, C.-.C., Wu, T.-.C., 2016. Community sustainability and resilience: similarities, differences and indicators. Tourism Geographies 18 (1), 18–27.

Lexico.com, 2021. (Definitions of *resilience* and *sustainable*.)

Lockeretz, W., 1988. Open questions in sustainable agriculture. American Journal of Alternative Agriculture 3 (4), 174–181.

Marchese, D., Reynolds, E., Bates, M.E., Morgan, H., Clark, S.S., Linkov, I., 2018. Resilience and sustainability: similarities and differences in environmental management applications. The Science of the Total Environment 613-614, 1275–1283.

Marion, B.W., 1986. The organization and performance of the U.S. food system. Lexington Books, Lexington, MA.

Mason, P., Lang, T., 2017. Sustainable diets: How ecological nutrition can transform consumption and the food system. Routledge, London, pp. 1–21.

Meadows, D.H., Meadows, D.L., Randers, J., Behrens III, W.W., 1972. *The limits to growth*. A Report for the Club of Rome's Project on the Predicament of Mankind. Universe Books, New York.

Mensah, J., 2019. Sustainable development: meaning, history, principles, pillars and implications for human action: literature review. Cogent Social Sciences 5, 1–21.

Meuwissen, M.P.M., Feindt, P.H., Spiegel, A., Termeer, C., Mathijs, E., de Mey, Y., … Reidsma, P., 2019. A framework to assess the resilience of farming systems. Agricultural Systems 176, 1–10.

Miller, M., Holloway, W., Perry, E., Zietlow, B., Kokjohn, S., Lukszys, P., … Morales, A., 2016. Regional food freight: lessons from the Chicago region. Center for Integrated Agricultural Systems. Project report for USDA-AMS.. Transportation Division, City is Madison, WI.

Monat, J.P., Gannon, T.F., 2015. What is systems thinking? A review of selected literature plus recommendations. American Journal of Systems Science 4 (1), 11–26.

Müller, A., Sukhdev, P., 2018. Chapter 2: why eco-agri-food systems can only be understood with a systems perspective. The economics of ecosystems and biodiversity (TEEB). Measuring what matters in agriculture and food systems: A synthesis of the results and recommendations of TEEB for Agriculture and Food's scientific and economic foundations report. UN Environment, Geneva.

National Academies of Sciences, Engineering and Medicine., 2019. Science Breakthroughs to Advance Food and Agricultural Research by 2030. The National Academies Press, Washington, DC.

National Sustainable Agriculture Coalition. No date. What is sustainable ag? Available at: https://sustainableagriculture.net/about-us/what-is-sustainable-ag/.

Nicholson, C.F., Kopainsky, B., Stephens, E.C., Parsons, D., Jones, A.D., Garrett, J., Phillips, E., 2020. Conceptual frameworks linking agriculture and food security. Nature Food 1, 541–551.

Nicholson, C.F., Gómez, M.I., Gao, O.H., 2011. The costs of increased localization for a multiple-product food supply chain: dairy in the United States. Food Policy 36 (2), 300–310.

Peterson, C.A., Eviner, V.T., Gaudin, A.C.M., 2018. Ways forward for resilience research in agroecosystems. Agricultural Systems 162, 19–27.

Pinstrup-Andersen, P., Watson II, D.D., 2011. Food policy for developing countries: The role of government in global, national and local food systems. Cornell University Press, Ithaca, NY.

Purvis, B., Mao, Y., Robinson, D., 2019. Three pillars of sustainability: in search of conceptual origins. Sustainability Science 14, 681–695.

Resilience Alliance, 2015. Key concepts. Available at http://resalliance.org/resilience.

Rogers, K., 2021. What is systems thinking? bigThinking. Available at bigthinking.io/what-is-systems-thinking?/.

Richmond, B., 2018. The "thinking" in systems thinking: how can we make it easier to master?. Available at: Syst Thinker, 6pp.

Satterthwaite, D., 2006. Barbara Ward and the origins of sustainable development. International Institute for Environment and Development, London.

Siebrecht, N., 2020. Sustainable agriculture and its implementation gap—Overcoming obstacles for implementation. Sustainability 12 (9), 27pp.

Southern Region Sustainable Agriculture Research and Education, 2009. Common Ground, 2009. Perspectives on Systems Research. Autumn, Athens, GA.

Stock, P., Burton, R.J.F., 2011. Defining terms for integrated (multi-inter-trans-disciplinary) sustainability research. Sustainability 3, 1090–1113.

Szostak, R., 2007. How and why to teach interdisciplinary research practice. Journal of Research Practice 3 (2), 16pp.

Szostak, R., 2015a. Defining "interdisciplinary. University of Alberta, Department of Economics, Edmonton. Available at https://sites.google.com/a/ualberta.ca/rick-szostak/research/about-interdisciplinarity/definitions/defining-instrumental-interdisciplinarity .

Szostak, R., 2015b. History of disciplines and interdisciplinarity. Available at: https://sites.google.com/a/ualberta.ca/rick-szostak/research/about-interdisciplinarity/history-of-disciplines-and-interdisciplinarity

Tendall, D.M., Joerin, J., Kopainsky, B., Edwards, P., Shreck, A., Le, Q.B., ... Six, J., et al., 2015. Food system resilience: defining the concept. Glob Food Sec 6, 17–23.

Thompson, J.L., 2009. Building collective communication competence in interdisciplinary research teams. Journal of Applied Communication Research 37 (3), 278–297.

Thompson, P., 2010. The agrarian vision: Sustainability and environmental ethics. The University Press of Kentucky, Lexington, KY.

Ungar, M., McRuer, J., Liu, X., Theron, L., Blais, D., Schnurr, M.A., 2020. Social-ecological resilience through a biocultural lens: a participatory methodology to support global targets and local priorities. Ecology and Society 25 (3), 12pp.

United States Department of Agriculture–National Institute of Food and Agriculture (USDA-NIFA), 2015. Integrated programs application information: Thursday, February 5, 2015. Available at: https://nifa.usda.gov/resource/integrated-programs-application-information

United States National Commission on Food Marketing, 1966. Food from farmer to consumer. U.S. GPO, Washington D.C.

Vadrevu, K.P., Cardina, J., Hitzhusen, F., Bayoh, I., Moore, R., Parker, J., ... Hoy, C., 2008. Case study of an integrated framework for quantifying agroecosystem health. Ecosystems 11, 283–306.

van Berkum, S., Dengerink, J., Ruben, R., 2018. The food systems approach: Sustainable solutions for a sufficient supply of healthy food. Wageningen Economic Research, Wageningen.

Walker, B., Holling, C.S., Carpenter, S.R., Kinzig, A., 2004. Resilience, adaptability and social-ecological systems. Ecology and Society 9 (2), 9pp.

Williams, B., 2018. All methods are wrong. Some methods are useful. Systems Thinker, 22pp. Thesystemsthinker.com.

Science Education Resource Center at Carleton College (SERC), 2018. What Is Systems Thinking? InTeGrate (Interdisciplinary Teaching about Earth for a Sustainable Future). Available at https://serc.carleton.edu/190020.

Wilson, K.K., Morren, G.E.B., 1990. Systems Approaches for Improvement in Agriculture and Resource Management. MacMillan Publishing Company, New York.

Woodhill, J., 2019. The Dynamics of Food Systems—A Conceptual Model. Foresight4Food. International Collaborative Initiative, 5pp.

World Commission on Environment and Development (WCED), 1987. Our Common Future. United Nations, Oxford.

Zhang, W., Gowdy, J., Bassi, A.M., Santamaria, M., DeClerck, F., Adegboyega, A., ... Wood, S.L.R., 2018. Systems Thinking: an Approach for Understanding 'Eco-Agri-Food Systems. TEEB for Agriculture & Food: Scientific and Economic Foundations. UN Environment, Geneva, pp. 17–55.

# Life cycle assessment of food systems and diets

*Greg Thoma[a], Nicole Tichenor Blackstone[b],*
*Thomas Nemecek[c] and Olivier Jolliet[d]*

[a]Ralph E. Martin Department of Chemical Engineering, University of Arkansas, Fayetteville, AR, USA [b]Friedman School of Nutrition Science and Policy, Tufts University, Boston, MA, USA [c]LCA Research Group, Agroscope, Switzerland [d]Environmental Health Sciences, School of Public Health, University of Michigan, Ann Arbor, MI, USA

## 3.1 Introduction

The Anthropocene, the current period in which human activities are significantly affecting conditions on the planet is a time when the adage "you manage what you measure" is increasingly important. Our manipulation of and reliance on the natural environment for the provision of basic human needs like food are increasingly stressed by the growing demands of humanity driven by a combination of factors including population growth and an expanding and increasingly affluent global middle class. Many of the systems we have developed for provision of food are highly complex interconnected global networks. At each stage of the supply network, resources are used (e.g., water, materials, land, energy), substances are emitted to the environment (e.g., greenhouse gas emissions to the air, nutrient runoff to water), and waste is produced (e.g., packaging, uneaten food). We can account for all the resources used, pollution, and waste produced across a supply chain or life cycle of a food product to estimate its potential environmental impacts using life cycle assessment.

Life cycle assessment is a framework for quantifying the impacts of complex systems providing goods and services. As a tool, it enables evaluating system performance in support of better management towards the goal of sustaining food systems into the future.

In this chapter, we will introduce life cycle assessment (LCA), which focuses on the environmental impacts of products; Chapter 9 presents an alternate computational framework known as environmentally extended input output (EEIO) LCA, which focuses on whole economic sectors. We will use strawberry yogurt as a case study to illustrate major steps in an LCA

*Food Systems Modelling.*
DOI: https://doi.org/10.1016/B978-0-12-822112-9.00004-7

and to discuss both the capabilities and limitations of LCA as a framework for evaluating and comparing sustainability characteristics of foods and diets. We will also explore the two major LCA paradigms, attributional and consequential.

> **LCA has three main characteristics:**
>
> It considers the whole life cycle from the extraction of resources to disposal of any waste generated ("cradle-to-grave").
>
> It considers all relevant environmental impact categories, thus highlighting trade-offs and potential burden shifting burdens from one impact category to another. This distinguishes LCA from footprint methodologies, such as carbon footprint, ecological footprint, or water footprint, which only consider one impact category.
>
> The environmental impacts are set in relation to a unit of product, the so-called functional unit (see below).

## 3.2 A brief history of life cycle assessment

Life cycle assessment dates to the late 1960s when the Coca-Cola Corporation conducted an environmental evaluation of packaging, although the report was not made public. However, it was in the early 1990s that the field began moving towards evaluation of agriculture and food systems and efforts to standardize the practice were also initiated. These efforts were led by the Society for Environmental Toxicology and Chemistry (SETAC) and the United Nations Environment Program, various European research projects as well as the U.S. Environmental Protection Agency (USEPA). These organizations remain active today, with the Life Cycle Initiative hosted by the UN as a leading example (Koellner et al., 2013; Teixeira et al., 2016; UNEP, 2011; Verones et al., 2017). Standardization of the methodology was not formalized till publication of the International Organization for Standardization (ISO) 14,040 standard in 1997 (with subsequent updates) (ISO, 2006a, 2006b). More recently, other guidelines, largely based on the ISO series of standards, have been produced, notably the Product Environmental Footprint (PEF) guidelines (European Commission, 2013a), PAS 2050 (British Standards Institution, 2011) for greenhouse gas emissions, and a series focused on agriculture by the Food and Agriculture Organization of the United Nations (LEAP, 2018;2016;2015).

LCA has become widely regarded as one of the principal frameworks for evaluating the environmental sustainability characteristics of goods and services. It has been adopted by the European Union's PEF program as the basis for communicating sustainability metrics to consumers (European Commission, 2013b). In addition, a wide variety of industries from construction to agriculture have established LCA as the backbone of their environmental sustainability efforts. These efforts are often based in companies' sustainability reporting and serve multiple goals ranging from: internally identifying supply chain hotspots that can be targeted for improvement of environmental and potentially economic performance (e.g., opportunities to reduce energy use), establishing benchmarks to gauge future improvements against, and informing marketing efforts. Some policy analysts are beginning to consider how to incorporate environmental sustainability characteristics into dietary guidelines (Blackstone, El-Abbadi, McCabe, Griffin, and Nelson, 2018; Fischer et al., 2016; van Dooren, Marinussen, Blonk, Aiking, and Vellinga, 2014).

# Life Cycle of Strawberry Yogurt

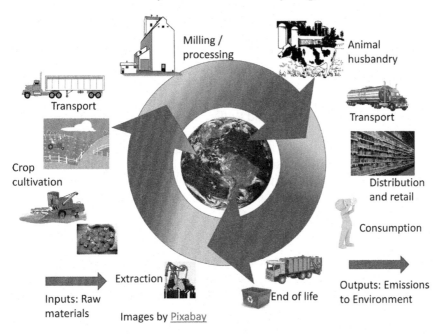

FIGURE 3.1    Schematic of life cycle stages included in an LCA.

## 3.3 The four phases of LCA

As mentioned, life cycle assessment is guided by a series of international standards, the ISO 14,040 series (ISO, 2006a, 2006b). The standards identify four phases of performing an LCA. These are 1) goal and scope definition, 2) life cycle inventory, 3) life cycle impact assessment, and 4) interpretation.

### 3.3.1  Phase 1: goal and scope

The goal of an LCA identifies its intended application, rationale, intended audience and whether a comparative assertion of superior performance (e.g., a product marketing claim) will be made. The scope of an LCA defines the system boundaries, functional unit, and impact categories to be considered. System boundaries are a set of criteria defining which activities or unit processes are part of the product system that delivers or provides the functional unit. The most comprehensive boundary is cradle-to-grave. Another common system boundary in LCA is cradle-to-gate, which sets the downstream boundary at either the farm or processing gate and excludes distribution, retail, consumption, and end-of-life activities (Fig. 3.1). A complete system boundary definition should also specify any ancillary activities which are expressly excluded from the system. Activities which are commonly excluded from the system are, for example, employee commuting, executive air travel, or financial services.

The functional unit is a key aspect that sets LCA apart from other methods. The function of a system describes what the system does or is for. This is often difficult to define for food systems research (see box). The functional unit is intended to represent the function of a system in quantitative terms and may be distinct from the reference flows of materials/energy required to produce the function (see box). Study results are scaled by the functional unit, and the functional unit provides the basis establishing a fair comparison across the systems being studied.

Appropriate functional units for food systems studies will vary depending on the goal and scope of the research (Heller, Keoleian, and Willett, 2013). For example, if the goal of the study is to compare dietary patterns, a functional unit of daily or standardized intake (2000 kcal per capita per day) may be appropriate. If the goal of the study is to compare disparate food items, the basis for a fair comparison across foods could be an index of nutrient quality (see nutrition and LCA section). If the goal of the study is to compare versions of a food made with different agricultural production or processing methods, such as our strawberry yogurt case, a mass or volume based functional unit is sufficient.

### Functional unit

As an example, the function of exterior paint is protecting the surface from the elements (with an added characteristic of aesthetics). Different quality paints may protect for different lengths of time; thus, the requirement of functional equivalence necessitates specification of the length of time of the protection and may require different amounts (reference flows) of different paints. Suppose two coats of paint requires one liter to cover the surface. If the low-quality paint's lifetime is five years and the high-quality paint's lifetime is 10 years, the reference flows necessary to provide the function of 10 years of protection would be one liter and for the high-and low-quality paints, respectively.

Food is more complex to characterize – is its function satiety? Nutrition? Enjoyment? Since no two foods are entirely equivalent, the challenge of comparative evaluation is significant.

Defining the life cycle impact categories relevant to the system under study is another aspect of the goal and scope definition phase of an LCA. Here, the practitioner should identify the classes of environmental impact which are most relevant in the context of both the system under study and the target audience's intended use of the results. It is important to note that the ISO standards do not support "cherry picking" of impact categories to avoid reporting on aspects that may not be favorable for the system under study. There are several impact assessment methods which have been critically evaluated by the LCA community and are generally considered to be suitable for use in life cycle assessment studies (European Commission and Joint Research Centre, 2010).

## 3.3.2 Phase 2: life cycle inventory (LCI)

### 3.3.2.1 *Unit process and databases*

The life cycle inventory (LCI) phase consists of data collection for the major contributing activities or stages of the supply chain. For a typical LCA study, we need hundreds or thousands of datasets. Our case study, for example, uses more than 13,000 datasets to calculate the resources used and emissions released over the whole life cycle. Why do we need so many? This is due to the manifold interrelationships in supply chains. To grow sugar beets, we need a tractor, machinery, fertilizers, fuels, pesticides, seed, and more. To produce mineral fertilizer, we need raw materials, such as phosphate rock or potassium chloride, fossil fuels, a factory, and electricity. To produce electricity, we need a power plant, built with cement, steel and other materials. Furthermore, all inputs need to be transported. This results in a complex network of interrelationships. To sum up all the natural resources consumed, and emissions released to the environment for a product's supply chain, we need comprehensive databases to provide datasets for the required inputs and processes across the full supply chain.

## Yogurt case study: goal and scope

The goal of our case study is to compare the potential environmental impacts of consuming one serving of strawberry yogurt, produced under five different scenarios. The intended audience for this case study is readers of this textbook. The results will not be used to support a comparative assertion to the public regarding which strawberry yogurt option has superior environmental performance.

The five strawberry yogurt production systems are as follows: Yogurt A includes milk sourced from dairies where the cows are primarily grain fed, the strawberries are produced in an open field farming system, and beet sugar is used as a sweetener. We have chosen a recipe with a relatively high fruit content of 13 percent and sugar content of 9 percent to accentuate the impacts of these ingredients for illustrative purposes. Yogurt B simply replaces the beet sugar in Yogurt A with cane sugar as the sweetener. Our third alternative (Yogurt C) replaces the grain-fed milk in Yogurt A with milk produced from a pasture-based dairy. Yogurt D replaces the open field strawberries in Yogurt A with strawberries grown in a heated greenhouse. This production method is included primarily for illustrative purposes, as greenhouse strawberries would not be commonly processed into a yogurt ingredient. Finally, we consider an additional scenario which uses a smaller quantity of strawberry (3.5 percent) and less sugar (6.1 percent) for the purpose of demonstrating the effect of composition on human health outcomes (Yogurt E).

We define our system boundary as cradle-to-grave. This includes all the necessary extractions of resources from nature, production of all inputs such as fertilizer, fuel, and electricity, necessary for agricultural production of yogurt ingredients, followed by processing and packaging for retail distribution, consumption, and disposal of packaging. The functional unit is the consumption of one serving (170 g) of strawberry yogurt.

We adopt the ReCiPe methodology which includes both midpoint and endpoint impact categories (described in more detail in a subsequent section).

For each input and process, the natural resources used, and emissions generated are calculated. These exchanges between the economic system (technosphere) and the environment (ecosphere) are called flows and can be goods, services, or natural resources. Goods and services (output of economic activities) consumed as inputs to an activity are often termed technosphere flows, while direct exchanges with the environment are termed elementary flows. Resources are the elementary flow inputs from the environment and emissions are the elementary flow outputs to the environment. Several hundreds or thousands of elementary flows can be recorded in an LCI at this stage. Examples of resources (elementary flow inputs) are mining activities, land- and water-use. Examples of emissions (elementary flow outputs) are the greenhouse gases (GHGs) such as $CO_2$, $CH_4$ and $N_2O$, and other emissions like ammonia ($NH_3$), nitrate ($NO_3^-$), phosphorus containing substances, toxic pollutants, and more. We distinguish emissions to the environmental compartments air, soil, and water, with sub-compartments such as surface freshwater (rivers and lakes), groundwater, and ocean.

Each activity that uses inputs and produces outputs in the supply chain is known as a unit process. Unit processes are linked together as a model of the supply chain by the flows of products that connect them. The final process produces the functional unit for the study. In the strawberry yogurt case, an example of these connections would be where the product of the dairy farm, milk, is an input into the yogurt manufacturing process. This set of linked unit processes is known as a life cycle inventory model and is constructed with the use of databases for the numerous inputs and processes (e.g., transportation, electricity production, waste treatment, etc.).

After more than two decades of development, we now have several comprehensive LCI databases. There are generic databases like ecoinvent, which cover the major economic sectors and provide some data for the agri-food sector. Examples of databases specific to the food sector are the World Food Life Cycle database (WFLDB), Agri-Footprint and the French Agri-Balyse database. Construction and maintenance of these databases is time consuming and expensive. Despite the existence of these databases, due to the multitude of food products and the various geographical regions and production systems, many gaps still exist, and techniques are needed to fill these gaps, as introduced in the following section.

### 3.2.2 Emission modeling

Estimating emissions in agricultural systems is challenging because of the strong dependence on natural resources, the high variability of soil, climate, topography and production systems, and the large number of often small production units (farms). We need data, models, and efficient tools for the calculation of emissions. An emission can be calculated by a simple emission factor. An example are the emissions of nitrous oxides ($N_2O$) from N fertilizers: the amount of N applied is multiplied by a default emission factor of 1 percent (Ogle et al., 2019), and subsequently converted from N to $N_2O$ by stoichiometric calculation. Such emission factors can also be differentiated by soil, climate, or production system. As an example, IPCC (Ogle et al., 2019) gives the values of 1.6 percent, 0.6 percent and 0.5 percent for synthetic fertilizer in wet climates, other N inputs in wet climates, and N inputs in dry climates, respectively. However, there are frequently situations where these simple factors are not sufficient to adequately describe the environmental mechanisms. Therefore, LCA practitioners often use complex, process-based models to calculate emissions. For example, the Integrated Farm System Model simulates dairy and beef production systems and includes simulation of all crop production activities (field preparation through harvest), biogeochemical cycling in the soil, ration formulation and animal performance (milk production and growth). It provides a detailed accounting of energy consumption, greenhouse gas emissions (from crop cultivation, animals, and manure), water consumption and reactive nitrogen losses (Kim et al., 2019; Veltman et al., 2018). The choice of model complexity depends on the goal and scope. Mechanistic models can simulate environmental processes; however, they often require a large number of input variables, which might not be available in many situations, or require adoption of default values that may not be fully representative of the situation under study. In LCA practice, we often use models of medium complexity (such as Tier 2 IPCC models), which are a good compromise between simple emission factors and complex process models.

### 3.3.2.3 *Data quality*

Even with the comprehensive databases, it is frequently not possible to find the specific dataset for some ingredient(s) from the country or region of origin and the given production system. Therefore, we need to simplify and approximate our data identification efforts to a certain extent. We need to consider three main data quality criteria:

- Temporal: do we have recent or outdated inventories?
- Spatial: do we have datasets from the country or region under consideration?
- Technological: do we have data for the technology under consideration? (e.g., organic vs. conventional, irrigated vs. rainfed, intensive vs. extensive, open field vs. greenhouse)

The data quality must be sufficient to satisfy the goal and scope of the study and answer our research question. For an ingredient, which is used only in very small quantities, rough approximations might be adequate, while we need high-quality data to model impacts of the main ingredients, which contribute most significantly to the environmental impacts. Database developers necessarily make decisions regarding the structure of the database, including decisions regarding multi-functionality and selection of inventory data. It is very important to note that when datasets from different databases are used, care must be taken to ensure their consistency; a harmonization effort is very likely to be needed to avoid unintended data quality effects due to different inventory modeling assumptions between databases. If, for example, we chose unit processes from different databases that each consume electricity in the same region, a harmonization check would require that the background electricity grid is the same for both selected processes – which could be different due to, for example, something as simple as the age of the databases. Using different background electricity introduces artificial differences between the processes. The practitioner should harmonize the datasets to remove this bias from the analysis.

## 3.3.3 The problem of multi-functionality

Unit processes frequently produce more than one product, yet it is desirable to distribute the impacts to the individual products. This is the problem of multi-functionality. Broadly speaking, two paradigms for solving multi-functionality exist: consequential and attributional. Consequential modeling intends to account for both direct and indirect environmental burdens from the activity. The indirect burdens are accounted through market-mediated effects in the broader economy, typically using system expansion, described below (UNEP, 2011). Attributional modeling allocates the burdens arising from multi-functional processes to individual products based on some attribute of the activity, such as revenue generated by each product or its energy content. The ISO standards provide an accounting hierarchy for managing multi-functionality. First is system separation followed by system expansion, then allocation based on physical causality, and finally allocation based on product characteristics (ISO, 2006a).

**System separation** relies on the ability to independently identify inputs and emissions from each individual product in a multifunctional unit process. This is not always possible (distiller's grains cannot be produced without also producing ethanol) and often not feasible (data are frequently only available for a full facility).

**System expansion** solves multi-functionality by substituting (mathematically taking a credit) for displacing the primary production of a product based on the amount of co-product made by the unit process; that is, the system under study is "expanded" to include separate processes that produce the co-products as their main product. This approach is sometimes also called "avoided impacts". Dairy farming, a multi-functional system that produces milk and meat, can be used as an example. Assuming the product of interest is milk, system expansion would give the dairy farming system a credit for the avoided production of meat. A fundamental consideration in selection of the credit is that the external market response will be through the marginal, not average production activity (Weidema, 2000; Weidema and Schmidt, 2010).

**Allocation** divides the total impact of a multi-functional unit process among the products based on physical causality or product attributes. This is an attributional approach and can take several forms. To illustrate, let us consider how we could allocate the environmental burdens of dairy farming between milk and meat. Thoma, Jolliet, and Wang (2013) and Nemecek and Thoma (2020) have proposed an approach based on the physiological requirements of cows to produce milk and body mass (meat) as an example of physical causality-based allocation. Other approaches to allocation between milk and meat on a dairy farm include simple product mass, protein or other nutrient contents, or revenue-based approaches.

The overarching goal of multifunctional accounting is to ensure that a mathematical solution to the system of Eqs. representing the supply chain exists. Although both consequential and attributional modeling solve the mathematical problem, and therefore, mathematically, can be applied simultaneously to different multi-functional processes in a product system, the conceptual basis for the two solutions is fundamentally different. The two approaches should not be combined in a single LCA to avoid complications in interpretation of the results. Nonetheless the literature is replete with examples of this kind of mixed modeling.

### 3.3.4 Phase 3: life cycle impact assessment (LCIA)

The impact assessment (LCIA) phase of LCA is typically performed with software tools designed to calculate the cumulative resource use and emissions across all the activities in the life cycle inventory model and characterize these emissions in terms of a smaller number of impact categories, such as climate change or eutrophication. The ISO standards require that life cycle impact assessment methods be based on a causal chain. For example, greenhouse gas emissions trap heat in the atmosphere, contributing to global warming, which can decrease

FIGURE 3.2   Midpoint and endpoint categories in the ReCiPe impact assessment framework. Arrows represent causal pathways; inputs to midpoint categories are the life cycle inventory (Huijbregts et al., 2017).

agricultural productivity and increase malnutrition (Fig. 3.2). In LCA, the impact categories we can analyze correspond to three areas of protection (AoP): human health, ecosystem quality, and resource conservation. Along the causal chain, there are different impact indicators, which are known as midpoint and endpoint impacts (Fig. 3.2). Midpoint impacts are assessed at some point in the causal chain between the emission or resource use and the final damage(s) they cause. Endpoint impacts correspond to the final damages for the three AoP.

The impact assessment phase consists of up to five steps - two required and three optional. First, in the *classification* step, the resources and emissions in the inventory are sorted into their corresponding impact categories. This stage is essentially assigning the flows of emissions and resources into the impact buckets to which they belong. For example, carbon dioxide ($CO_2$), nitrous oxide ($N_2O$) and methane ($CH_4$) all trap heat in the atmosphere and are therefore sorted into the climate change/global warming potential category.

**TABLE 3.1**  Ingredient list for yogurt varieties evaluated in the chapter case study. Quantities required to product 100 g of the final product, accounting for processing losses.

| Composition | A) Yogurt, standard high fruit content | B) Yogurt, with cane sugar | C) Yogurt, milk from grass-fed cows | D) Yogurt, with greenhouse strawberries | E) Yogurt, standard low fruit content |
|---|---|---|---|---|---|
| Milk, maize-fed | 80.7 | 80.7 | | 80.7 | 93.1 |
| Milk, grass-fed | | | 80.7 | | |
| Beet sugar | 9.0 | | | 9.0 | 6.1 |
| Cane sugar | | 9.0 | | | |
| Strawberries, open field, macrotunnels | 13.0 | 13.0 | 13.0 | | 3.6 |
| Strawberries, heated greenhouse | | | | 13.0 | |
| Total | 102.7 | 102.7 | 102.7 | 102.7 | 102.7 |

## Yogurt case study: life cycle inventory

For our case study of strawberry yogurt, the life cycle inventory encompasses all the inputs and outputs of the production of feed for cows, the dairy operation including animal husbandry, milking and manure management followed by processing of raw milk into yogurt, strawberry cultivation, and introduction to the yogurt, along with additional ingredients of packaging and then distribution to consumers. Food loss in the supply chain is also quantified and included. The processing stage inventory (yogurt recipe) for each of the five varieties of yogurt is presented in Table 3.1. The upstream production stage data, such as the rations consumed by dairy cows, milk and strawberry yields, and all other inputs are also included in the inventory (supplementary material). The multifunctionality of the milk production stage was adopted directly from the Agri-Balyse data sets for milk production which was based on a physical causality model. Other allocation in the model is based on the background datasets from Ecoinvent, which is based on revenue generation as the key.

Second, midpoint *characterization* quantifies the magnitude of impact. In the example of global warming potential, 1 kg of $CH_4$ is equivalent to 28 kg of $CO_2$ and 1 kg of $N_2O$ is equivalent to 265 kg $CO_2$ (Myhre et al., 2013). For each impact category, characterization converts the classified emissions or resource use flows to a common unit (e.g., $CO_2$-equivalent for global warming potential), and sums them up. The indicator units are different for most midpoint impact categories; therefore, we can compare only within an environmental category, but not between. An analogy is that if we had different currencies, but we did not know the exchange rates, we could not make price comparisons of commodities in different countries. Recent LCIA methods also include endpoint characterization factors, which aggregate all midpoint impact categories that contribute damage to a given area of protection, such as damage to human health shown in Fig. 3.2 (Huijbregts et al., 2017). Damages to the same AoP can be summed up or directly compared, but not across AoPs (as noted by the color coding in Fig. 3.2). Finally, it is important to note that classification and characterization occur within LCA software tools; LCA practitioners do not commonly compute these steps by hand.

The next three steps in impact assessment are optional due to the additional uncertainty and subjectivity that is introduced. *Grouping* is not commonly applied and means that similar impact categories are grouped according to certain criteria.

*Normalization* typically compares the environmental impact of the food product to the environmental impact of an average person from a given area during a specified time period. This allows expressing the different impact categories with a common unit typically as a fraction of the per capita impact for that impact category, which makes them comparable. However, these results still do not tell us which impact category is the most important when it comes to decision-making, but they can indicate which impact categories are relative hotspots.

The final step is *weighting*, which can be applied to indicate the relative importance of impact categories at the midpoint or endpoint levels. In the impact assessment method used in our case study (ReCiPe (H) 2016), there are three endpoint impacts, which have different weights that can be applied in the default method: 40 percent for ecosystem quality, 40 percent for human health, and 20 percent for resource availability. Weight sets can be generated using different approaches, such as surveying a panel of experts or using distance to science-based targets (Pizzol et al., 2017). Weighting is based on value choices and is the most subjective LCIA step. Normalized and weighted results can be combined to produce a single score, which can be helpful for decision-making and applications like eco-labeling. It is important to note, however, that an LCA that includes these optional LCIA steps should also provide raw results (i.e., unnormalized, unweighted) and/or normalized or weighted results still differentiated by impact category (as discussed below) for transparency, according to ISO standards.

In addition to the method adopted for this case study (ReCiPe 2016), there are numerous other LCIA methods commonly used by practitioners (Bulle et al., 2019; European Commission et al., 2011; Frischnecht and Jolliet, 2016; Jolliet et al., 2003; Verones et al., 2020). Additional information for impact assessment of water and land use is provided chapters 4 and 5 of this book.

### 3.3.5 Phase 4: interpretation of the assessment

The interpretation of the impact assessment includes identification of hotspots in the supply chain, that is, activities which are particularly impactful. Interpretation can be used to recommend improvements, identify trade-offs, establish a baseline benchmark for a service or product and support policy recommendations. It should also include an uncertainty analysis (see below). In short, LCA provides a quantitative view of the system providing our food and can identify resource use, pollution and waste considering the full production and consumption life cycle.

Providing quantitative estimates of systems' performance also reduces reliance on intuition. As an example, many consumers believe that packaging is a significant driver of the sustainability of food products. In fact, for a large majority of foods, packaging plays a relatively minor role in terms of direct contribution to environmental impacts (Williams and Wikström, 2011) (Fig. 3.3), but plays an important beneficial role in terms of preservation (reducing food loss) and safety. However, for some food categories like beverages, packaging can be quite relevant. Many consumers also believe that transportation has a major contribution to foods'

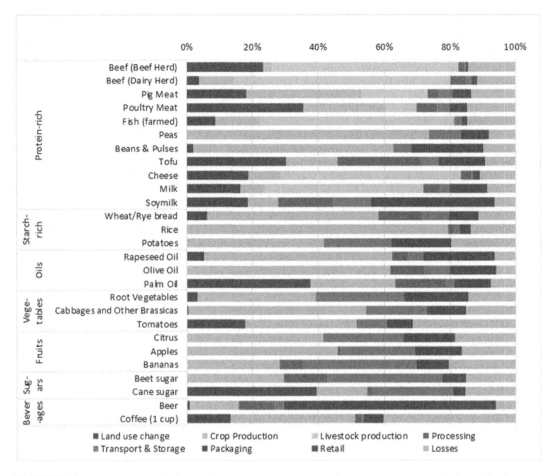

**FIGURE 3.3**  Contribution of different life cycle stages to the greenhouse gas emissions of selected food groups (Meyer et al., 2020). Reproduced with permission.

environmental impacts. While this is true for air freight and fruits and vegetables transported over longer distances, transport is of low importance for food groups like meat. The impacts of transportation are furthermore dependent on the mode of transport and decrease in the order of air freight, lorry, railway, inland water transport, and transoceanic freight ship. Therefore, domestic products are not necessarily more environmentally friendly than imported food; frequently the impacts are more dependent on the production system.

## 3.4  Yogurt case study: LCIA result and interpretation example at midpoint

Returning to our strawberry yogurt case study, we will provide some examples of impact assessment results and their interpretation. Fig. 3.4 shows how the production system con-

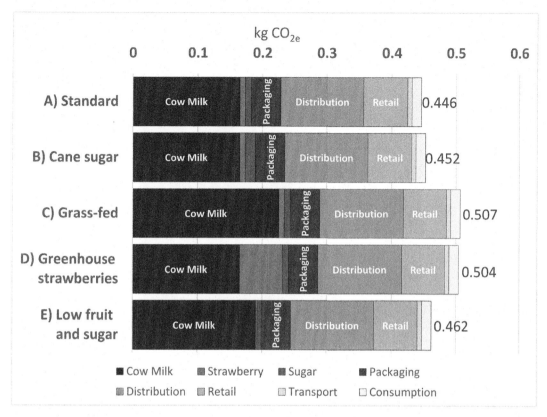

**FIGURE 3.4** Global warming potential for 100 years (IPCC 2013 methodology) for alternative strawberry yogurt products.

tributes to global warming potential impact. There are several interesting comparisons to note. First, we can explore the influence of sugar sourcing. Although sugar is not a major contributor to the overall impact, in our example, the climate change impacts of cane sugar are almost twice as high as of beet sugar (in Yogurt B versus A). Why might that be? We know from our LCI research that sugar cane is primarily produced in tropical regions. While the crop is very productive, in some cases sugar cane plantations have been established on land previously covered by tropical rain forest and cleared to grow crops (see also "Land use change" in Fig. 3.3). In such situations, the emissions from land clearing and the subsequent decomposition of organic matter cause greenhouse gas emissions ($CO_2$ and $N_2O$) that are much higher than the emissions from the sugar production itself. This is a major driver of the differences in climate impact between beet and cane sugar sourcing.

Let us next consider the alternate production of milk from pasture-based systems (Yogurt C versus A). Again, from our LCI research, we know that pasture-based cows have typically lower milk yields. In addition, feeding roughage tends to produce more enteric methane emissions compared to feeding grains. These effects typically lead to higher GHG emissions per kg milk for pasture-based production.

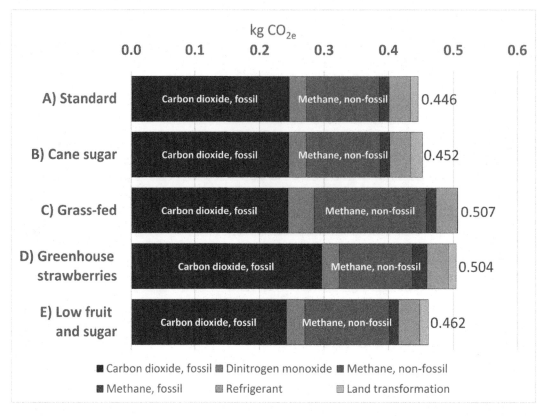

FIGURE 3.5    Contribution of various greenhouse gases across the full supply chain.

Across all yogurt systems studied, processing adds some emissions, but the distribution phase is a quite significant driver of impact, due to refrigerated transport and storage. Consumer transport is included in the consumption phase but is less important in terms of impact in this example than in-home refrigeration.

Now, let us take a different view of contributions to global warming potential impact for the five yogurt systems. We can explore which greenhouse gasses are the major contributors to global warming impact across the yogurt supply chains (Fig. 3.5). We can see that carbon dioxide is the dominating GHG contributing 55 percent in Yogurt A, followed by $CH_4$ (29 percent), refrigerants (7 percent) and $N_2O$ (6 percent). In Yogurt B the share of $CO_2$ from land transformation is slightly increased, due to deforestation for sugar cane production, while it is decreased in system C because the grass-fed milk does not include soy products from South America. Non-fossil methane increases in Yogurt C, due to higher enteric emissions in the pasture-based system, while $CO_2$ becomes even more important in system D, due to heating of the strawberry greenhouse with fossil fuels.

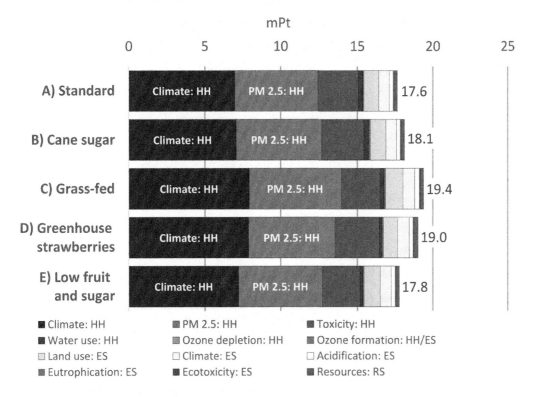

**FIGURE 3.6** Weighted endpoint impact per serving of strawberry yogurt.

## 3.5 Yogurt case study: LCIA results and interpretation at endpoint

We have explored in Fig. 3.4 and 3.5 the global warming scores as an example of midpoint impact results for our yogurt case study. We would also be able to analyze multiple impact categories side by side (e.g., global warming potential, eutrophication potential, and land use) at the midpoint level to identify tradeoffs and support decision-making. Our analysis could - and in many cases in food LCAs, does - stop there. Here, we take the results a few steps further. We also present LCIA results at the endpoint level, which have been normalized, weighted, and combined into a single score (Fig. 3.6). Remember, normalization and weighting are optional LCIA steps that can be helpful for decision-making.

Fig. 3.6 presents a normalized and weighted single score for each yogurt system. The score is reported in milli-points. One point represents the fractional contribution of one EU citizen in one year to the combined endpoint categories. The contribution of each weighted impact category to the single score is shown in the legend. As a reminder, these endpoint impacts correspond to damages to one or more Areas of Protection (AoP). Climate change, for example, corresponds to the AoP of human health (Climate: HH, in Fig. 3.6) and ecosystem quality (Climate: ES, in Fig. 3.6). The correspondence of the impact categories to each AoP is also shown in the Y axis of Fig. 3.7.

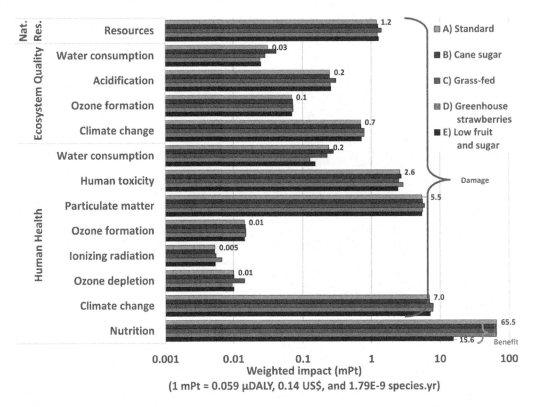

FIGURE 3.7    Endpoint contribution to natural resource, ecosystem quality, and human health impacts. Comparison with nutritional benefits (bottom bars) per serving of yogurt with high and low fruit content. The dietary health benefits are notably larger than the adverse impacts shown for all the other categories (remaining bars). Note the axis is logarithmic.

In the legend, HH = human health; ES = ecosystems; RS = resource availability.

Comparing the overall results for our five yogurt systems, Yogurt C has the highest overall environmental impact, while Yogurt A has the lowest (Fig. 3.6). At the same time, the differences between all systems are relatively small; they may not be significantly different once we consider the uncertainty of our results (see uncertainty in LCA section). The major factors contributing to the larger impact of Yogurt C are the climate change and particulate matter effects on human health followed by land use impact on ecosystems. This follows directly from the lower milk yield from pasture-based milk production.

For each yogurt system, the weighted impact of climate change (CC) and particulate matter on human health (Fig. 3.6) are the greatest contributors to overall system impacts. The red bar (ozone formation) separates the human health related impacts from ecosystem and resource depletion impacts in the single score. Concerning the impacts to ecosystems, land use is the largest, followed by climate change and terrestrial acidification. The standard yogurt with high fruit content (A) has the lowest impact on human health (excluding nutrition which is discussed below) while the grass-fed and greenhouse strawberry scenarios (C/D) are the highest, due to enteric methane and more fossil fuel consumption contributing to climate change and particulate matter formation. Yogurt (A) also has marginally lower impacts on ecosystems compared to the grass-fed milk-based yogurt (C) which has higher land use impacts. There is relatively little difference for the resource use categories. Yogurt (A) and the cane sugar yogurt (B) have the lowest impacts, while yogurt (D) as the highest resource consumption, from fossil fuel consumption to heat the greenhouse.

Since the uncertainty of the endpoint categories shown in Fig. 3.7 is high, it is in general preferable to report the weighted results differentiated by impact category. Comparing endpoint impacts across the five yogurt scenarios shows that the overall impact profiles are quite similar across all categories. Fig. 3.7 shows us the differences between scenarios for each impact category. We observe patterns of impact that are more similar across scenarios than are the differences between impact categories. For example, climate change and particulate matter impacts to human health are the top two contributing categories for all scenarios.

## 3.6 Including nutritional benefits and impacts in LCA

While human health is one of the three AoP of focus in LCA, paradoxically, the human health impacts and benefits of consuming foods (part of the use or consumer phase of an LCA) have not historically been included in food LCAs. Recent research has changed this, allowing us to incorporate these nutritional impacts into LCA. Building on the Global Burden of Disease (GBD) studies in nutrition epidemiology that have identified 15 risk factors contributing to detrimental or beneficial effects of foods on health, Stylianou et al. (2016), 2021) proposed an approach to translate the GBDs population-based impacts to the level of individual, complex foods. Based on the composition of individual foods coupled with the 15 GBD dietary risk factors (Fulgoni, Wallace, Stylianou, and Jolliet, 2018), nutrition impacts have been calculated with units of micro-Disability Adjusted Life Years (DALYs) per serving for 5000+ items found in the US and the Swiss diet (Ernstoff et al., 2020).

1 DALY is approximately equivalent to half a minute, and benefits and impacts can thus be expressed as minutes of life gained or lost per serving size via the Health Nutritional Index (HENI).

Fig. 3.8 illustrates the HENI index scores for Yogurt A (high strawberry content of 13 percent) versus Yogurt E (low strawberry content of 3.5 percent). Three risk factors contribute to most of the nutritional impacts and benefits. On the detrimental side, the sodium content of 0.10 g sodium/yogurt serving leads to 0.75 min of healthy life lost. This is largely compensated by the calcium (+0.7 min. gained per serving) and the fruit (+2.1 min. gained per serving for high strawberry yogurt A), yielding a net HENI score of 2.05 min of healthy life gained per serving for Yogurt A. Reducing the fruit content to 3.5 percent reduces the net benefit per serving down to only 0.5 min gained.

Since these scores can also be expressed in DALY, they can be combined as weighted, normalized endpoint damages and benefits (note that the nutritional benefit in Fig. 3.7 is presented as the absolute). In Fig. 3.7, the bottommost two bars represent the net nutritional benefit for the standard and low fruit/sugar alternates. The details of the specific components of the dietary contribution leading to a net health benefit is presented in Fig. 3.8. This shows that the direct nutritional impacts are substantial, exceeding environmental impacts by over a factor of 3 for the high fruit content options (65.5 mPt benefit vs. an average of 18.3 mPt damage from environmental impact), while the dietary benefit from the low-fruit yogurt (E)

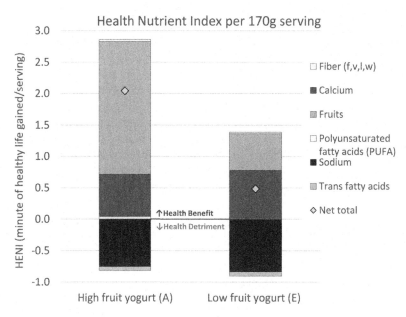

FIGURE 3.8  Human health burden by dietary risk in minutes of healthy life per serving of strawberry yogurt. Purple diamonds represent the net Health Nutritional Index (HENI) per serving.

nearly offsets the adverse environmental impacts. It is therefore crucial to consider nutrition in LCA when analyzing trade-offs between impact categories and scenarios over the entire life cycle.

### 3.6.1  Which yogurt is better?

Among these five options, which yogurt is the preferred choice from a life cycle perspective? It is not sufficient to base our analysis only on climate change (Fig. 3.4). As we stated, one of the primary strengths of LCA is that it provides a multidimensional picture of systems' performance. With LCA, we can (and should!) estimate multiple environmental impacts of systems (see Fig. 3.6). By including multiple impact categories and a life cycle perspective, the LCA framework enables an understanding of trade-offs between relevant impacts and across supply chain stages. Although, in this case study, there are not significant tradeoffs, by quantifying trade-offs we can identify and avoid burden shifting and other unintended consequences.

## 3.7  Uncertainty in LCA

The challenge of constructing a representative life cycle inventory model from multiple sources (measured, estimated or modeled) is generally the most time consuming and difficult stage of performing an LCA. There are several considerations regarding the accuracy of the data quality and model construction which are directly relevant for evaluating the robustness of conclusions from any LCA study. Uncertainties in LCA modeling arise from the following sources:

- Inherent variability of activities (e.g., inter-annual differences in milk or strawberry yields)
- Epistemic uncertainty, or the absence of knowledge or data for an activity known to contribute (an unknown amount of a pesticide used in strawberry cultivation) or missing a contributing activity altogether (exclusion of a pesticide which is used).
- Uncertainty in the causal chain defining the midpoint and endpoint impacts, that is uncertainty in the impact assessment method's characterization factors.

LCA is frequently used for benchmarking, product comparison, and setting policy. The uncertainty inherent in LCA modeling requires that practitioners consider the potential for reaching an incorrect conclusion. The most common approaches used in the field are a combination of uncertainty and sensitivity analyses, which together can provide insight regarding which activities require higher quality data and whether the conclusions are sufficiently robust to support the goal and scope of the study.

## 3.8 Sensitivity analysis

From a practical perspective, it is not feasible to conduct a sensitivity evaluation of all the parameters contributing to all the activities in a complete supply chain. Thus, one approach is to perform a one at a time evaluation of the contributing activities in the foreground system based on the fractional change in an impact category resulting from a specified fractional change in the input parameter value. Some authors choose a standard plus or minus 10 percent variation in the input value; however, a better approach is to base this on the expected range of variation in the input parameter. In the strawberry yogurt case, because milk represents a significant contribution to system impacts, it is logical that an error in the inventory (i.e., the quantity used) of milk used in the yogurt will lead to a significant error in the result, thus milk quantity is a sensitive parameter for which uncertainty must be reduced. In this instance, simple inspection is sufficient to identify milk as highly sensitive input parameter; in other cases, formal testing, as described above is needed.

### 3.8.1 Monte carlo analysis

Many life cycle inventory databases have included estimates of inventory uncertainties which can be used to support Monte Carlo Simulations (MCS) - a diagnostic tool to assess robustness of conclusions from an LCA study. More specifically, MCS is a method estimating the uncertainty of the results of a complex model based on known or estimated variability of input data. For example, the quantity of electricity consumed in many of the unit processes contributing to a supply chain will not be known with high precision (for example, the electricity usage of in-home refrigeration may have a mean value of 5 kWh and coefficient of variation of 20 percent). MCS randomly selects an input from the distribution (for the electricity and potentially thousands of other inventory flows), and in successive calculations for the functional unit, will provide the range of calculated impacts (for example, climate change) which can then be used for statistical inference tests.

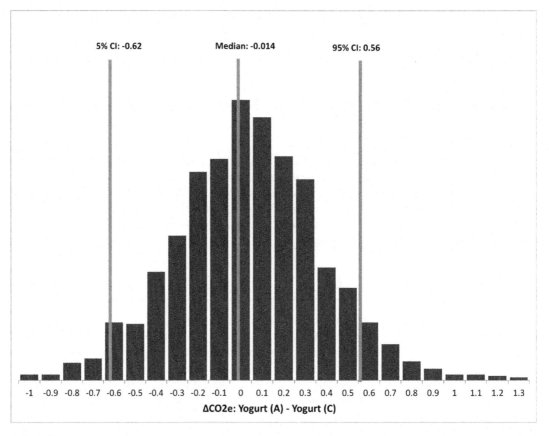

FIGURE 3.9  Pairwise comparative climate change impact of yogurt A minus yogurt C. The negative median indicates that C has slightly larger climate change impact, however it is not a statistically significant difference given the confidence intervals (CI).

Most standard LCA software provide a mechanism for Monte Carlo simulation. Each of the activities for which uncertainty information is known, or can be estimated, is assigned a mean value and a probability distribution. The software will then perform the specified number of complete simulations for the product or products to produce a distribution of impact values from which the statistical likelihood of a difference between systems can be inferred. In making product comparisons or longitudinal comparisons against a benchmark, it is imperative that shared background processes are matched in each of Monte Carlo simulations. This helps to prevent the inflation of variance between systems and provides for a more robust comparison.

Fig. 3.9 presents the results of 1500 Monte Carlo simulations comparing Yogurt A and Yogurt Cs climate change results. Each simulation selects the same background LCI data so that the difference between the two systems' impacts is not artificially inflated because of the different choices of, for example, electricity emissions in the upstream supply chain. In this case, although the median is slightly negative, the wide confidence interval indicates that the difference is not statistically significant at the 5 percent confidence level, meaning that these alternative yogurt systems yield very similar carbon footprint results.

The combination of sensitivity and Monte Carlo analysis is used to test the robustness of conclusions in comparative studies. If the uncertainty in the input data do not obscure the differences in the output results, the conclusions are considered robust.

## 3.9 Gaps and further research needs

LCA is a widely adopted framework for evaluation of food system sustainability characteristics, but there are limitations to the methodology that should be understood. One important point to understand is that most impact assessment frameworks do not have geospatially or temporally explicit characterization factors. This means, for example, that phosphorous emissions from fertilizer or manure application to crops from two different regions will be assigned the same potential eutrophication (water quality) impact – even if one emission source is near a sensitive waterbody and the other is not; the "true" impact will not be the same at the local level. This is a consequence of the geospatial scale (continental to global) for most characterization factors commonly used in life cycle assessment. There are two notable exceptions. The first is the Aware method for water scarcity which provides both geospatially (watershed level) and temporally explicit (seasonal) characterization that accounts for the local relative availability of water. The second is a biodiversity impact methodology proposed by Chaudhary, Pfister, and Hellweg (2016) and Chaudhary and Brooks (2018). The Aware LCIA method (Boulay et al., 2015 Boulay, Hoekstra, and Vionnet, 2013) is an alternative approach to the water footprint method presented in Chapter 4. The biodiversity method is linked to land use and land use change which is the topic of Chapter 5.

Additional impact categories are currently the subject of significant research in the field. These include ecosystem services, biodiversity, antibiotic resistance, and soil quality (Bakshi and Small, 2011; Chaplin-Kramer et al., 2017; Chaudhary and Brooks, 2018; Chaudhary et al., 2016; Cowell and Clift, 2000; De Laurentiis et al., 2019; He et al., 2016). While there are some initial and directionally correct estimates for several of these categories, more work is needed.

There are also erroneous extensions of life cycle assessment results, for example, if results obtained for a product or service are applied at a scale beyond that chosen for the study. This is because of a foundational assumption of life cycle assessment that the systems modeled are marginal in a macro-economic sense; specifically, the effects modeled are not large enough to affect the background economy. Thus, one cannot extrapolate the large-scale consequences of, for example, eliminating dairy products from the US diet based on a study comparing a serving of a plant-based beverage versus fluid milk. Elimination of the dairy sector would have far reaching effects that are not captured by a simple product LCA. In such cases, LCA should be coupled with general equilibrium modeling where such large shocks to the economy can be simulated providing a completely new background economy in which to contextualize the conclusions.

# 3.10 Conclusions

LCA provides a systems framework for evaluating and comparing food systems. It is designed to quantify potential impacts and identify tradeoffs between actors in the supply chain as well as among different facets of environmental concern. We have shown that even for simple foods there can be notable differences arising from different production systems. In addition, determining an appropriate function for food remains a challenge. Some researchers propose inclusion of nutritional quality in the functional unit (Saarinen, Fogelholm, Tahvonen, and Kurppa, 2017), while others argue it should be considered in impact assessment, as we have done in this chapter (Weidema, 2018). Further there are shifts in sources of emissions which may seem to be intuitively beneficial but may not provide the expected result after careful quantification using the LCA framework. As a simple example, it is possible that environmentally friendly reduction in food packaging material quantity may in fact result in larger food loss since a primary function of packaging is food preservation leading to larger negative impacts from the loss of the food.

LCA can also provide the basis for product or system comparison and benchmarks for longitudinal studies demonstrating continual improvement. However, in these applications caution in interpretation is needed because there are pitfalls in performing LCA which can lead to the results from different studies/authors not being directly comparable – for example, mixed system boundaries or background databases can cause incomparability issues.

LCA is rapidly becoming a prominent tool in the assessment of sustainable food systems and holds great potential, particularly when used with other modeling tools described in this text. While the capability of LCA is not limited in principle, as discussed in the uncertainty section, the robustness of conclusions, especially comparative assertions of superior performance must be evaluated in the context of the available data.

## Acknowledgements

We thank Katerina Stylianou for providing the underlying data to calculate the Health Nutritional Index.

## References

Bakshi, B., Small, M.J., 2011. Incorporating Ecosystem Services Into Life Cycle Assessment. Journal of Industrial Ecology 15, 477–478. https://doi.org/10.1111/j.1530-9290.2011.00364.x.

Blackstone, N.T., El-Abbadi, N.H., McCabe, M.S., Griffin, T.S., Nelson, M.E., 2018. Linking sustainability to the healthy eating patterns of the Dietary Guidelines for Americans: a modelling study. The Lancet Planetary Health 2, e344–e352. https://doi.org/10.1016/S2542-5196(18)30167-0.

Boulay, A.-.M., Bare, J., De Camillis, C., Döll, P., Gassert, F., Gerten, D., et al., 2015. Consensus building on the development of a stress-based indicator for LCA-based impact assessment of water consumption: outcome of the expert workshops. International Journal of Life Cycle Assessment 20, 577–583. https://doi.org/10.1007/s11367-015-0869-8.

Boulay, A.-.M., Hoekstra, A.Y., Vionnet, S., 2013. Complementarities of Water-Focused Life Cycle Assessment and Water Footprint Assessment. Environmental Science & Technology 47, 11926–11927. https://doi.org/10.1021/es403928f.

British Standards Institution, (2011). PAS 2050:2011: Specification for the assessment of the life cycle greenhouse gas emissions of goods and services

Bulle, C., Margni, M., Patouillard, L., Boulay, A.-.M., Bourgault, G., De Bruille, V., et al., 2019. IMPACT World+: a globally regionalized life cycle impact assessment method. International Journal of Life Cycle Assessment 24, 1653–1674. https://doi.org/10.1007/s11367-019-01583-0.

Chaplin-Kramer, R., Sim, S., Hamel, P., Bryant, B., Noe, R., Mueller, C., et al., 2017. Life cycle assessment needs predictive spatial modelling for biodiversity and ecosystem services. Nature Communications 8, 15065–15073. https://doi.org/10.1038/ncomms15065.

Chaudhary, A., Brooks, T.M., 2018. Land Use Intensity-Specific Global Characterization Factors to Assess Product Biodiversity Footprints. Environmental Science & Technology 52, 5094–5104. https://doi.org/10.1021/acs.est.7b05570.

Chaudhary, A., Pfister, S., Hellweg, S., 2016. Spatially Explicit Analysis of Biodiversity Loss Due to Global Agriculture, Pasture and Forest Land Use from a Producer and Consumer Perspective. Environmental Science & Technology 50, 3928–3936. https://doi.org/10.1021/acs.est.5b06153.

Cowell, S.J., Clift, R., 2000. A methodology for assessing soil quantity and quality in life cycle assessment. Journal of Cleaner Production 8, 321–331. https://doi.org/10.1016/S0959-6526(00)00023-8.

De Laurentiis, V., Secchi, M., Bos, U., Horn, R., Laurent, A., Sala, S., 2019. Soil quality index: exploring options for a comprehensive assessment of land use impacts in LCA. Journal of Cleaner Production 215, 63–74. https://doi.org/10.1016/j.jclepro.2018.12.238.

Ernstoff, A., Stylianou, K.S., Sahakian, M., Godin, L., Dauriat, A., Humbert, S., et al., 2020. Towards Win–Win Policies for Healthy and Sustainable Diets in Switzerland. Nutrients 12, 2475–2499. https://doi.org/10.3390/nu12092745.

European Commission, 2013a. Building the Single Market for Green Products: facilitating better information on the environmental performance of products and organisations. Communication From the Commission to the European Parliament and the Council. CELEX:52013DC0196.

European Commission, 2013b. Recommendations on the use of common methods to measure and communicate the life cycle environmental performance of products and organisations. Official Journal of the European Union 56.

European Commission, 2010. Joint Research Centre. ILCD handbook: General guide for life cycle assessment : Detailed guidance. Publications Office of the European Union, Luxembourg.

European Commission, Joint Research Centre, Institute for Environment and Sustainability, 2011. International reference life cycle data system (ILCD) handbook general guide for life cycle assessment: Provisions and action steps. Publications Office, Luxembourg.

Fischer, C.G., Garnett, T., 2016. Plates, pyramids, and planets: Developments in national healthy and sustainable dietary guidelines : A state of play assessment. Food and Agriculture Organization of the United Nations, University of Oxford, Food Climate Research Network.

Frischknecht, R., Jolliet, O., 2016. Global Guidance for Life Cycle Impact Assessment Indicators, 1. UNEP, Paris, France No. Vol..

Fulgoni, V., Wallace, T., Stylianou, K., Jolliet, O., 2018. Calculating Intake of Dietary Risk Components Used in the Global Burden of Disease Studies from the What We Eat in America/National Health and Nutrition Examination Surveys. Nutrients 10, 1441–1457. https://doi.org/10.3390/nu10101441.

He, L.-.Y., Ying, G.-.G., Liu, Y.-.S., Su, H.-.C., Chen, J., Liu, S.-.S., et al., 2016. Discharge of swine wastes risks water quality and food safety: antibiotics and antibiotic resistance genes from swine sources to the receiving environments. Environment International 92, 210–219. https://doi.org/10.1016/j.envint.2016.03.023.

Heller, M.C., Keoleian, G.A., Willett, W.C., 2013. Toward a Life Cycle-Based, Diet-level Framework for Food Environmental Impact and Nutritional Quality Assessment: a Critical Review. Environmental Science & Technology 47, 12632–12647. https://doi.org/10.1021/es4025113.

Huijbregts, M.A.J., Steinmann, Z.J.N., Elshout, P.M.F., Stam, G., Verones, F., Vieira, M., et al., 2017. ReCiPe2016: a harmonised life cycle impact assessment method at midpoint and endpoint level. International Journal of Life Cycle Assessment 22, 138–147. https://doi.org/10.1007/s11367-016-1246-y.

ISO, 2006a. Environmental management — Life cycle assessment — Requirements and guidelines (No. ISO 14044:2006(E)), Environmental Management. International Organization for Standardization, Geneva.

ISO, 2006b. Environmental management — Life cycle assessment — Principles and framework. International Organization for Standardization, Geneva, Switzerland.

Jolliet, O., Margni, M., Charles, R., Humbert, S., Payet, J., Rebitzer, G., et al., 2003. IMPACT 2002+: a new life cycle impact assessment methodology. Int J LCA 8, 324–330. https://doi.org/10.1007/BF02978505.

Kim, D., Stoddart, N., Rotz, C.A., Veltman, K., Chase, L., Cooper, J., et al., 2019. Analysis of beneficial management practices to mitigate environmental impacts in dairy production systems around the Great Lakes. Agricultural Systems 176, 102660–102672. https://doi.org/10.1016/j.agsy.2019.102660.

Koellner, T., de Baan, L., Beck, T., Brandão, M., Civit, B., Margni, M., et al., 2013. UNEP-SETAC guideline on global land use impact assessment on biodiversity and ecosystem services in LCA. International Journal of Life Cycle Assessment 18, 1188–1202. https://doi.org/10.1007/s11367-013-0579-z.

LEAP, (2018). Environmental performance of pig supply chains: guidelines for assessment

LEAP, 2016. Environmental performance of large ruminant supply chains: gudelines for Assessment, Livestock Environmental Assessment and Performance Partnership. Food and Agriculture Organization of the United Nations. Rome, IT.

LEAP, 2015. Environmental performance of animal feeds supply chains: Guidelines for assessment. Food and Agriculture Organizaiton of the United Nations, Rome, Italy.

Myhre, G., Shindell, D., Bréon, F.-.M., Collins, W., Fuglestvedt, J., Huang, J., et al., 2013. Anthropogenic and Natural Radiative Forcing, in: climate Change 2013: the Physical Science Basis. Contribution of Working Group I to the Fifth Assessment Report of the Intergovernmental Panel on Climate Change. Intergovernmental Panel on Climate Change 8, 659–740.

Nemecek, T., & Thoma, G. (2020). Allocation between milk and meat in dairy LCA: critical discussion of the International Dairy Federation's standard methodology 4.

Ogle, S.M., Wakelin, S.J., Buendia, L., Mc Conkey, B., Baldock, J., Akiyama, H., Kishimoto-Mo, A.W., Chirinda, N., Bernoux, M., Bhattacharya, S., et al., 2019. Cropland-Chapter 5 in: 2019 Refinement to the 2006 IPCC Guidelines for National Greenhouse Gas Inventories, Chapter 5. IPCC, Hayama. p 1–102.

Pizzol, M., Laurent, A., Sala, S., Weidema, B., Verones, F., Koffler, C., 2017. Normalisation and weighting in life cycle assessment: quo vadis? International Journal of Life Cycle Assessment 22, 853–866. https://doi.org/10.1007/s11367-016-1199-1.

Saarinen, M., Fogelholm, M., Tahvonen, R., Kurppa, S., 2017. Taking nutrition into account within the life cycle assessment of food products. Journal of Cleaner Production 149, 828–844.

Stylianou, K.S., Heller, M.C., Fulgoni, V.L., Ernstoff, A.S., Keoleian, G.A., Jolliet, O., 2016. A life cycle assessment framework combining nutritional and environmental health impacts of diet: a case study on milk. International Journal of Life Cycle Assessment 21, 734–746. https://doi.org/10.1007/s11367-015-0961-0.

Stylianou, K.S., Fulgoni, V.L., Jolliet, O., 2021. Small targeted dietary changes can yield substantial gains for human and environmental health. Nat. Food 2, 616–627. https://doi.org/10.1038/s43016-021-00343-4.

Teixeira, R.F.M., Maia de Souza, D., Curran, M.P., Antón, A., Michelsen, O., Milà i Canals, L., 2016. Towards consensus on land use impacts on biodiversity in LCA: UNEP/SETAC Life Cycle Initiative preliminary recommendations based on expert contributions. Journal of Cleaner Production 112, 4283–4287. https://doi.org/10.1016/j.jclepro.2015.07.118.

Thoma, G., Jolliet, O., Wang, Y., 2013. A biophysical approach to allocation of life cycle environmental burdens for fluid milk supply chain analysis. International Dairy Journal 31, S41–S49. https://doi.org/10.1016/j.idairyj.2012.08.012.

UNEP, 2011. Global Guidance Principles for Life Cycle Assessment Databases A Basis for Greener Processes and Products. United Nations Environment Programme. Shonan, JP.

van Dooren, C., Marinussen, M., Blonk, H., Aiking, H., Vellinga, P., 2014. Exploring dietary guidelines based on ecological and nutritional values: a comparison of six dietary patterns. Food Policy 44, 36–46. https://doi.org/10.1016/j.foodpol.2013.11.002.

Veltman, K., Rotz, C.A., Chase, L., Cooper, J., Ingraham, P., Izaurralde, R.C., et al., 2018. A quantitative assessment of Beneficial Management Practices to reduce carbon and reactive nitrogen footprints and phosphorus losses on dairy farms in the US Great Lakes region. Agricultural Systems 166, 10–25. https://doi.org/10.1016/j.agsy.2018.07.005.

Verones, F., Bare, J., Bulle, C., Frischknecht, R., Hauschild, M., Hellweg, S., et al., 2017. LCIA framework and cross-cutting issues guidance within the UNEP-SETAC Life Cycle Initiative. Journal of Cleaner Production 161, 957–967. https://doi.org/10.1016/j.jclepro.2017.05.206.

Verones, F., Hellweg, S., Antón, A., Azevedo, L.B., Chaudhary, A., Cosme, N., et al., 2020. LC-IMPACT: a regionalized life cycle damage assessment method. Journal of Industrial Ecology 24, 1201–1219. https://doi.org/10.1111/jiec.13018.

Weidema, B., 2018. Nutrition: function or Impact. In: Presented at the 2018 LCA Food Conference. Bangkok, Thailand October 2018.

Weidema, B., 2000. Avoiding Co-Product Allocation in Life-Cycle Assessment. Journal of Industrial Ecology 4, 11–33. https://doi.org/10.1162/108819800300106366.

Weidema, B.P., Schmidt, J.H., 2010. Avoiding Allocation in Life Cycle Assessment Revisited. Journal of Industrial Ecology 14, 192–195. https://doi.org/10.1111/j.1530-9290.2010.00236.x.

Williams, H., Wikström, F., 2011. Environmental impact of packaging and food losses in a life cycle perspective: a comparative analysis of five food items. Journal of Cleaner Production 19, 43–48. https://doi.org/10.1016/j.jclepro.2010.08.008.

# Water Footprint Assessment: towards water-wise food systems

*Joep F. Schyns[a], Rick J. Hogeboom[a,b] and Maarten S. Krol[a]*

[a]Multidisciplinary Water Management Group, Faculty of Engineering Technology, University of Twente, the Netherlands [b]Water Footprint Network, the Netherlands

## 4.1 Introduction

Freshwater is both a vital agricultural input to grow crops and raise animals as well as a sink for agricultural outflows, such as nutrient or pesticide runoff, that affect water quality. As such, food systems put significant pressure on global freshwater resources, both in terms of water quantity and water quality (Hoekstra and Mekonnen, 2012; Mekonnen and Gerbens-Leenes, 2020).

Freshwater is a renewable resource, but its availability is limited. Freshwater availability stems from precipitation over land, which differentiates into a blue water flow – runoff via groundwater and surface water – and a green water flow – rainfall that infiltrates the soil or is intercepted by vegetation and eventually flows back to the atmosphere as evapo(transpi)ration (Schyns, Hoekstra, Booij, Hogeboom, and Mekonnen, 2019). Every year, people claim parts of these blue and green water flows for use at home (blue), in industry (blue), in agriculture (mostly green; also blue) and forestry (mostly green; also blue). These claims should not exceed local annual replenishment rates. After all, the use of blue and green water flows subtracts from the freshwater availability to ecosystems in water and on land, which also require minimum environmental flows for their subsistence. Therefore, there are limits to the sustainable freshwater use by humanity which are lower than the annual replenishments rates of blue (Vörösmarty et al., 2010) and green water (Rockström and Gordon, 2001; Schyns et al., 2019).

### 4.1.1 The water footprint concept

To measure humanity's pressure on freshwater resources, the water footprint (WF) concept was coined by the late professor Arjen Y. Hoekstra in 2002 (Hoekstra, 2003). The concept is

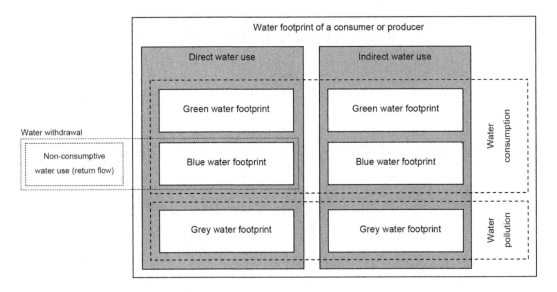

FIGURE 4.1  Schematic representation of the components of a water footprint. It shows that the non-consumptive part of water withdrawals (the return flow) is not part of the water footprint. It also shows that, contrary to the measure of 'water withdrawal', the 'water footprint' includes green and grey water and the indirect water-use component. Credit: From Hoekstra et al. (2011), Copyright © Water Footprint Network 2011. Reproduced with permission of Taylor & Francis Group through PLSclear.

inspired by the concept of 'ecological footprint' (Wackernagel & Rees, 1996) and is built upon a number of important notions (Hoekstra, 2017): (i) freshwater is a *global* resource, because goods produced with freshwater can be traded, resulting in virtual water trade, such that people can benefit from freshwater resources elsewhere (Allan, 1998); (ii) to comprehensively measure human pressure on freshwater systems, one must consider both *green* and *blue* water consumption (Falkenmark, 2000) and water *pollution* (Postel, Daily, and Ehrlich, 1996); (iii) *supply-chain thinking* is needed to link human consumption patterns to impacts on freshwater resources. Subsequently, the concept has been further developed into a multidimensional indicator of human appropriation of freshwater consumption and pollution in volumetric terms, which is embedded into a broader assessment framework: Water Footprint Assessment (WFA) (Hoekstra, Chapagain, Aldaya, and Mekonnen, 2011). The WF concept and WFA methodology have experienced wide uptake and led to the evolution of a new research field (Hoekstra, 2017).

The WF is an indicator of environmental pressure: it measures the human appropriation of freshwater. The WF has three components – green, blue, and grey – that are each specified geographically and temporally (Fig. 4.1). The green and blue WF measure consumptive use of green and blue water flows, respectively. Consumptive use in WF terms refers to water that is "lost" from the system, and that therefore cannot be used for other purposes at that particular time at that particular location (Hogeboom, 2020). These "losses" include the evaporation of water, water contained in products, or water that is transferred to another system, such as another river basin. The grey WF refers to the volume of water that is needed to assimilate

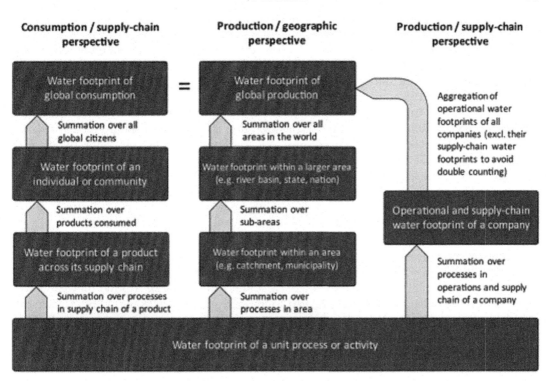

FIGURE 4.2   Water footprints of single processes or activities form the basic building blocks for the water footprint of a product, consumer, or producer or for the footprint within a certain geographical area. The footprint of global consumption is equal to the footprint of global production. Credit: Reprinted from Hoekstra (2017).

pollutants associated with a particular activity to meet ambient water-quality standards (more in Section 4.3.1.4). The WF thus measures both the appropriation of freshwater as a natural resource, via the green and blue WF, and the appropriation of freshwater as an agent to assimilate waste, via the grey WF (Hogeboom, 2020). The WF is not restricted to direct water use only, but also includes indirect water use in the supply chain. As such, the WF uncovers the link between the consumption of goods in one place and water use in another place.

WF accounts can be made for a range of different entities, such as the WF of a product, a consumer, a business, or within a certain geographic region. The building blocks of any WF account are the WFs of single processes or activities, which are mutually exclusive (see Fig. 4.2). WF accounts give spatiotemporally explicit information on how water is appropriated for various human purposes, which can feed the discussion on sustainable, efficient and equitable water use and allocation. To facilitate this, the WFA framework follows four main phases (Hoekstra et al., 2011): (i) setting the goals and scope of the assessment; (ii) WF accounting, in which data are collected and accounts are developed; (iii) WF sustainability assessment, in which the environmental sustainability, efficiency and equitability of the accounted WF are evaluated; and (iv) WF response formulation, in which response options, strategies or policies are developed.

### 4.1.2 The position of this chapter in WFA literature

A rich body of literature on WFA exists. The core consists of the WFA manual (Hoekstra et al., 2011), which is a technical reference containing the Global Water Footprint Standard, with definitions and calculation methods. The book *'Water Footprint of Modern Consumer Society'* (Hoekstra, 2020) is described by its author as a companion to the WFA manual. Hoekstra (2020) uses improved conceptual frameworks and insights from scientific publications to describe global risks of freshwater consumption and pollution and possible response options, taking a holistic perspective. For grey WF accounting, the flagship report by Franke, Boyacioglu, and Hoekstra (2013) provides additional guidelines to those laid out in the WFA manual. Next to these key references, there is a large and growing body of scientific literature that apply the methods and terminology of WFA (Hoekstra, 2017; Mubako, 2018; Zhang, Huang, Yu, and Yang, 2017).

This chapter focuses on the WF of food systems, which play a dominant role in the WF of many economies. Agricultural production contributes 92 percent to the WF of humanity (Hoekstra and Mekonnen, 2012). Trade in commodities, which have a WF associated with their production, results in virtual water flows. These international virtual water flows more than doubled from 1986 to 2008 (Carr, D'Odorico, Laio, and Ridolfi, 2012). Around the year 2000, 88 percent of these virtual flows were related to trade in agricultural commodities (Hoekstra and Mekonnen, 2012). The WF related to the consumption pattern of individuals is largely determined by the food they consume (Hoekstra and Mekonnen, 2012; Vanham, Mekonnen, and Hoekstra, 2013b). For the average consumer in the European Union, 74 percent of their consumptive WF lies in the food sector, 26 percent in the energy sector, and only a minor fraction in the water sector itself, relating to direct use of water at home (Vanham, Medarac, Schyns, Hogeboom, and Magagna, 2019).

This chapter is structured according to the major phases identified by the WFA framework (Section 4.1.1), so that Section 4.2 focuses on WF accounting, Section 4.3 on WF sustainability assessment, and Section 4.4 on WF response formulation. In Section 4.2, we describe how to account the green and blue WF of growing a crop. In Section 4.3, we summarize methods to assess the environmental sustainability, efficiency and equitability of the WF of food systems. Lastly, in Section 4.4, we discuss possible response strategies of different actors in the society to transform current food systems towards more water-wise food systems.

## 4.2 Accounting the consumptive water footprint of growing a crop

This section focuses on accounting the WF of growing a crop and omits the WF of other stages in the food chain, such as the processing, packaging and distribution. Studies on the WF of specific food and beverage products have shown that the majority of the WF is in the stage of crop production (Aldaya & Hoekstra, 2010; Ercin, Aldaya, & Hoekstra, 2011, 2012). Also for farm animal products (Mekonnen and Hoekstra, 2012) and fish from aquaculture production (Pahlow, van Oel, Mekonnen, and Hoekstra, 2015), the largest part of the WF relates to growing the crops which are used as animal and fish feed. Furthermore, the WF of growing a crop is an essential building block in calculating, for example, the WF of the agricultural sector in a nation (Hoekstra and Mekonnen, 2012) or the WF of alternative diets (Vanham, Comero, Gawlik, & Bidoglio, 2018; Vanham, Hoekstra, & Bidoglio, 2013a; Vanham, Mekonnen, & Hoekstra, 2013b).

The consumptive WF accounts for the water lost to the atmosphere in the process of growing a crop. It includes all (green plus blue) freshwater that evaporates from the crop field or transpires through the crop's leaves and is, therefore, no longer available to be consumed for other purposes within that area and part of the year. The WF of growing a crop differs largely from place to place, depending on differences in climate, agricultural practices, and the type of crop. For example, the global average green plus blue WF of growing a crop increases from sugar crops (182 m$^3$ of water per tonne of crop), vegetables (237 m$^3$/tonne), roots and tubers (343 m$^3$/tonne), fruits (874 m$^3$/tonne), cereals (1,460 m$^3$/tonne), oil crops (2,243 m$^3$/tonne), pulses (3,321 m$^3$/tonne), spices (6,616 m$^3$/tonne) to nuts (8,383 m$^3$/tonne) (Mekonnen and Hoekstra, 2011).

In layman's terms, the consumptive WF measures the inverse of "crop per drop." It is calculated as the sum of the green and blue crop water use (*CWU*, in m$^3$ of water per ha) divided by the harvestable crop yield (*Y*, in tonne of crop per ha) and measured in m$^3$ of water per tonne of crop (Hoekstra et al., 2011). Crop water use refers to the accumulation of daily actual evapotranspiration from the crop field (*ET*, in mm per day) over the length of the growing period (*lgp*, in days). See Eqs. (4.1) and (4.2) below, in which the factor 10 is used to convert mm per day into m$^3$ per ha (Hoekstra et al., 2011).

$$WF = \frac{CWU}{Y} \qquad (4.1)$$

$$CWU = 10 \times \sum_{d=1}^{lgp} ET \qquad (4.2)$$

Technically, the water content in the harvested crop is also part of the WF of a crop, but in practice this part is often neglected since its order of magnitude is typically less than one percent of the crop water use (Hoekstra et al., 2011).

The consumptive WF of a crop is often modeled rather than measured. Although *ET* from a crop field can be measured in the field this process is laborious and costly, and therefore unusual. This is especially true when the scope of interest is larger than a number of specific locations or time frames for which measuring equipment can be installed and monitored, which is the case for many research questions in WFA. For example, how does the WF of a crop vary across the globe? (Mekonnen and Hoekstra, 2011, 2014); how has the WF in a region changed over time? (Zhuo et al., 2016b; Zhuo, Mekonnen, Hoekstra, and Wada, 2016c); how does the WF of a crop respond to a range of different agricultural management practices? (Chukalla, Krol, and Hoekstra, 2017). Therefore, *ET* is generally estimated with models that account for variables and relations in the atmosphere-plant-soil continuum. Crop yields are commonly measured in the field and are reported on at the level of administrative units, such as the province or country. Still, modeling of crop yields is often desired, for example when exploring the effects of climate change or alternative agricultural management strategies.

We describe here how to estimate the consumptive WF of a crop with the use of soil water balance models (Section 4.2.1) or crop models (Section 4.2.2), which has become the new standard in WFA. Also, we explain why and how to distinguish between the green and blue WF of a crop (Section 4.2.3).

## 4.2.1 The use of soil water balance models for crop water footprint accounting

In the early stages of the development of the field of WFA (2002–14), studies estimated the green and blue WF of growing a crop according to the guidelines by Allen, Pereira, Raes, and Smith (1998) using a soil water balance model and crop coefficients (Hoekstra et al., 2011), and leveraging the strong empirical relationship between evapotranspiration and crop yield. The approach can be summarized in three steps. Step 1 estimates the potential $ET$ of a crop given unlimited access to water. Step 2 determines the actual $ET$ based on actual soil water availability during the growing season. Step 3 calculates crop yield based on the difference between actual and potential $ET$ and the crop's yield response to water shortage (Doorenbos & Kassam, 1979). In this approach, the crop factor and the rooting depth vary over the growing season, but are pre-described inputs to the soil water balance model. Actual crop growth is not simulated by the model itself. The approach has been applied in the key publication on the WF of crops and derived crop products by Mekonnen and Hoekstra (2011), which has served as a basis for many subsequent studies.

## 4.2.2 The use of crop models for crop water footprint accounting

Alternatively, the consumptive WF of growing a crop can be estimated with crop models that simulate both crop growth and the soil water balance. Such crop models are more process-based, meaning that higher level responses (like crop transpiration and biomass growth) are determined by lower level simulated processes (like root growth and rate of photosynthesis). These processes represent biophysical relationships between the atmosphere, the crop and the soil, and often influence each other via feedback loops. Therefore, crop models more accurately represent reality and are better suited to analyze particular responses, like the yield response to irrigation water applied, compared to the soil water balance approach. Another advantage is that these models often include possibilities to simulate the effect of alternative irrigation (irrigation scheduling and application technique) and field (mulching, tillage, bunds) management options on the WF of a crop. It is for these reasons that in recent years, crop modeling has become the new standard in WFA (Chouchane, Krol, & Hoekstra, 2018b; Chukalla, Krol, & Hoekstra, 2015, 2017, 2018b; Gobin et al., 2017; Hogeboom & Hoekstra, 2017; Karandish, Hoekstra, & Hogeboom, 2018; Masud, McAllister, Cordeiro, & Faramarzi, 2018; Masud, Wada, Goss, & Faramarzi, 2019; Nouri, Stokvis, Galindo, Blatchford, & Hoekstra, 2019; Zhuo, Mekonnen, Hoekstra, & Wada, 2016c). The disadvantages of crop models are the higher input data requirements and computational demands, which makes their use for high spatial resolution modeling at large geographic scales (countries, river basins, global) challenging.

Crop models can be categorized in three groups, depending on the main mechanism that drives biomass accumulation (Steduto, 2006) namely: carbon-driven, solar-driven, and water-driven crop models. Carbon-driven models, for which growth is based on the carbon assimilation by the photosynthetic process, are highly mechanistic and simulate all the main biophysical processes and feedback loops that translate incoming solar radiation into biomass (Steduto, 2006). Solar-driven models simulate biomass accumulation directly from the intercepted solar radiation through a single coefficient, the radiation use-efficiency, which synthetically incorporates the underlying biophysical processes (Steduto, 2006). Water-driven models simulate biomass accumulation based on accumulated crop transpiration through a single coefficient, the water-use efficiency or biomass water productivity (Steduto, 2006). Although biomass is produced through photosynthesis, the relation between transpiration and photosynthesis is so close that it can be captured in a single coefficient when normalized for climatic conditions (Steduto, Hsiao, Raes, and Fereres, 2009).

The choice for a crop model to use will depend on one's objective and available means. Models differ in input data requirements and computational resource needs. In the field of WFA, the main interest is often the crop's response to water. For this purpose, the water-driven crop growth model type is particularly suited, because the crop's response to water is at the core of the model and the relationship between transpiration on biomass growth has proven to be robust, even under water stress conditions (Steduto, 2006). For this reason, the AquaCrop model (Steduto et al., 2009) has often been applied in WFA studies. The downside of the AquaCrop model is that it does not simulate a nutrient cycle, which impairs its use to assess the trade-offs between the consumptive (green and blue) WF and the grey WF related to fertilizer or pesticide application. For such assessments, other models such as the APEX model are needed (Chukalla et al., 2018b; Chukalla, Krol, and Hoekstra, 2018a).

### 4.2.3 Distinguishing between green and blue crop water use

Because green and blue water differ in terms of possibilities for storage and use, it has become common practice to distinguish between green and blue crop water use (Hoekstra, 2019a). Making this distinction is non-trivial, because it is not 'visible' which part of modeled $ET$ originates from rainwater and which part from irrigation water or capillary rise. Hoekstra (2019a) describes a generic and physically-based method to differentiate green and blue soil evaporation ($E$) and crop transpiration ($T$) that is depicted in Fig. 4.3. The method demands a systematic accounting of the fractions of green and blue water in the different soil and vegetation (referring to water intercepted by vegetation) layers on a daily basis, as a basis to estimate green and blue water fractions in all fluxes leaving each layer. The input for this method – the daily water fluxes in each layer – is often the output of a crop model simulation, such that the green-blue accounting can be done in a post-processing step (Chukalla, Krol, & Hoekstra, 2015; Zhuo, Mekonnen, Hoekstra, & Wada, 2016c). Details of the method are provided by Hoekstra (2019a).

When the green-blue accounting method by Hoekstra (2019a) cannot be followed, an alternative, less accurate method exists to distinguish between green and blue crop water use. This method, which has been practiced in the past (Hoogeveen, Faurès, Peiser, Burke, & van de Giesen, 2015; Liu & Yang, 2010; Mekonnen & Hoekstra, 2010, 2011; Siebert & Döll, 2010), is to estimate blue $ET$ as the difference between $ET$ under irrigated conditions and $ET$ under rainfed conditions. Note that this method thus requires two simulations for the same crop field: an irrigated and a rainfed case. Hoekstra (2019a) mentions a couple of drawbacks of this method. First, the rooting depth of crops is different under rainfed versus irrigated conditions, which affects the water uptake by plants and consequently the partitioning of green versus blue $E$ and $T$. This issue can be partially resolved by simulating the rainfed case with a rooting depth as it would be under irrigated conditions. However, this approach is still inaccurate

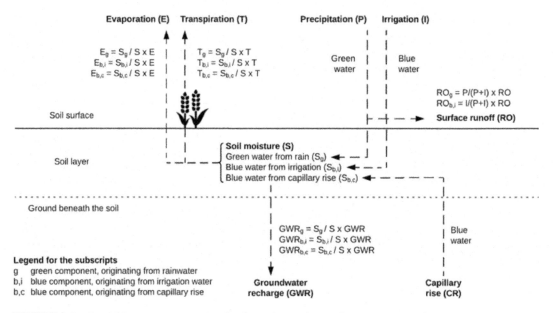

**FIGURE 4.3**    Green-blue water accounting of soil moisture and water fluxes entering and leaving the soil moisture in case of one soil layer. The 'colour codes' in the form of subscripts refer to the origin of the water. Credit: Reprinted from Hoekstra (2019a) with permission from Elsevier.

because irrigation affects the soil moisture dynamics over time, such that the partitioning of green versus blue $E$ and $T$ is different under irrigated versus rainfed conditions. Furthermore, a rainfed case is not always available, which is the case in many (arid) regions where rainfed cropping is not a viable alternative to irrigated agriculture. In these regions, a crop model will not be able to simulate crop development (the crop will hardly develop and/or die-off early in the season) and soil moisture dynamics under rainfed conditions that are realistically representative for the green $E$ and $T$.

## 4.3  Environmental sustainability, efficiency and equitability of the water footprint of food systems

The purpose of WFA is to feed a debate on wise freshwater allocation. Relevant questions in the domain of food systems are for example: in which places is food production associated with unsustainable water abstractions or water pollution, and which consumers benefit from this food? Can we produce the same crop with a smaller water footprint? Or can we replace it for other foodstuffs with equivalent nutritional value but a smaller water footprint? Given that the global freshwater availability is limited, what are fair shares of water use per community to meet their (food) demand?

To shed light on such conversations, WFA goes beyond WF accounting to sustainability assessment. We follow the three pillars under wise freshwater allocation as postulated by

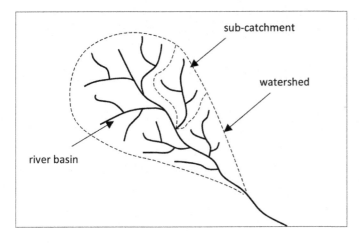

FIGURE 4.4 A river basin is a catchment area in which the precipitation is discharged by a network of streams, with one common outflow point. A river basin is a typical unit for analysis of freshwater availability and use. A river basin can be schematized into sub-catchments. The watershed is the dividing line between different river basins. The watershed will generally follow the highest mountain ridges: precipitation on the one side of the ridge will run off in one direction and finally end up in one river, while precipitation on the other side of the ridge will run off in the other direction and finally end up in another river. Credit: Reprinted from Hoekstra (2019b) with permission from University of Twente.

Hoekstra (2014): assessing the environmental sustainability of a WF (Section 4.3.1), the efficiency of a WF (Section 4.3.2), and the equitability of a WF (Section 4.3.3). The methods we discuss in this section step away from the narrower focus on growing a crop (as per Section 4.2), to encompass food production, consumption and trade in a wider sense.

## 4.3.1 Environmental sustainability

In the field of WFA, the sustainability of a WF of a certain entity (an activity, a producer, or a consumer) is addressed by analyzing *total* WFs in the context of *maximum sustainable* WFs for a certain geographic area, often a river basin (see Fig. 4.4), and a certain period of time. The reasoning is that one cannot label a single WF in itself as 'sustainable' or 'unsustainable', but each WF contributes to a state at the system level which is sustainable or unsustainable, considering how the WF of all activities in the system altogether compare to the maximum sustainable WF of the system (Hoekstra, 2015). So, sustainability of an activity is always conditional on other activities present in its context. Note that this approach differs from the typical life-cycle analysis (see Chapter 3 of this book) approach to assessing the sustainability of an activity or product, which is to compare the potential environmental impacts of substitutable products. In other words, WFA tries to answer whether a WF is part of a system that is sustainable or unsustainable (absolute approach), while life-cycle analysis tries to answer whether a product is more sustainable than other substitutable products (comparative approach) (Hoekstra, 2015). In the following sub sections, we describe the concept of the maximum sustainable footprint and how this concept is used to assess the environmental sustainability of the blue, green, and grey components of a WF.

### 4.3.1.1 *The maximum sustainable water footprint*

Freshwater availability in a river basin stems from precipitation over land. Part of the precipitation evapotranspires from the land surface (green water flow), the rest generates runoff via surface and groundwater (blue water flow) towards the sea. Not all green and blue

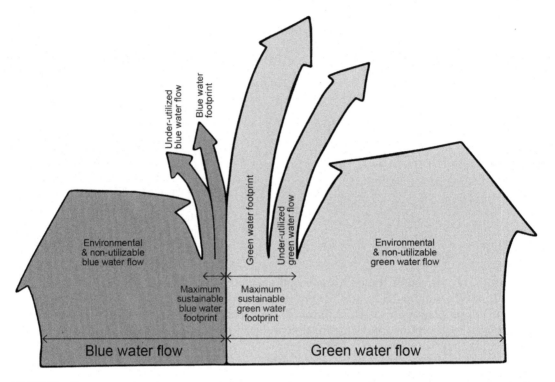

**FIGURE 4.5** The partitioning of precipitation over land into blue and green water flows. Both flows further partition into environmental and non-utilizable (or non-accessible) flows, flows allocated to human activities (i.e. water footprint) and under-utilized flows below the maximum sustainable level. Credit: Reprinted from Schyns (2018).

water flows are available for consumption, and fractions of these flows need to be reserved for the functioning of terrestrial (for green water) and aquatic (for blue water) ecosystems, the so-called environmental flow requirements (see Fig. 4.5). The maximum sustainable green WF, or 'green water availability', is equal to the total green water flow minus environmental green water requirements. The maximum sustainable blue WF, or 'blue water availability', is equal to the total blue water flow minus environmental blue water requirements. The maximum sustainable green and blue WF reflect environmental limits to the consumptive use of green and blue water, respectively. The grey WF is defined in volumetric terms and also represents a pressure on blue water. The effect of the total grey water footprint in a river basin depends on the available blue water flow available to assimilate the waste (Hoekstra et al., 2011). The maximum sustainable grey WF, or 'waste assimilation capacity', is therefore equal to the actual runoff, which equals the natural runoff minus the blue WF.

### 4.3.1.2 Sustainability of the blue water footprint

To evaluate whether the blue WF of a particular activity, producer or consumer is sustainable or not requires placing that blue WF in the context of the degree of blue water scarcity experienced in the geographic area in which the WF is located. This is done using the blue

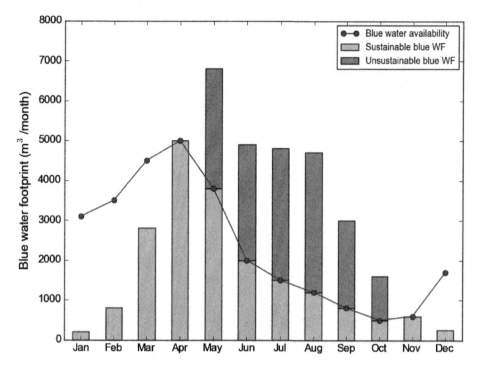

**FIGURE 4.6** Theoretical example of the monthly sustainable (light gray) and unsustainable (dark gray) components of the blue water footprint. Credit: Reprinted from Mekonnen and Hoekstra (2020b) with permission from Elsevier.

water scarcity index, defined as the ratio of the total blue WF over the blue water availability (Hoekstra et al., 2011). A blue WF located in a place in which the total blue WF exceeds blue water availability (blue water scarcity index > 1) is labelled as unsustainable, because it contributes to the infringement of environmental flows (Fig. 4.6). This method has for example been used to assess the sustainability of the blue WF of crop production (Mekonnen and Hoekstra, 2020b) and of national consumption (Mekonnen and Hoekstra, 2020a).

Blue water scarcity typically manifests within particular parts of the year – because both blue water availability and blue WF vary strongly within the year (Fig. 4.6) – in particular parts of a river basin (often downstream). Therefore, the assessment of the environmental sustainability of the blue WF is ideally done on a month-by-month basis on the level of the river basin or sub-catchments (Fig. 4.4) (Hoekstra, Mekonnen, Chapagain, Mathews, & Richter, 2012; Mekonnen & Hoekstra, 2016).

### 4.3.1.3 Sustainability of the green water footprint

Like the blue WF, evaluating the sustainability of a particular green WF requires placing that green WF in the context of the degree of local green water scarcity. Green water scarcity is a bit harder to grasp than blue water scarcity. It refers to the competition over limited green water flows, which can either support a natural ecosystem or the production of biomass for various purposes in the human economy (Schyns, Hoekstra, and Booij, 2015). A green water

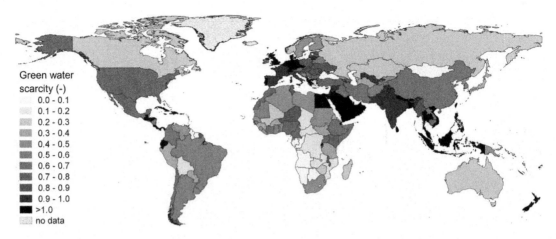

FIGURE 4.7   Green water scarcity per country, expressed as the ratio of the national aggregate green water footprint to the national aggregate green water availability. Credit: Reprinted from Schyns et al. (2019) with permission from the author.

scarcity index has been proposed (Hoekstra, Chapagain, Aldaya, & Mekonnen, 2011; Schyns, Hoekstra, & Booij, 2015) and recently developed (Schyns et al., 2019) that measures the ratio of the total green WF to green water availability (see Fig. 4.7) similar to the blue water scarcity index. Green water availability considers the agroecological suitability and accessibility of land, biophysical constraints to intensifying land use, and the need to set aside part of the land and green water flows for biodiversity conservation (Schyns et al., 2019). Countries with green water scarcity equal to 1 have fully allocated their sustainably available green water flow to human activities (or overshoot of the maximum sustainable level is canceled out by remaining potential in another part of the country).

In contrast to blue water scarcity, which is best estimated per month per sub-catchment, green water scarcity should be assessed on a (multi-)annual scale and larger spatial entities, such as countries, large river basins, or ecoregions. Green water is indirectly allocated through land-use decisions, which are generally made over long planning horizons and do not change on a monthly basis. Land use decisions tend to be made by local land owners, but they impact the biodiversity of the larger ecosystems in which the land resides. Thus, to estimate green water availability, one should consider a fairly large spatial region to determine which places within that region have the most biodiversity value and are most in need of conservation.

### 4.3.1.4 Sustainability of the grey water footprint

While the blue and green WF are measures of water use, the grey WF is a measure of water pollution. The grey WF of an activity is not a measure of the actual polluted volume (which can never exceed the actual volume of water), but an indicator of the severity of water pollution, expressed in terms of the water volume required to assimilate the load of pollutants that the activity adds to a freshwater body (Hoekstra et al., 2011). To evaluate whether a particular grey WF is sustainable or not requires assessing whether or not the waste assimilation capacity in the geographic area in which the WF is located is already fully consumed. For this, the water pollution level is used as indicator. The water pollution level is defined as the ratio of the

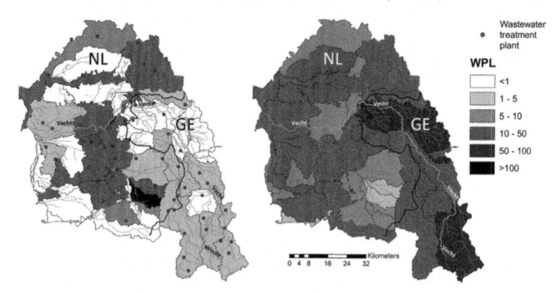

**FIGURE 4.8**  Annual average water pollution level (WPL) in the Vecht catchment – located in Germany (GE) and the Netherlands (NL) – resulting from the maximum grey water footprint of human (left) and veterinary (right) pharmaceutical use, resulting from the substances ethinylestradiol and amoxicillin, respectively. Credit: Reprinted from Wöhler et al. (2020b).

total grey WF in an area to the waste assimilation capacity. The total grey WF in an area is determined by the pollutant that is most critical, i.e. the pollutant with the largest associated grey WF.

Since the waste assimilation capacity depends on the blue water runoff, the assessment of the sustainability of the grey WF is (similar to the blue WF, see Section 4.3.1.2) best determined for each month of the year separately, and at the level of the river basin or sub-catchments (Fig. 4.4).

An example of an assessment of the environmental sustainability of the grey WF is provided in Fig. 4.8. This study by (Wöhler et al., 2020b) shows the water pollution level within the Vecht catchment as a result of the grey WF of pharmaceutical use by livestock and humans in the catchment. Another notable example is by Aldaya et al. (2020), who assessed the grey WF of diffuse nitrogen pollution in an agricultural region in Spain. Just like blue and green water scarcity (Mekonnen & Hoekstra, 2016; Schyns et al., 2019), the water pollution levels related to anthropogenic loads of nitrogen (Mekonnen and Hoekstra, 2015) and phosphorus (Mekonnen and Hoekstra, 2018) have also been mapped with global coverage.

## 4.3.2  Efficiency

The efficiency of a WF can be addressed from three perspectives (Hoekstra, 2020): a production (Section 4.3.2.1), a geographic (Section 4.3.2.2), and a consumption perspective (Section 4.3.2.3).

### 4.3.2.1 *Efficiency from a production perspective*

A production perspective on the efficiency of a WF addresses the question of how we can produce a certain process or product with less water (Hoekstra, 2020). In other words, how can we reduce the WF (volume per unit of product) or increase the water productivity (unit of product per volume of water, also called the water-use efficiency). This question can be answered by comparing the actual WF of a crop in a certain location and time period to a benchmark value. Such a benchmark represents a reasonable WF per unit of crop. There are two main methods that have been applied to derive a benchmark for the WF of growing a crop.

One method is to look at the spatial differences in the WF per unit of crop and derive a benchmark by ranking the WF achieved by producers under similar circumstances. A benchmark can then be set at the WF achieved by the top-10 percent or top-25 percent most efficient producers, for example. This method thus compares and ranks producers that operate under similar circumstances. So the key question is: what producers can we compare? Recent studies have grouped and compared producers that operate in similar climates (Karandish et al., 2018; Mekonnen, Hoekstra, Neale, Ray, and Yang, 2020; Zhuo, Mekonnen, and Hoekstra, 2016a) and/or on similar soils (Zhuo et al., 2016a), and by separating rainfed and irrigated agricultural systems (Mekonnen et al., 2020). However, no matter what spatial scope one takes in grouping producers, within that scope there will still be variability from place to place. Rainfall, for example, shows strong spatial variability over short distances, such that a few producers in a larger area simply had more favourable local circumstances. Therefore, one can always question the comparability of producers that operate in different locations and the WFs they achieve.

Alternatively, to avoid this drawback, one can derive a benchmark for the WF of a crop in a certain location by estimating the WF of the crop in the *same* location under a best practice scenario (Chukalla et al., 2015, 2017). These studies modeled crop WFs under the same conditions (location, year) under alternative management packages, to evaluate which package (the best practice) results in the smallest WF. Management packages are combinations of irrigation techniques (application methods, such as flooding or sprinkler systems), irrigation strategies (when and how much water to apply) and mulching practices that cover part of the soil to reduce soil evaporation. This method does not suffer from the drawback of comparability, but it can be laborious. One needs to have a crop model (or field measurements, which is arguably even more laborious) that can simulate agricultural management packages. Moreover, deciding on the exact setup of alternative potential 'best management packages' is not straightforward, because options are practically unlimited.

A final note on benchmarking is that comparing a WF in a particular location against a benchmark indicates whether or not the WF of that process or product can be reduced (and by how much). However, it does not indicate whether using the water in that location for that purpose is a wise decision. Better alternative uses of the water may exist in that location or upstream of that location. For example, consider a case in which a river flows through a fertile landscape into marginal land. A farmer on marginal land can use the water to irrigate crops in an efficient way such that the farmer achieves the local benchmark. However, it is probably wiser to not grow crops on the marginal land, but instead on the fertile lands upstream where more food can be produced with the same amount of water due to the more favourable circumstances.

### 4.3.2.2 Efficiency from the geographic perspective

The geographic perspective on the efficiency of a WF addresses the question of where we can best produce what to save water and reduce water scarcity (Hoekstra, 2020). Across the globe, there is large spatial variability in the WF of growing a crop (Liu and Yang, 2010; Mekonnen and Hoekstra, 2011), indicating potential to save water by changing spatial cropping patterns and international trade in agricultural goods, along with their accompanying virtual water flows. Indeed, a number of studies have assessed changes in national, regional or global cropping patterns and associated water savings (Chouchane, Krol, & Hoekstra, 2020; Davis, Rulli, Seveso, & D'Odorico, 2017a; Davis, Seveso, Rulli, & D'Odorico, 2017b; Schyns & Hoekstra, 2014; Ye et al., 2018). Recently, Chouchane et al. (2020) optimized global cropping patterns to minimize national blue water scarcity. They found that with a minimum allowance ($\leq$10 percent) of expansion of irrigated and rainfed harvested area per crop, it is possible to decrease the global blue water footprint of crop production by 21 percent and decrease the global total harvested and irrigated areas by 2 percent and 10 percent, respectively. Consequently, the blue water scarcity in the world's most water-scarce countries can be greatly reduced (Chouchane et al., 2020).

### 4.3.2.3 Efficiency from the consumption perspective

The consumption perspective on the efficiency of a WF addresses the question of how we can best fulfill certain consumer needs with less water (Hoekstra, 2020). Studies taking this perspective on the efficiency of the WF of food systems have focused on reducing food losses (Jalava et al., 2016; Karandish, Hoekstra, and Hogeboom, 2020; Kummu et al., 2012; Mekonnen and Fulton, 2018) and changing diets (Harris et al., 2020; Jalava, Kummu, Porkka, Siebert, & Varis, 2014; Kim et al., 2020; Mekonnen & Fulton, 2018; Vanham, Comero, Gawlik, & Bidoglio, 2018; Vanham, Hoekstra, & Bidoglio, 2013a; Vanham, Mekonnen, & Hoekstra, 2013b). The question is how we can fulfill energy and protein needs in a water efficient manner. Reductions in food waste directly translate into reduced freshwater needs to grow the food in the first place. Meat and dairy generally have significantly larger WFs per kcal of energy and per gram of protein compared to plant-based alternatives (Mekonnen and Hoekstra, 2012). Therefore, the water efficiency of certain diets can be increased by replacing (part of) meat and/or dairy by suitable amounts of plant-based alternatives, while maintaining the overall nutritional value of the diet. A recent global study has shown that, for countries around the world, changes in diets (to diets with less red meat or pescetarian, lacto-ovo vegetarian or vegan diets) have significant effects on the both the blue and total consumptive (green plus blue) WFs per capita (Kim et al., 2020). They found the largest potential reductions in per capita consumptive WFs from shifting to vegan diets in countries where animal products account for a large share of the current (baseline) consumption pattern.

## 4.3.3 Equitability

The central question when assessing the equitability of a WF is: which communities finally benefit from the water used (Hoekstra, 2020)? This question is at the root of the WFA

methodology which – by incorporating supply-chain thinking into water resources management – provides insight into what end-purposes and communities benefit from freshwater
use. While a WFA can reveal what end-purposes benefit from freshwater used (cf. Zhuo
et al. (2019)), it does not provide quantitative measures to answer the question whether that
freshwater allocation is fair or equitable. Rather, indicators on the sustainability and efficiency
of a WF – as discussed in previous sections – feed into a debate on what is equitable water
sharing.

Hoekstra (2020) sheds light on the topic of equitable water use from three perspectives:
the consumption, production and geographic perspective. The consumption perspective addresses the question of how we can fairly share the world's limited freshwater resources.
Approaches to assess the efficiency of a WF from the consumption perspective feed into the
debate on fair shares of water use per community. The production perspective raises the question of how we can produce more commodities to be shared amongst communities, without
increasing the pressure on freshwater resources. Solutions to increase the water-use efficiency
of food production are part of this debate. The geographic perspective is concerned with how
we can fairly share water given large differences in the water availability per capita across
nations. This relates to trade in water-intensive commodities (which encompasses practically
all agricultural goods) in solidarity with people in water-poor countries. Research in this
direction has for example assessed the need for increased export of staple crops to countries
with decreasing water availability per capita (Chouchane, Krol, and Hoekstra, 2018a).

Compared to assessing the environmental sustainability and efficiency of WFs, the topic
of assessing equitable water sharing is still relatively underdeveloped. As such, it remains an
area for further research.

## 4.4 Towards water-wise food systems

It is difficult enough to make sure that complex and geographically dispersed food systems
reach goals for food security, let alone achieving goals for water security (Hogeboom et al.,
2021). The usual suspects to prevent harmful water outcomes and ignite positive change are
farmers and governments: the former produce food and use water directly, while the latter
typically oversee water allocation to users. However, other actors that may not be readily
associated with water-for-food discourses can and should play a role too. Think of investors,
civil society groups, or consumers. While it is true that such actors mainly carry an indirect or
supply chain responsibility, their power to influence food and water outcomes should not be
underestimated. To the contrary, there is not one actor that can transform current food systems
into water-wise food systems by itself. This transformation is a shared responsibility, one that
requires inclusion of a broad spectrum of actors (UN-WWAP, 2006).

So what can or should these actors do? It is clear there is no silver bullet, but what set
of measures might lead us in the right direction? Without holding to the illusion of being
complete and conclusive, this section highlights possible responses across the landscape of
actors: governments, citizens, food sector companies, investors, international organizations,
civil society, media, and academia.

A summary overview of response actions for each of these actor groups is provided
in Table 4.1. Note that these examples are not exclusively derived from the WF concept,

**TABLE 4.1**   Examples of response options per actor to transform food systems into water-wise food systems.

| Actor | Response options |
|---|---|
| Governments | *Regulatory instruments* |
| |     Set WF caps |
| |     Formulate WF benchmarks |
| |     Mandatory product WF labeling |
| |     Certification |
| |     Institute a WF permit system |
| |     Ensure minimum water rights |
| |     Enforcement and monitoring |
| |     Water neutral spatial development |
| | *Economic instruments* |
| |     (Full cost) water pricing |
| |     Taxing pollution |
| |     Taxing water-intensive commodities |
| |     Subsidizing Best Available Technologies |
| |     Stop subsidizing ill-performing practices and technologies |
| | *Institutional arrangements* |
| |     Harmonize policies across policy domains |
| |     Take responsibility for external WF of consumption |
| |     Make companies take responsibility for their supply chains |
| | *International agreements* |
| |     Water-sensitive trade |
| |     Fair shares of WF of consumption |
| | *Standards* |
| |     WF disclosure |
| |     WF performance |
| | *Awareness raising* |
| |     Campaigning |
| |     Water in educational programmes |
| |     Promoting water transparency |
| |     Promoting water-sustainable diets |
| Citizens | *As consumer* |
| |     Substitute water-intensive foodstuffs in diet |
| |     Reduce overall caloric intake |
| |     Reduce meat and animal product intake |
| |     Buy local and organic |
| |     Reduce/avoid food waste |
| |     Avoid water-intensive non-food products that compete with foodstuffs over water (e.g., biofuels, cotton) |
| | *As voter* |
| |     Vote for parties that are mindful of water resources |
| |     Vote not for parties that harm or neglect water resources |
| | *As investor* |
| |     Invest in companies/pension funds that are mindful of water resources |
| |     Invest not in companies/pension funds that harm or neglect water resources |
| |     Choose a water-aware bank |
| | *As activist* |
| |     Hold opinions on water issues |
| |     Demand product transparency from companies |
| |     Demand government regulation concerning water issues |

*(continued on next page)*

**TABLE 4.1**   Examples of response options per actor to transform food systems into water-wise food systems—cont'd

| Actor | Response options |
|---|---|
| (Food sector) Companies | *Reduce operational WF*<br>  Substitute water-intensive for water-extensive crops, products, and processes<br>  Use Best Available Technologies and Best Available Practices for water<br>  Set and meet WF reduction targets<br>*Reduce supply chain WF*<br>  Supplier WF agreements<br>  Change (non-compliant) suppliers<br>  Set and meet WF reduction targets<br>*Corporate water stewardship*<br>  Measure and monitor WFs in operations<br>  Measure and monitor WFs in supply chain<br>  Certify sites for water performance<br>  Comply with (WFA) standard<br>  Actively engage with (basin) communities<br>  Label products<br>  Disclose corporate water performance |
| Investors | *Investment water policy*<br>  Have one<br>  Require WF disclosure/compliance of investees<br>  Adjusted interest rates for water-wise investment activities<br>  Increase share of water-wise companies/products in portfolio<br>  Screen companies on water performance<br>  Divest from water harming companies or water-scarce locations<br>  Set portfolio WF reduction targets<br>*Corporate water stewardship*<br>  Measure and monitor WFs in operations<br>  Measure and monitor WFs in portfolio<br>  Disclose portfolio water dependency and risk<br>  Adopt responsible investment principles<br>  Share best investment practices |
| International organisations | Agenda setting<br>Leverage stakeholder convening power<br>Invest in WF reduction programs<br>Water-informed trade agreements<br>Advocate a 'Kyoto Protocol' for water<br>Advocate minimum water rights<br>Develop an international water pricing protocol<br>Develop international nutrient housekeeping agreements<br>Develop international water standards and taxonomies |
| Civil society groups | Awareness raising<br>Agenda setting<br>Promoting transparency<br>Lobbying for water sustainable food systems |
| Media | Investigative journalism<br>Maintaining private and public sector accountability<br>Monitoring transparency |
| Academia | Providing water solutions (technical and non-technical)<br>Evaluating water policies |

nor is their applicability limited to water considerations in food systems alone. Rather, we deduce insights that emerge from the conceptual framework of WFA to explore perhaps unconventional measures and solution pathways (Hoekstra, 2020).

### 4.4.1 What can governments do?

Governments come in many shapes and sizes. They are a collective of many different governmental bodies that each has its own policy mandates and jurisdictions. The most direct responsibility to manage water resources wisely in their local context lies with local authorities, such as municipalities, utilities, and waterboards. These are the bodies that can, for instance, set a WF cap at the catchment level. Capping WFs entails earmarking sufficient water for nature, so that what remains can be considered sustainably allocable to human activities; the sum of all WF permits issued in that catchment should remain below that agreed upon WF cap (Hogeboom et al., 2020).

More challenging is the fact that other governmental bodies without direct mandates related to water or food hold important pieces of the water-for-food puzzle as well. For example, Ministries of Agriculture can promote installing efficient irrigation technologies and growing drought-resistant crop varieties. Ministries of Environment can define ecological boundary conditions within which total water consumption should remain (viz. define a 'safe operating space', (cf Hogeboom, de Bruin, Schyns, Krol, & Hoekstra (2020)). Ministries of Infrastructure can maintain and cover irrigation canals and ponds to prevent leakage or evaporation (cf. Haghighi, Madani, and Hoekstra (2018)). Ministries of Energy should steer clear of solving the carbon crisis at the expense of worsening water crises, which may happen through for instance the production of biofuels that compete with food systems over water resources (cf. Holmatov, Hoekstra, and Krol (2019), Vanham et al. (2019)). Ministries of Health can enforce protocols for disposing of pharmaceuticals to avoid water pollution, rendering this water useless for further use (cf. Wöhler, Hoekstra, Hogeboom, Brugnach, and Krol (2020)). Ministries of Economic Affairs can stop subsidizing water-wasting farming practices and water-polluting industries and many more examples could be mentioned.

For some solutions, multilateral cooperation between governments is needed. For example, to reduce their internal WF, national governments can save water domestically by importing more food (and thus virtual water) from other countries. The risk with such a trade strategy is that it solves water issues in the importing country at the cost of the exporting countries to which the water issues are being externalized. This need not be the case, though. If WFs per unit of product in the exporting countries are lower than domestic unit WFs, there will be a net saving associated with this trade flow. National governments could therefore investigate differences in unit WFs across countries, and select trade partners on the basis of their comparative advantage in terms of water productivities (Hoekstra and Chapagain, 2008). Such trade strategies create win-win situations in which not only national, but also global water savings are possible.

### 4.4.2 What can citizens do?

For all governments can do, they cannot legislate virtue nor force consumers to make water-wise food decisions. Hence, citizens have a responsibility too. Given there are almost 8 billion of us on this planet, citizens in the role of consumers are arguably the most powerful actor group to solve our collective water issues. Adopting a vegetarian diet, for example, reduces the typical WF by 40 percent (Vanham et al., 2018). Reducing food waste goes a long way too (Karandish et al., 2020). But if you are standing in the grocery store, contemplating tonight's dinner, how will you know if you are choosing the best product from a water perspective? A major obstacle for consumers to making an informed decision is that many products do not come with the necessary transparency on water performance (nor on other sustainability aspects, for that matter). If you are of the activist type, pressuring companies to provide this information might thus be a sensible thing to do. You might be surprised to see how sensitive companies can be to feedback that may harm their reputation. But beyond your own diet and idealistic activism, have you thought about promoting water sustainability through the ballot box? And how about your personal investments? Do you hold stocks in water-wasting companies? Is your bank mindful about water sustainability or will they lend your savings to water-polluting businesses? Where does your pension fund invest your retirement savings?

### 4.4.3 What can companies do?

Actions to reduce WFs that can be taken by food sector companies, including fishers, fertilizer producers, processors, traders, retailers, transporters, markets, tech companies, and others, again cover a wide spectrum of solutions. In general, though, we postulate that each can and should adopt Best Available Practices in terms of water for their respective business processes and products. What is particularly important to note is that, although reducing water use and pollution in their direct operations is imperative, the bulk of the WF is often in the supply chain, especially in agricultural products (Linneman, Hoekstra, and Berkhout, 2015). WFs in supply chains thus need to be addressed too. Fortunately, there are increasingly better water stewardship programmes and tools that help guide companies on their water journey. The WFA Global Standard (Hoekstra et al., 2011), for example, provides a systematic, science-based methodology to measure WFs along corporate value chains and provides guidelines to help formulate quantitative WF reduction targets within and beyond the company fence.

### 4.4.4 What can investors do?

Financial institutions, including asset owners (banks, pension funds, insurance companies), asset managers, rating agencies, and regulators, are a powerful yet enigmatic actor group that to date has hardly wielded their influence in water-for-food discourses. Lagging behind companies, investors by and large turn a blind eye to the effect that the activities they finance may have on (the future state of) water resources. This is unfortunate, as investing in, say, a large-scale avocado plantation in a water-scarce area today will have implications for consumption of local water resources for decades to come. A recent assessment by (Hogeboom, Kamphuis, & Hoekstra (2018)) confirmed that there is still a lot to be done to build investor awareness on the reality and potential impact of water crises, their role therein, and the development of sound, water-informed investment policies.

## 4.4.5 What can international organizations do?

The example above in Section 4.4.1 on national and global water saving strategies already illustrated that there is a global dimension to water management beside the traditional local one (Vörösmarty, Hoekstra, Bunn, Conway, and Gupta, 2015). In fact, one could argue that the global market dictates to a large extent which crops and food products are produced. Therefore, international organizations, such as the United Nations, multilateral development banks, and regional collaborations such as the European Union, seem logical actors to address water considerations associated with supra-national trade flows. Trade agreements through the World Trade Organization, for example, include a so-called non-discrimination clause, meaning countries cannot differentiate between similar products in trade (Hoekstra, 2020). Consequently, countries cannot favor trade partners on the basis of differences in water performance of products either: according to WTO rules, a tomato is a tomato, regardless the WF or impact on water resources that the cultivation of that tomato has. Trade agreements could thus be renegotiated to accommodate for differentiation in products based on water and/or other sustainability criteria.

Another potential action by international organizations is to negotiate a water version of the Kyoto Protocol. Such a protocol for water could define fair shares for WFs of consumption per capita, coupled to an international WF pricing scheme (Hoekstra, 2020).

## 4.4.6 What can civil society, media, and academia do?

A final, broad category of actors that can take meaningful water action in the food domain are civil society groups, the media, and academia. Civil society groups can be particularly effective in putting water-for-food discussions higher on political and corporate agendas. Media, in turn, have an important role to play in unearthing detrimental water performance by other actors, for example by diving deeply into often hidden supply chains of companies or investigative journalism of water impact of governmental policies. Lastly, academics can contemplate, research and develop both technical and non-technical (behavioural, governance, political) solutions to transforming food systems into water-wise food systems.

## References

Aldaya, M.M., Hoekstra, A.Y., 2010. The water needed for Italians to eat pasta and pizza. Agricultural Systems 103, 351–360.

Aldaya, M.M., Rodriguez, C.I., Fernandez-Poulussen, A., Merchan, D., Beriain, M.J., Llamas, R., 2020. Grey water footprint as an indicator for diffuse nitrogen pollution: the case of Navarra, Spain. The Science of the Total Environment 698, 134338.

Allan, J.A., 1998. Virtual water: a strategic resource global solutions to regional deficits. Groundwater 36, 545–546.

Allen, R.G., Pereira, L.S., Raes, D., Smith, M., 1998. Crop evapotranspiration: Guidelines for computing crop water requirements. FAO Irrigation and drainage paper no, 56. FAO, Rome, Italy.

Carr, J.A., D'Odorico, P., Laio, F., Ridolfi, L., 2012. On the temporal variability of the virtual water network. Geophysical Research Letters 39, L06404.

Chouchane, H., Krol, M.S., Hoekstra, A.Y., 2018a. Expected increase in staple crop imports in water-scarce countries in 2050. Water Research X 1, 100001.

Chouchane, H., Krol, M.S., Hoekstra, A.Y., 2018b. Virtual water trade patterns in relation to environmental and socioeconomic factors: a case study for Tunisia. The Science of the Total Environment 613-614, 287–297.

Chouchane, H., Krol, M.S., Hoekstra, A.Y., 2020. Changing global cropping patterns to minimize national blue water scarcity. Hydrology and Earth System Sciences 24, 3015–3031.

Chukalla, A.D., Krol, M.S., Hoekstra, A.Y., 2015. Green and blue water footprint reduction in irrigated agriculture: effect of irrigation techniques, irrigation strategies and mulching. Hydrology and Earth System Sciences 19, 4877–4891.

Chukalla, A.D., Krol, M.S., Hoekstra, A.Y., 2017. Marginal cost curves for water footprint reduction in irrigated agriculture: guiding a cost-effective reduction of crop water consumption to a permit or benchmark level. Hydrology and Earth System Sciences 21, 3507–3524.

Chukalla, A.D., Krol, M.S., Hoekstra, A.Y., 2018a. Grey water footprint reduction in irrigated crop production: effect of nitrogen application rate, nitrogen form, tillage practice and irrigation strategy. Hydrology and Earth System Sciences 22, 3245–3259.

Chukalla, A.D., Krol, M.S., Hoekstra, A.Y., 2018b. Trade-off between blue and grey water footprint of crop production at different nitrogen application rates under various field management practices. The Science of the Total Environment 626, 962–970.

Davis, K.F., Rulli, M.C., Seveso, A., D'Odorico, P., 2017a. Increased food production and reduced water use through optimized crop distribution. Nature Geoscience 10, 919–924.

Davis, K.F., Seveso, A., Rulli, M.C., D'Odorico, P., 2017b. Water savings of crop redistribution in the United States. Water (Basel) 9, 83.

Doorenbos, J., Kassam, A.H., 1979. Yield response to water. FAO Irrigation and Drainage Paper 33. Food and Agriculture Organization of the United Nations, Rome, Italy.

Ercin, A.E., Aldaya, M.M., Hoekstra, A.Y., 2011. Corporate water footprint accounting and impact assessment: the case of the water footprint of a sugar-containing carbonated beverage. Water Resources Management 25, 721–741.

Ercin, A.E., Aldaya, M.M., Hoekstra, A.Y., 2012. The water footprint of soy milk and soy burger and equivalent animal products. Ecological Indicators 18, 392–402.

Falkenmark, M., 2000. Competing freshwater and ecological services in the river basin perspective. Water International 25, 172–177.

Franke, N.A., Boyacioglu, H., Hoekstra, A.Y., 2013. Grey water footprint accounting: tier 1 supporting guidelines. Value of Water Research Report Series No. 65. UNESCO-IHE, Delft, the Netherlands.

Gobin, A., Kersebaum, K.C., Eitzinger, J., Trnka, M., Hlavinka, P., Takáč, J., et al., 2017. Variability in the water footprint of arable crop production across European regions. Water (Basel) 9, 93.

Haghighi, E., Madani, K., Hoekstra, A.Y., 2018. The water footprint of water conservation using shade balls in California. Nature Sustainability 1, 358–360.

Harris, F., Moss, C., Joy, E.J.M., Quinn, R., Scheelbeek, P.F.D., Dangour, A.D., et al., 2020. The water footprint of diets: a global systematic review and meta-analysis. Advances in nutrition 11, 375–386.

Hoekstra, A.Y., 2003. Virtual water: An introduction. In: HOEKSTRA, A.Y. (Ed.), Virtual water trade: proceedings of the international expert meeting on virtual water trade, Value of Water Research Report Series No.12. UNESCO-IHE, Delft, the Netherlands, pp. 13–23.

Hoekstra, A.Y., 2014. Sustainable, efficient, and equitable water use: the three pillars under wise freshwater allocation. Wiley Interdisciplinary Reviews: Water 1, 31–40.

Hoekstra, A.Y., 2015. The sustainability of a single activity, production process or product. Ecological Indicators 57, 82–84.

Hoekstra, A.Y., 2017. Water footprint assessment: evolvement of a new research field. Water Resources Management 31, 3061–3081.

Hoekstra, A.Y., 2019a. Green-blue water accounting in a soil water balance. Advances in water resources 129, 112–117.

Hoekstra, A.Y., 2019b. Lecture notes 'Water'. Faculty of Engineering Technology. Enschede. University of Twente, the Netherlands.

Hoekstra, A.Y., 2020. The water footprint of modern consumer society, Second edition Routledge, London, UK.

Hoekstra, A.Y., Chapagain, A.K., 2008. Globalization of water: Sharing the planet's freshwater resources. Blackwell Publishing, Oxford.

Hoekstra, A.Y., Chapagain, A.K., Aldaya, M.M., Mekonnen, M.M., 2011. The water footprint assessment manual: Setting the global standard. Earthscan, London.

Hoekstra, A.Y., Mekonnen, M.M., 2012. The water footprint of humanity. Proceedings of the National Academy of Sciences of the United States of America 109, 3232–3237.

Hoekstra, A.Y., Mekonnen, M.M., Chapagain, A.K., Mathews, R.E., Richter, B.D., 2012. Global monthly water scarcity: blue water footprints versus blue water availability. Plos One 7, e32688.

Hogeboom, R.J., 2020. The water footprint concept and water's grand environmental challenges. One Earth 2, 218–222.

Hogeboom, R.J., Borsje, B.W., Deribe, M.M., van der Meer, F.D., Mehvar, S., Meyer, M.A., et al., 2021. Resilience meets the water–energy–food nexus: mapping the research landscape. Frontiers in Environmental Science 9, 630395.

Hogeboom, R.J., de Bruin, D., Schyns, J.F., Krol, M.S., Hoekstra, A.Y., 2020. Capping human water footprints in the world's river basins. Earth's Future 8, e2019EF001363.

Hogeboom, R.J., Hoekstra, A.Y., 2017. Water and land footprints and economic productivity as factors in local crop choice: the case of silk in Malawi. Water (Basel) 9, 802.

Hogeboom, R.J., Kamphuis, I., Hoekstra, A.Y., 2018. Water sustainability of investors: Development and application of an assessment framework. Journal of Cleaner Production 202, 642–648.

Holmatov, B., Hoekstra, A.Y., Krol, M.S., 2019. Land, water and carbon footprints of circular bioenergy production systems. Renewable & Sustainable Energy Reviews 111, 224–235.

Hoogeveen, J., Faurès, J.M., Peiser, L., Burke, J., van de Giesen, N., 2015. GlobWat – a global water balance model to assess water use in irrigated agriculture. Hydrology and Earth System Sciences 19, 3829–3844.

Jalava, M., Guillaume, J.H.A., Kummu, M., Porkka, M., Siebert, S., Varis, O., 2016. Diet change and food loss reduction: what is their combined impact on global water use and scarcity? Earth's Future 4, 62–78.

Jalava, M., Kummu, M., Porkka, M., Siebert, S., Varis, O., 2014. Diet change—A solution to reduce water use? Environmental Research Letters 9, 074016.

Karandish, F., Hoekstra, A.Y., Hogeboom, R.J., 2018. Groundwater saving and quality improvement by reducing water footprints of crops to benchmarks levels. Advances in water resources 121, 480–491.

Karandish, F., Hoekstra, A.Y., Hogeboom, R.J., 2020. Reducing food waste and changing cropping patterns to reduce water consumption and pollution in cereal production in Iran. Journal of hydrology 586, 124881.

Kim, B.F., Santo, R.E., Scatterday, A.P., Fry, J.P., Synk, C.M., Cebron, S.R., et al., 2020. Country-specific dietary shifts to mitigate climate and water crises. Global Environmental Change 62, 101926.

Kummu, M., de Moel, H., Porkka, M., Siebert, S., Varis, O., Ward, P.J., 2012. Lost food, wasted resources: global food supply chain losses and their impacts on freshwater, cropland, and fertiliser use. The Science of the Total Environment 438, 477–489.

Linneman, M.H., Hoekstra, A.Y., Berkhout, W., 2015. Ranking water transparency of Dutch stock-listed companies. Sustainability 7, 4341–4359.

Liu, J., Yang, H., 2010. Spatially explicit assessment of global consumptive water uses in cropland: green and blue water. Journal of hydrology 384, 187–197.

Masud, M.B., McAllister, T., Cordeiro, M.R.C., Faramarzi, M., 2018. Modeling future water footprint of barley production in Alberta, Canada: implications for water use and yields to 2064. The Science of the Total Environment 616-617, 208–222.

Masud, M.B., Wada, Y., Goss, G., Faramarzi, M., 2019. Global implications of regional grain production through virtual water trade. The Science of the Total Environment 659, 807–820.

Mekonnen, M.M., Fulton, J., 2018. The effect of diet changes and food loss reduction in reducing the water footprint of an average American. Water International 43, 860–870.

Mekonnen, M.M., Gerbens-Leenes, W., 2020. The water footprint of global food production. Water (Basel) 12, 2696.

Mekonnen, M.M., Hoekstra, A.Y., 2010. A global and high-resolution assessment of the green, blue and grey water footprint of wheat. Hydrology and Earth System Sciences 14, 1259–1276.

Mekonnen, M.M., Hoekstra, A.Y., 2011. The green, blue and grey water footprint of crops and derived crop products. Hydrology and Earth System Sciences 15, 1577–1600.

Mekonnen, M.M., Hoekstra, A.Y., 2012. A global assessment of the water footprint of farm animal products. Ecosystems 15, 401–415.

Mekonnen, M.M., Hoekstra, A.Y., 2014. Water footprint benchmarks for crop production: a first global assessment. Ecological Indicators 46, 214–223.

Mekonnen, M.M., Hoekstra, A.Y., 2015. Global gray water footprint and water pollution levels related to anthropogenic nitrogen loads to fresh water. Environmental Science & Technology 49, 12860–12868.

Mekonnen, M.M., Hoekstra, A.Y., 2016. Four billion people facing severe water scarcity. Science Advances 2, e1500323.

Mekonnen, M.M., Hoekstra, A.Y., 2018. Global anthropogenic phosphorus loads to freshwater and associated grey water footprints and water pollution levels: a high-resolution global study. Water Resources Research 54, 345–358.

Mekonnen, M.M., Hoekstra, A.Y., 2020a. Blue water footprint linked to national consumption and international trade is unsustainable. Nature Food 1, 792–800.

Mekonnen, M.M., Hoekstra, A.Y., 2020b. Sustainability of the blue water footprint of crops. Advances in water resources 143, 103679.

Mekonnen, M.M., Hoekstra, A.Y., Neale, C.M.U., Ray, C., Yang, H.S., 2020. Water productivity benchmarks: the case of maize and soybean in Nebraska. Agricultural Water Management 234, 106122.

Mubako, S.T., 2018. Blue, green, and grey water quantification approaches: a bibliometric and literature review. Journal of Contemporary Water Research & Education 165, 4–19.

Nouri, H., Stokvis, B., Galindo, A., Blatchford, M., Hoekstra, A.Y., 2019. Water scarcity alleviation through water footprint reduction in agriculture: the effect of soil mulching and drip irrigation. The Science of the Total Environment 653, 241–252.

Pahlow, M., van Oel, P.R., Mekonnen, M.M., Hoekstra, A.Y., 2015. Increasing pressure on freshwater resources due to terrestrial feed ingredients for aquaculture production. The Science of the Total Environment 536, 847–857.

Postel, S.L., Daily, G.C., Ehrlich, P.R., 1996. Human appropriation of renewable fresh water. Science 271, 785–788.

Rockström, J., Gordon, L., 2001. Assessment of green water flows to sustain major biomes of the world: implications for future ecohydrological landscape management. Physics and Chemistry of the Earth, Part B: Hydrology, Oceans and Atmosphere 26, 843–851.

Schyns, J.F., 2018. Sustainable and efficient allocation of limited blue and green water resources [PhD thesis]. University of Twente, Enschede, the Netherlands.

Schyns, J.F., Hoekstra, A.Y., 2014. The added value of water footprint assessment for national water policy: a case study for Morocco. Plos One 9, e99705.

Schyns, J.F., Hoekstra, A.Y., Booij, M.J., 2015. Review and classification of indicators of green water availability and scarcity. Hydrology and Earth System Sciences 19, 4581–4608.

Schyns, J.F., Hoekstra, A.Y., Booij, M.J., Hogeboom, R.J., Mekonnen, M.M., 2019. Limits to the world's green water resources for food, feed, fiber, timber, and bioenergy. Proceedings of the National Academy of Sciences of the United States of America 116, 4893–4898.

Siebert, S., Döll, P., 2010. Quantifying blue and green virtual water contents in global crop production as well as potential production losses without irrigation. Journal of hydrology 384, 198–217.

Steduto, P., 2006. Biomass water-productivity: Comparing the growth-engines of crop models. WUEMED training course "Integrated approaches to improve drought tolerance in crops". Bologna, Italy.

Steduto, P., Hsiao, T.C., Raes, D., Fereres, E., 2009. AquaCrop-the FAO crop model to simulate yield response to water: I. concepts and underlying principles. Agronomy Journal 101, 426–437.

UN-WWAP, 2006. The United Nations World Water Development Report 2006 - Water: A shared responsibility. United Nations Educational, Scientific and Cultural Organization (UNESCO), Paris, France.

Vanham, D., Comero, S., Gawlik, B.M., Bidoglio, G., 2018. The water footprint of different diets within European sub-national geographical entities. Nature Sustainability 1, 518–525.

Vanham, D., Hoekstra, A.Y., Bidoglio, G., 2013a. Potential water saving through changes in European diets. Environment International 61, 45–56.

Vanham, D., Medarac, H., Schyns, J.F., Hogeboom, R.J., Magagna, D., 2019. The consumptive water footprint of the European Union energy sector. Environmental Research Letters 14, 104016.

Vanham, D., Mekonnen, M.M., Hoekstra, A.Y., 2013b. The water footprint of the EU for different diets. Ecological Indicators 32, 1–8.

Vörösmarty, C.J., Hoekstra, A.Y., Bunn, S.E., Conway, D., Gupta, J., 2015. Fresh water goes global. Science 349, 478–479.

Vörösmarty, C.J., McIntyre, P.B., Gessner, M.O., Dudgeon, D., Prusevich, A., Green, P., et al., 2010. Global threats to human water security and river biodiversity. Nature 467, 555–561.

Wackernagel, M., Rees, W., 1996. Our ecological footprint: reducing human impact on the Earth, Gabriola Island. New Society Publishers, Gabriola Island, Canada, 176p.

Wöhler, L., Hoekstra, A.Y., Hogeboom, R.J., Brugnach, M., Krol, M.S., 2020a. Alternative societal solutions to pharmaceuticals in the aquatic environment. Journal of Cleaner Production 277, 124350.

Wöhler, L., Niebaum, G., Krol, M., Hoekstra, A.Y., 2020b. The grey water footprint of human and veterinary pharmaceuticals. Water Research X 7, 100044.

Ye, Q., Li, Y., Zhuo, L., Zhang, W., Xiong, W., Wang, C., et al., 2018. Optimal allocation of physical water resources integrated with virtual water trade in water scarce regions: a case study for Beijing, China. Water Research 129, 264–276.

Zhang, Y., Huang, K., Yu, Y., Yang, B., 2017. Mapping of water footprint research: a bibliometric analysis during 2006–2015. Journal of Cleaner Production 149, 70–79.

Zhuo, L., Liu, Y., Yang, H., Hoekstra, A.Y., Liu, W., Cao, X., et al., 2019. Water for maize for pigs for pork: an analysis of inter-provincial trade in China. Water Research 166, 115074.

Zhuo, L., Mekonnen, M.M., Hoekstra, A.Y., 2016a. Benchmark levels for the consumptive water footprint of crop production for different environmental conditions: a case study for winter wheat in China. Hydrology and Earth System Sciences 20, 4547–4559.

Zhuo, L., Mekonnen, M.M., Hoekstra, A.Y., 2016b. The effect of inter-annual variability of consumption, production, trade and climate on crop-related green and blue water footprints and inter-regional virtual water trade: a study for China (1978–2008). Water Research 94, 73–85.

Zhuo, L., Mekonnen, M.M., Hoekstra, A.Y., Wada, Y., 2016c. Inter- and intra-annual variation of water footprint of crops and blue water scarcity in the Yellow River basin (1961–2009). Advances in water resources 87, 29–41.

# Land use modeling: from farm to food systems

## H.H.E. van Zanten[a], A. Muller[b,c] and A. Frehner[a,b]

[a]Farming Systems Ecology group, Wageningen University and Research, Wageningen, Netherlands [b]Department of Socioeconomics, Research Institute of Organic Agriculture FiBL, Frick, Switzerland [c]Institute of Environmental Decisions, Federal Institutes of Technology Zurich ETHZ, Zurich, Switzerland

## 5.1 Introduction

Land is a strictly limited resource and therefore affects nutrition security. Furthermore, land use is a central concern as it is associated with critical processes affecting the living space for human societies, such as climate change, biosphere integrity and biochemical flows. Worldwide, roughly about 4.8 billion ha is used for agriculture (FAOstat 2018). Of the total agricultural area, about one third is crop land - of which about a third is used to produce feed - while grazing land occupies the other two-thirds (FAOstat 2018).

Assuming the world population grows to 10 billion people by mid-century, then just 0.16 ha of cropland will be available per person in 2050. Production of an average vegan diet already requires about 0.13 ha of arable land per person per year, thus this exceeds the 0.16 ha of arable land when we account for current food waste and losses (Van Zanten et al., 2018). Currently, a considerable amount of agricultural production is lost or wasted, resulting in increased land requirements beyond the areas needed to provide a sufficient diet for all people. Although in recent decades, agricultural production has become increasingly efficient - resulting in less land needed to produce the same amount of food - we also see an increased pressure for sourcing more land due to fertility losses, erosion, etc. on cropland areas that have been used intensively for decades. Expanding the area for crop production further will lead to losses of grazing areas or deforestation in the tropics, resulting in loss of biodiversity and increased carbon emissions (Foley et al., 2011; IPCC 2019). These factors demonstrate the urgency to better understand the dynamics, impacts, potentials and challenges of future land use and land use change. To do this, we apply land various use modeling approaches.

So how do we model land use? Envision an area of agricultural land. What do you see? A natural grazing area with sheep or goats? Or do you see a crop land with grains or soybeans?

We know that each area of land is different - think about soils or climate conditions – which defines their current uses. The question, therefore, is how can we compare land use within food systems when our land is so highly diverse? This question becomes even more complex when we aim to allocate a certain land use to specific food items. For example, soybeans; the cultivation of soybeans can yield soybean oil for human consumption and soybean meal used as animal feed. So, how should we allocate the land used between the oil and the meal?

The questions addressed above are those faced by researchers when assessing the land use of our food system. This is, logically, a highly debated subject. Especially when we talk about the land use related to animal products, researchers come to contradictory conclusions - for example regarding which types of animal source food (ASF) should be reduced to minimize land use or sustainability impacts in general, and by how much (Ridoutt, Hendrie, and Noakes, 2017; Schader et al., 2015; Van Zanten et al., 2018). It is also unclear what optimal production systems should look like to achieve these improvements (Muller et al., 2017; Poore and Nemecek, 2018). Some studies conclude that ASF from monogastric animals (mainly pork and poultry) should be favored over ruminants. Others argue that production systems operating with low-opportunity-cost biomass (LCB), such as grassland-based ruminant production, can most sustainably contribute to food security. The latter is based on the argument that these systems avoid competition with resources that could be directly used for human food production. The differing results are mainly caused by the chosen methodologies, reflecting differences in time-horizon (current versus future), scale (farm versus country assessments), scope (narrow focus on single products and direct upstream value chains versus broad focus on systemic aspects including co- and by-product flows and their use), reference unit (per unit product impacts: efficiency measures versus total impacts), and demographics (individual versus population) (Frehner et al., 2020).

Understanding why these differences occur is highly important as they can lead to confusion and hinder the decision-making process and therefore the transition towards more sustainable food systems. In this chapter, we explain the main differences between the methods available, the differences in results, and their advantages and disadvantages. We address three commonly applied methods: Life cycle assessment (LCA) (section 5.2), the Land Use Ratio (LUR) (section 5.3) and a Food Systems Approach (section 5.4). These have been selected as they illustrate how conclusions can differ based on the chosen methodology due to, for example, differences in system boundaries (from farm to food systems). Each section describes the methodology, its relevance and provides examples of results. We end this chapter with an overview of the pros and cons of the different methodologies and, in section 5.5, we provide recommendations. This chapter focuses on land use issues related to the production of animal source food commodities, as these are the hotly debated commodities regarding their role in sustainable healthy diets.

## 5.2 Life cycle assessment

### 5.2.1 Methodology and relevance

LCA is an internationally accepted and standardized method for evaluating the environmental impact of a product through the entire production chain. To assess land use, LCA studies generally focus on the land used in m$^2$ per year to provide a certain amount of output, expressed as a functional unit (**FU**). The FU can be a kg of harvested grain, but it can also be a kg of protein from egg production. The boundaries of the system can range from a farm assessment to assessing a complete value chain. In chapter 3, the methodological aspects of an LCA are discussed in more detail. However, in this chapter we highlight one methodological aspect crucial to understanding the contradictory results between different studies related to the land use of food items: the allocation method between multi-functional processes.

A multi-functional process is one that yields more than one output. Examples of multi-functional processes are sugar production, where beet pulp is produced as a byproduct in addition to the main product, sugar, or wheat flour production, where middlings are produced as a byproduct. Although it is recognized by ISO that the allocation of the environmental impact between the different multi-functional processes should be avoided – for example by using expanding the system - it still is commonly done. When applying allocation, the environmental impact of a multi-functional process is allocated between the outputs based on their underlying physical relationship, such as the relative weight or calorie contents of each output, or on their relative economic value, the so-called "economic allocation". Most LCA studies providing a fixed impact intensity of agricultural products are attributional LCA (**ALCA**) studies based on economic allocation. The land use of a certain food item, therefore, also depends on its economic value at a certain moment in time.

## 5.2.2 Example results

LCA can be applied at a product level, focusing on the environmental impact of the production process (**production-oriented**) or at a diet level, focusing on the environmental impact of a diet (**diet-oriented**). LCAs focusing on the production process quantify the land needed to produce one kg of a certain product. To produce one kg of ASF, LCAs implicitly combine information about crop productivity (yield per ha) and animal productivity (feed efficiency along the chain, including breeding, rearing, and producing animals). The LCA results are often used to compare different food items (Fig. 5.1) or to explore solutions to reduce the environmental impact along the production chain.

Fig. 5.1 shows that ASF products have a higher environmental impact, including land use, than plant-based products. However, there are large differences between ASF products, as beef in general has a much higher land use compared to, for example, eggs. We furthermore see a higher variation per kg of beef compared to kg of eggs or chicken meat. This high variation is caused by the variation in production systems, for example outdoor grazing versus feedlot. As LCAs often do not differentiate between arable land use and grasslands, high variations can be found between different ruminant production systems based on higher or lower shares of grassland use as shown in Fig. 5.1.

Solutions often address increasing environmental efficiency along the chain (Van Zanten et al., 2018). They include strategies to increase crop yield per input of natural resource or per amount of pollutants lost to the environment, strategies to increase feed efficiency or life-time-productivity, or to combat disease. They also cover strategies to reduce enteric methane emissions from ruminants (such as feed additives) or to change feed composition to reduce the impact of feed (replacing feed ingredients with a high impact, e.g., soybean meal), strategies

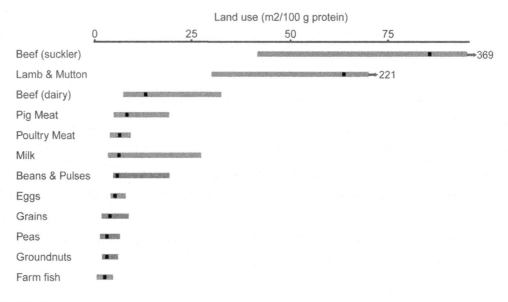

**FIGURE 5.1**   Land use per food item based on data from Poore and Nemecek, 2018.

to reduce emissions from manure in stables and storage systems, and those that reduce waste losses along the chain. These solutions favor, for example, a transition from pasture based-beef production to producing beef in feedlots, or more generally from grazing systems to mixed crop-livestock systems, and breeding of high yielding, feed use efficient, and robust animals. Thus, in general, these strategies all aim at increasing the land use efficiency of the single product and the underlying processes relevant for its production, which, in total, indicates the need for intensifying production systems along several key dimensions related to product quantities. On the downside, this strategy tends to neglect performance regarding other impact measures not related to production volumes but related to regional ecosystem boundaries, such as nutrient surplus per hectare.

LCAs focusing on human diets are based on the same metrics as the production-oriented studies but have a different focus. While production-oriented strategies focus on reducing footprints of single commodities, consumption-oriented dietary strategies compare footprints of different diets and focus on reducing the whole diet's footprint by changing its commodity composition, without addressing potential changes of the footprints of the various commodities themselves. Consumption-oriented dietary strategies are mostly assessed by scenario analysis, ranging from **predefined scenarios** to **optimized scenarios** (Frehner et al., 2020). Examples of predefined scenarios are assessments of vegan or vegetarian diets (Scarborough et al., 2014) as well as of dietary guidelines (Springmann et al., 2020). Examples of optimized scenarios are minimizations of dietary costs or certain environmental impacts (Benvenuti, De Santis, Di Sero, and Franco, 2019; Ferrari et al., 2020), or of the difference from current diets, while fulfilling certain environmental goals such as reducing land use or GHG emissions (Gazan et al., 2018). This form of scenario can indicate short-term improvements that do not deviate too much from current diets. These consumption-oriented dietary solutions are

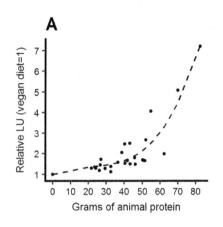

**FIGURE 5.2**  Relative reduction in land use per g of protein from animal source food (ASF). Each dot represents a dietary scenario. Dietary scenarios were derived from peer reviewed studies that assessed the environmental impact of different diets. Excerpted from Fig. 1 of Van Zanten et al. (2018). Licensed under Creative Commons BY 4.0.

based on the following approach. Studies first define existing or desired consumption patterns, ideally with similar nutritional value. The environmental impact of each diet is subsequently determined by multiplying the amount of each food product consumed with its associated footprint. Thus, LCAs are often employed to assess environmental impacts of aggregate food consumption or individual diets (Chukalla et al., 2018b; Chukalla, Krol, and Hoekstra, 2018a). For the former, food consumption of a certain region – typically a country – is estimated based on food supply data. Possibly some data conversion is still required, such as to derive food consumption values from food supply values via the consideration of food waste fractions. By multiplying these food consumption values with footprints of food products, total environmental impacts of the respective region can be estimated. Different methods are used to assess environmental impacts for individual dietary habits. An example is dietary recalls, where people recall what they ate over a specific period (usually 24 h). These values can, again, be multiplied with environmental footprints to derive the total footprint of individual diets. It is important to note that these assessments of both aggregate food consumption and individual diets mostly relate environmental impacts to the macronutrient protein or to calories, and hence do not capture the full nutritional function of the different foods including amino acid profiles, vitamins, and mineral contents. This is further complicated by interaction effects between different nutrients and foods, and other aspects such as lifestyle factors.

In sum, the results of the consumption-oriented dietary studies focus on changing consumption patterns by reducing or avoiding consumption of ASF, or shifting from ASF with a higher environmental impact, such as beef, to ASF with a lower environmental impact, such as chicken meat or fish (Hallström, Carlsson-Kanyama, and Börjesson, 2015). Most consumption-oriented dietary studies focus on affluent countries, and therefore usually address a reduction in consumption of ASF. Fig. 5.2 summarizes results of consumption-oriented dietary studies that evaluated at least two diets differing in the percentage of ASF, including one vegan diet. Results illustrate that increasing the amount of ASF in a diet exponentially increases land use. Overall, all LCA consumption-oriented dietary studies conclude that a vegan diet will always result in the lowest land use.

## 5.3 Land use ratio

LCA studies assess land use based on the total area needed to produce one unit of a commodity. The methodology, therefore, does not account for differences in consumption of humanly-edible products by livestock or differences in suitability of land used for feed production to directly cultivate food crops. In other words, it does not account for the competition between humans and animals for land. For our future food supply, it might be better not to use highly productive croplands to produce animal feed because, regardless of how efficiently animals are raised, using these land resources to produce biomass streams for direct human consumption is always more land use efficient (Foley et al., 2011). Feed produced on land suitable for human food production, therefore, may result in competition for land between feed and food.

**Feed-food competition** can occur directly or indirectly. Direct competition occurs when biomass suitable for direct human consumption is fed to animals instead of humans (Wilkinson and Lee, 2018). Indirect competition occurs when feed is cultivated on areas where food for direct human consumption could be cultivated (Van Zanten et al., 2016). Thus, either directly or indirectly, land is allocated to produce biomass used to feed farm animals instead of humans.

One way to measure this competition for land is to compute humanly-edible energy and protein conversion ratios (Wilkinson, 2011). These represent the amount of energy or protein in animal feed that is potentially edible for humans compared to the amount of energy or protein in that animal product that is edible for humans. Ratios above 1 indicate that the animals produce less edible energy or protein than they consume (Wilkinson, 2011). A ratio below 1 means the animals produce more edible energy or protein than they consume and implies more efficient land use (meaning that direct feed-food competition is avoided). This is not true, however, if the non-edible feed is produced on land suited for direct human food production (the so called indirect competition). Thus, these conversion ratios do not immediately imply efficient land use in terms of global food security, because the ratios do not yet include, for example, that grass fed to dairy cows can also be produced on land suitable for the cultivation of human food crops. In other words, the opportunity costs of land for human food production are not considered (Van Zanten et al., 2016).

### 5.3.1 The methodology and relevance

Van Zanten et al. (2016) developed a land use efficiency assessment method called Land Use Ratio (**LUR**) to deal with feed-food competition. LUR is calculated by defining the maximum amount of human digestible protein (**HDP**) that could be derived from food-crops on all land used to cultivate the feed required to produce one kilogram ASF, over the amount of human digestible protein in that one kilogram ASF (Eq. 5.1).

$$LUR = \frac{\sum_{i=1}^{n}(LO_{ij} \times HDP_j)}{HDP \, of \, one \, kg \, ASF} \qquad (5.1)$$

where $LO_{ij}$ is the land area occupied for a year ($m^2 \, yr^{-1}$) to cultivate the amount of feed ingredient i ($i = 1,n$) in country j ($j = 1,m$) that is needed to produce one kg ASF, including breeding and rearing of young stock. $HDP_j$ is the maximum amount of human-digestible protein (HDP) that can be produced per $m^2$ $yr^{-1}$ by direct cultivation of food crops in country j. The denominator is the HDP of one kg ASF.

An LUR lower than 1.0 implies that livestock produce more HDP per $m^2$ than the food crops that could be grown there. Only if the LUR is zero, which can occur by solely feeding co-products and food-waste, for example, or by cows only grazing on marginal land, is feed-food competition completely avoided. In terms of global food supply, however, livestock systems with an LUR lower than 1 are more efficient than crops because they produce more HDP per $m^2$ than crops.

To compute the LUR of one kg ASF from a specific livestock system, four steps are required. First, quantify the land area occupied ($LO_{ij}$) to cultivate the amount of each feed ingredient ($i = 1,n$) in the different countries of origin of the feed ($j = 1,m$) needed to produce one kg ASF. Second, assess the suitability of each land area currently occupied to produce feed to directly grow human food crops, using a crop suitability index (such as the Global Agro-Ecological Zones) to estimate the percentage of feed crop land that is also suitable for growing food crops. By doing so, the different land classes including their yields are considered which is one of the major differences between the LUR and LCA studies. The land suitability is needed to calculate the potential amount of HDP from food crops which can be grown on the current area used to grow feed crops. Third, for each area of land suitable for direct cultivation of food crops ($LO_{ij}$), the maximum $HDP_j$ from cultivation of food crops need to be determined by combining information about crop yield per ha for each suitable crop, with its protein content, and human digestibility. The amount of HDP that can be produced on all land required for feed production is summed and used as numerator. Last, assess the amount of HDP in one kg ASF, the denominator (Van Zanten et al., 2016).

## 5.3.2 Example results

The LUR shows that in the current situation, ruminant systems have in general a lower LUR value compared to monogastric systems in European production systems (for example Van Zanten et al., 2016). Ruminant systems in general use land less suitable for food crop production such as marginal grasslands on peat soils (for example Van Zanten et al., 2016). Monogastric systems only derive a value below 1 if they are fed with food waste or co-products of a low value for other use. In general, we see that most current monogastric systems in the EU have an LUR >1 as their diets consist of large amounts of small grains, maize, and meals such as soybean meal or rapeseed meal and pulses. In the current situation, feed-food competition is therefore higher in monogastric production systems compared to the ruminant systems. Nevertheless, in cases where ruminants are mainly kept in confinement, their ratios exceed the values of monogastric production systems by far.

LUR results imply that a considerable amount of grassland could in principle be used for crop production. Bear in mind, however, such land use change would lead to adverse effects such as soil carbon and biodiversity losses, Therefore, this possible alternative use of grassland is often undesirable from a more holistic perspective that addresses environmental indicators beyond cropland use. On the other hand, areas of intensively used grasslands with low species diversity and high nutrient inputs and correspondingly low ecological value, could be transformed into croplands. Furthermore, the LUR also implicitly assumes that all currently available grassland on marginal lands should be used for grazing. A large proportion of this land, however, was previously covered by forests and to reach biodiversity conservation targets it is probable that grazing would need to cease in some places. As such, there is an opportunity cost entailed in rearing livestock. Not all biomass from grasslands can be considered 'free', without or with merely low opportunity costs for livestock production. To increase biodiversity, we recommend, to identify which grasslands can be considered 'available' for animals, which for crop production, and which should be left for rewilding or other conservation strategies (Van Zanten et al., 2018).

### 5.3.3 Allocation

The LUR proposed by Van Zanten et al. (2016) is based on the LCA framework, as most LCA studies use economic allocation for the LU between the multifunctional processes. The LU related to sugar beets, for example, is allocated to various products (sugar, molasses, and beet pulp) based on their relative economic values. The greater the relative economic value of the main product, sugar, or a co-product, the greater the portion of LU related to the cultivation of the underlying primary product allocated to this main or co-product. The economic value, however, does not represent the suitability of a product for human consumption. Van Hal, Weijenberg, De Boer, and Van Zanten (2019) therefore presented a novel allocation method that reflects the (un)suitability of feed products for human consumption. This food-based allocation assigns zero environmental impact to byproducts that humans cannot or do not want to consume, whereas the determining (food) product is given full allocation. For example, 100 percent of the environmental impact of producing one kg of sunflower seed is now allocated to the sunflower oil as this is the only edible end-product (Fig. 5.3). Although this food-based allocation is simplified and binary - a single product is allocated all the impact of cultivation and processing if suitable for human consumption, and none if unsuitable - it illustrates the complexity of feed-food competition within the food system and how different methodological assumptions can significantly influence results.

To evaluate the impact of accounting for feed-food competition on LCA results, Van Hal et al. (2019) compared economic and food-based allocation in an LCA of a novel egg production system - called Kipster - that only feeds chickens with products unsuitable or undesired for human consumption. Using economic allocation, the land use was 2.99 $m^2$, and the land use ratio 1.70 per kg egg, lower than that of free range or organic eggs. Accounting for feed-food competition with food-based allocation further reduced impacts per kg egg by 96 percent for LU to 0.11 $m^2$, and by 88 percent for LUR to 0.30.

The example given by Van Hal et al. (2019) illustrates the importance of accounting for feed-food competition. However, capturing the complete nutritional values within the allocation methods is complex. This requires implementing a measure expressing a nutritional value that includes multiple nutritional aspects, and not only protein or calorie contents such as the nutrient density score (Van Kernebeek et al., 2014). Allocation, therefore, remains a core challenge of studies that assess land use along the chain.

To conclude, the LCA approach provides insights in strategies to minimize the land-footprint while the LUR provides insights in strategies to minimize feed-food competition.

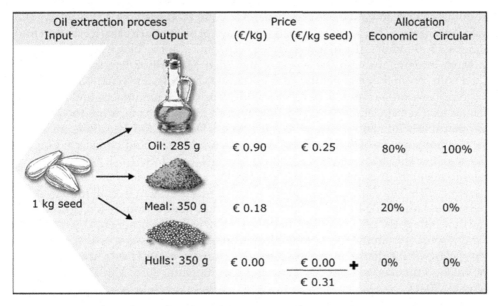

| Oil extraction process | | Price | | Allocation | |
| Input    Output | | (€/kg) | (€/kg seed) | Economic | Circular |
|---|---|---|---|---|---|
| 1 kg seed | Oil: 285 g | € 0.90 | € 0.25 | 80% | 100% |
| | Meal: 350 g | € 0.18 | | 20% | 0% |
| | Hulls: 350 g | € 0.00 | € 0.00 + | 0% | 0% |
| | | | € 0.31 | | |

**FIGURE 5.3** Environmental impact allocation of the co-products resulting from the multifunctional process sunflower seed crushing under economic and circular allocation. Figured derived from Van Hal et al., 2019.

Each approach makes assumptions about the nature of the problem and comparing the results of the two different optimization strategies provides a great opportunity to define strategies of sustainable land management.

## 5.4 Food systems approach – accounting for food system level interlinkages and circularity

Assessing land use along the chain falls short in addressing the complexity of food systems because interlinkages - such as the fact that we cannot produce milk without producing meat from the slaughtered calves and dairy cows or sugar without producing beet pulp - in the food system are not acknowledged. Although the LUR can capture feed-food competition, it is unable to consistently tackle the allocation issue. To redesign today's food system to make optimal use of the earth's natural resources, we need to move away from the current product footprint approach and start using a food-systems approach. In such an approach, the focus is not on efficiency measures, like land use per unit commodity or for a human diet, but on the total impact of the whole food system, derived from all underlying interlinked processes, mass and nutrient flows, only putting these total impacts in relation to single commodities or human diets at the end. Therefore, the food systems approach does not use footprints as key building blocks to derive which diets may be most sustainable, but as end results to be communicated along with other performance indicators. In the case of animal production systems this is crucial because if we aim to avoid feed-food competition and to reduce land use, the role of animals will be limited to the amount of non-competing feed-stuff available.

This implies a shift towards using animals in circular food systems. Let us elaborate a bit more on what a circular food system entails. The concept of circularity has its roots in multiple disciplines such as industrial ecology, ecological economics, agroecology, and general systems theory. More recently, through the work of the Ellen MacArthur Foundation, it has been combined with concepts of regenerative design, cradle-to-cradle, and biomimicry. In a circular food system, losses are prevented, or recovered for reuse, for remanufacturing, and recycling. In a circular food system, arable land is primarily used to produce nutritious foods from plant biomass that fulfils human nutritional requirements (De Boer and Van Ittersum, 2018; Van Zanten, Van Ittersum, and De Boer, 2019). Edible byproducts should be recycled in the food system – together with byproducts inedible for humans (e.g. straw, husk and human excreta) – in order to maintain or improve the soil and fertilize crops, or to feed animals (De Boer and Van Ittersum, 2018; Van Zanten et al., 2019). A circular food system overall implies searching for practices and technologies that minimize the input of finite resources (e.g. phosphate rock and land), encourage the use of regenerative ones (e.g. wind and solar energy), prevent leakage of natural resources from the food system (e.g. nitrogen (N), phosphorus (P)), and stimulate reuse/recycling of inevitable resource losses (e.g. human excreta) in a way that adds the highest value to the food system (De Boer and Van Ittersum, 2018; Van Zanten et al., 2019; Muscat et al., 2021).

By recycling byproducts, waste, and biomass from grasslands (further referred to as low-opportunity cost biomass (**LCB**)) into the food system, farm animals can play an important role in feeding humanity. These farm animals - further referred to as low-cost livestock - will no longer consume human-edible biomass, such as grains, but convert biomass that we cannot or do not want to eat into valuable food, manure, and other ecosystem services (Fig. 5.4). LCB streams consist of biomass from grassland and "leftovers": crop residues left over from harvesting food crops, co-products left over from industrial processing of plant-source and animal-source food, and losses and waste in the food system. By converting these LCB streams, low-cost livestock recycle nutrients back into the food system that otherwise would have been lost in food production. The availability of LCB and the type of animal production systems used largely determine the boundaries for animal production and consumption. Van Zanten et al., 2018 demonstrated that farm animals raised solely on these LCB streams could provide a significant part (9–23 g/per capita) of our daily protein needs (50–60 g/per capita) (Fig. 5.5). With this concept, competition for land for feed or food would be minimized (Van Zanten et al., 2018).

The current average global supply of terrestrial animal protein (excluding fish) per capita is 27 g per day (FAOSTAT, 2017), while there are large differences between countries. For example, the average European supply is 102 g of protein per capita per day, of which 51 g is terrestrial animal protein, while the average West African supply is 62 g of protein per day, of which just 8 g is terrestrial animal protein. Notably, the average European supply substantially exceeds protein requirements. As the amount of daily per capita ASF protein that could be sourced from low-cost livestock ranges globally from 9 to 23 g, it could potentially fulfill a useful part of our daily protein needs – while enabling arable land to be dedicated to the cultivation of food crops. Nevertheless, this still requires a high reduction in the consumption of ASF in high income-countries.

Diets containing animal protein from low-cost livestock use less arable land than a vegan diet and considerably less arable land than the current diets in high-income countries

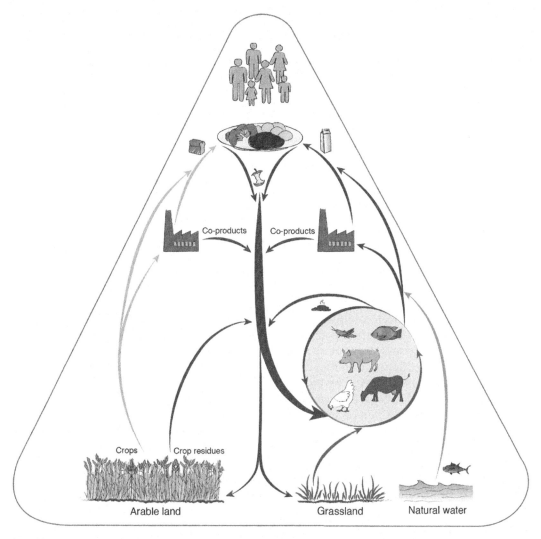

**FIGURE 5.4** The biophysical concept of circularity: arable land is primarily used for food production; biomass unsuited for direct human consumption is recycled as animal feed; byproducts and manure are used to maintain soil fertility. In this way, nutrients are recycled and animals contribute to a circular food system while sustainably feeding the future population. Fig. derived from Van Zanten et al., 2019).

(Röös et al., 2017; Schader et al., 2015; Van Kernebeek, Oosting, Van Ittersum, Bikker, and De Boer, 2016; Van Zanten et al., 2016). However, low-cost livestock diets require the use of grasslands for feed biomass, whereas grasslands remain unused by vegan diets. Hence, defining a 'biodiversity boundary' is essential to properly specifying a low-cost livestock system. The reduction in arable land largely depends on the plant-based foods we consume (refined grains versus whole grains) and the amount of food we waste. Reducing the amounts of co-products or food wasted will benefit the environment directly but will reduce the amount

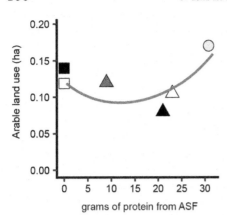

**FIGURE 5.5** Illustration of the theoretical relation between the percentage of animal protein in the human diet and the total amount of arable land used to produce human food. The Fig. is based on global studies assessing the principle of livestock with low-opportunity-costs: Schader et al. (2015), dark grey; Van Zanten et al. (2016), black; and, Röös et al. (2017) white. The squares represent a vegan diet; triangles represent the value of low-cost livestock; and the circle (light grey) represents a current diet. It clearly shows that arable land use is most efficient with a moderate consumption of protein from livestock with low-opportunity-costs. Adapted from Fig. 3 of Van Zanten et al. (2018), licensed under Creative Commons BY 4.0.

of ASF produced from low-cost livestock. This brings us back to the fundamental goal of food production, providing healthy food for humans. Current dietary habits in most high-income countries are a major risk factor for multiple non-communicable diseases such as diabetes and cancer (Afshin et al., 2019). To address this problem, research should focus on how to provide food compatible with dietary guidelines in a sustainable way.

Recently, the EAT-Lancet commission issued a healthy reference diet that respects the share of the planetary boundaries available for food systems (Willett et al., 2019). Moreover, Van Selm et al. (2021) assessed to what extent these healthy dietary guidelines are compatible with feed-food competition-free animal-source foods. In other words, is it possible to produce the recommended amount of ASF by only feeding the animals with the LCB from the plant-based part of the EAT-Lancet reference diet. The authors found that it is difficult to meet the exact composition of the different animal-source foods, in particular for the case of chicken versus beef. Chicken is seen as the most sustainable meat source while broilers seem to be less suitable to convert LCB. In a circular food system, most of the animal source proteins come from milk and beef related to the dairy sector (Frehner et al., 2021a, Van Selm et al., 2021). This example illustrates the importance of addressing food consumption and associated human health aspects in the analysis of land use in agriculture and food production systems and underlines the need for a food systems approach.

To conclude, the key difference between assessing the land use of ASF based on a chain approach (LCA and LUR) versus a food systems approach is that the food systems approach allows to define a land use boundary. In other words, it estimates a range in which animal source foods do not compete with direct production of plant-based foods.

## 5.5 Pros and cons and recommendations for land use modeling

In this section we aim to provide an overview of the differences between the three method-ologies - LCA, LUR and the food systems approach - described in this chapter (Table 5.1). We first presented the LCA methodology commonly used in consumption-oriented dietary changes. Consumption-oriented studies rely on fixed impact intensities derived from LCA studies that assess land use of a certain product at a certain time. The time component is

**TABLE 5.1** Overview of the difference between the life cycle assessment (LCA), land use ratio (LUR) and food systems approach (FSA).

|  | LCA | LUR | FSA |
| --- | --- | --- | --- |
| Boundary | Chain | Chain | Food system |
| Time-line | Current | Current | Future |
| Accounts for feed food competition | No | Yes | Yes |
| Avoids allocation per default | No | No | Yes |
| Accounts for land quality | No | Yes | Yes |
| Favored scenario for minimized land use | Vegan diet | Small amount of ASF | Small amount of ASF |
| Type of ASF favored | Chicken meat, milk, eggs, fish | All animals fed with LCB | Milk, beef from dairy cows, small amount of pork, fish and eggs, chicken meat from laying hens |
| Application | Identifying hotspots | Identifying hotspots | Food systems redesigns |

an important element as most LCAs studies use economic allocation reflecting the economic value of a certain product at a certain time. Examples of these scenario specifications are dietary guidelines, vegetarian scenarios, or the reduction of all red and processed meat. These studies mostly recommend greatly reducing beef and pork, while chicken, dairy, eggs, and fish can be reduced to a lesser extent – or, in some cases, even be increased and taken as substitute. The results of consumption-oriented dietary studies clearly show the current environmental hotspots within a static non-changing production context and, as such, provide a starting point to define strategies for future improvements.

Second, we presented the LUR methodology, showing that results depend on the suitability of land, and that animal production systems can play an important role in the food system by providing high quality proteins. With this, land use can be assessed while considering feed-food competition. The LUR shows how land use can be assessed by considering different utilization options which in turn are defined by the suitability of resources. Thereby, this assessment provides information on the maximum amount of humanly digestible protein that can be derived from the respective utilization. The results show that if we feed animals with products that we cannot or do not want to eat, or we keep animals on land unsuited for crop production, animal production systems can be more efficient in terms of land use compared to plant production systems, but only if absolute animal numbers stay within the bounds of these feeding methods. The results from LUR studies clearly illustrate that current monogastric production systems yield higher values of feed-food competition compared to ruminant productions systems in European production systems, especially ruminant systems grazing on marginal lands. The method however does not account for the interconnections within the food system and still relies on (economic or other) allocation to divide land use among multi-functional processes.

Last, we presented the food systems approach. These studies explicitly consider the intrinsic interlinkages between different products produced in multi-functional processes, such as milk and meat from dairy production. These studies thus avoid the need for allocation of environmental impacts of multi output products to the single products as a prerequisite for producing results. For ASF, this is particularly relevant for the relation milk – meat in dairy production systems, and likewise for eggs – meat in egg production systems. As an example, employing a vegetarian scenario on a population-level scale would result in inconsistencies, since the associated meat of the milk and egg production is not considered in the consumer-oriented dietary LCA studies. Accounting for interlinkages is especially important to identify future food system redesigns, as large food system changes affect the interlinkages.

To sum up, we would like to emphasize two main conclusions:

1. In spite of methodological differences, all methods conclude that a reduction in ASF is recommended for high-income (and high ASF consumption) countries to reduce land use. Studies based on the LCA methodology focusing on consumption-oriented dietary scenarios tend to suggest that beef should be reduced most followed by pork and chicken, whereas studies applying food system modeling approaches tend to recommend larger reductions for chicken and pork and beef (from beef and not from dairy cattle). Notably, in the former approaches, the appropriate role of ASF is based on assessing current production systems, while in the latter, the appropriate role of ASF is based on potential alternative systems in which animals are fed mainly low-opportunity-cost biomass. Within this limit, feed-food competition is avoided, and circularity principles can be considered.
2. Each method has its own strengths. LCA offers the best estimates of how we should change our diets at the margin in the near term. However, the food systems approach provides key insights for deciding what systems we want in the long term. Large-scale dietary changes will cause a cascade of effects and require cautious consideration of co-product links and suitability and scarcity of resources. To best assess these changes and propose suitable solutions, food system modeling studies should be applied, as fixed impact intensities derived from LCA studies are unable to capture the full consequences of such changes. Fixed impact intensities derived from LCA studies are, however, adequate to assess the impacts within the current food system or of relatively small systems, where changes have negligible effects only beyond narrowly chosen boundaries. Researchers can use fixed impact intensities derived from LCA studies to assess land use of new novel food or feed to gain insights into the land use of a production process with the aim of finding opportunities to optimize the operation in terms of land use. Researchers furthermore can use fixed impact assessments to identify land use hotspots within the current food system. These hotspots can be used to identify opportunities to reduce the current land use of the food system. Making these underlying assumptions explicit in future scientific work is key for improved transparency regarding dietary solutions, thus enabling effective communication and policy design.

# Acknowledgement

This chapter is an extensive synthesis of the work we have been conducting over the last decade (Van Zanten et al., 2018, 2019; Frehner et al., 2020, 2021 and Van Hal et al., 2019) together with many other researchers, which we hereby want to acknowledge for their valuable contributions.

# References

Afshin, A., Sur, P.J., Fay, K.A., Cornaby, L., Ferrara, G., Salama, J.S., et al., 2019. Health effects of dietary risks in 195 countries, 1990–2017: a systematic analysis for the Global Burden of Disease Study 2017. The Lancet 393 (10184), 1958–1972.

Benvenuti, L., De Santis, A., Di Sero, A., Franco, N., 2019. Concurrent economic and environmental impacts of food consumption: are low emissions diets affordable? Journal of Cleaner Production 236, 1–19.

De Boer, I.J., Van Ittersum, M.K., 2018. Circularity in agricultural production. Wageningen University & Research, Wageningen, the Netherlands.

Ferrari, M., Benvenuti, L., Rossi, L., De Santis, A., Sette, S., Martone, D., et al., 2020. Could Dietary Goals and Climate Change Mitigation Be Achieved Through Optimized Diet. The Experience of Modeling the National Food Consumption Data in Italy. Frontiers in nutrition. 7 (48), 1–13. http://doi.org/10.3389/fnut.

Foley, J.A., Ramankutty, N., Brauman, K.A., Cassidy, E.S., Gerber, J.S., Johnston, M., et al., 2011. Solutions for a cultivated planet. Nature 478 (7369), 337–342.

Frehner, A., Muller, A., Schader, C., De Boer, I.J.M., Van Zanten, H.H.E., 2020. Methodological choices drive differences in environmentally-friendly dietary solutions. Global Food Security 24, 1–8.

Frehner, A., Van Zanten, H.H.E., Schader, C., De Boer, I.J.M., Pestoni, G., Rohrmann, S., et al., (2021a). How food choices link sociodemographic and lifestyle factors with sustainability impacts. Journal of Cleaner Production 300, 1–14.

Frehner, A., Cardinaals, R.P.M., De Boer, I.J.M., Muller, A., Schader, C., Van Selm, B. et al., (2021b). Integrating circularity principles with animal-source food recommendations in national dietary guidelines: nutritional and environmental consequences. Submitted.

Gazan, R., Brouzes, C.M., Vieux, F., Maillot, M., Lluch, A., Darmon, N., 2018. Mathematical optimization to explore tomorrow's sustainable diets: a narrative review. Advances in nutrition 9 (5), 602–616.

Hallström, E., Carlsson-Kanyama, A., Börjesson, P., 2015. Environmental impact of dietary change: a systematic review. Journal of Cleaner Production 91, 1–11.

IPCC, (2019). IPCC special report on climate change, desertification, land degradation, sustainable land management, food security, and greenhouse gas fluxes in terrestrial ecosystems.

Muller, A., Schader, C., Scialabba, N.E.-H., Brüggemann, J., Isensee, A., Erb, K.-H., et al., 2017. Strategies for feeding the world more sustainably with organic agriculture. Nature communications 8 (1), 1–13.

Muscat, A., De Olde, E.M., Ripoll-Bosch, R., Van Zanten, H.H.E., Metze, T.A.P., Termeer, C.J.A.M., et al., 2021. Principles, drivers and opportunities of a circular bioeconomy. Nature Food (2), 561–566.

Poore, J., Nemecek, T., 2018. Reducing food's environmental impacts through producers and consumers. Science 360 (6392), 987–992.

Ridoutt, B.G., Hendrie, G.A., Noakes, M., 2017. Dietary strategies to reduce environmental impact: a critical review of the evidence base. Advances in nutrition 8 (6), 933–946.

Röös, E., Bajželj, B., Smith, P., Patel, M., Little, D., Garnett, T., 2017. Greedy or needy? Land use and climate impacts of food in 2050 under different livestock futures. Global Environmental Change 47, 1–12.

Scarborough, P., Appleby, P.N., Mizdrak, A., Briggs, A.D., Travis, R.C., Bradbury, K.E., et al., 2014. Dietary greenhouse gas emissions of meat-eaters, fish-eaters, vegetarians and vegans in the UK. Climatic Change 125 (2), 179–192.

Schader, C., Muller, A., Scialabba, N.E.-H., Hecht, J., Isensee, A., Erb, K.-H., et al., 2015. Impacts of feeding less food-competing feedstuffs to livestock on global food system sustainability. Journal of the Royal Society Interface 12 (113), 1–12.

Springmann, M., Spajic, L., Clark, M.A., Poore, J., Herforth, A., Webb, P., et al., 2020. The healthiness and sustainability of national and global food based dietary guidelines: modelling study. BMJ 370 (m2322), 1–16.

Van Hal, O., Weijenberg, A., De Boer, I.J.M., Van Zanten, H.H.E., 2019. Accounting for feed-food competition in environmental impact assessment: towards a resource efficient food-system. Journal of Cleaner Production 240, 8pp.

Van Kernebeek, H., Oosting, S., Feskens, E., Gerber, P., De Boer, I., 2014. The effect of nutritional quality on comparing environmental impacts of human diets. Journal of Cleaner Production 73, 88–99.

Van Kernebeek, H.R., Oosting, S.J., Van Ittersum, M.K., Bikker, P., De Boer, I.J., 2016. Saving land to feed a growing population: consequences for consumption of crop and livestock products. International Journal of Life Cycle Assessment 21 (5), 677–687.

Van Selm, B., Frehner, A., De Boer, I.J.M., Van Hal, O., Hijbeek, R., Van Ittersum, M.K., et al., 2021. The compatibility of animal-source food and circularity in healthy European diets. Accepted with revisions in *Nature Food*. Under open-review. http://doi.org/10.21203/rs.3.rs-147410/v1NatureFood.

Van Zanten, H.H.E., Meerburg, B.G., Bikker, P., Herrero, M., De Boer, I.J.M., 2016. Opinion paper: the role of livestock in a sustainable diet: a land-use perspective. Animal 21, 18–22.

Van Zanten, H.H.E., Mollenhorst, H., Klootwijk, C.W., Van Middelaar, C.E., De Boer, I.J.M., 2016. Global food supply: land use efficiency of livestock systems. International Journal of Life Cycle Assessment 21 (5), 747–758.

Van Zanten, H.H.E., Herrero, M., Van Hal, O., Röös, E., Muller, A., Garnett, T., et al., 2018. Defining a land boundary for sustainable livestock consumption. Global Change Biology 24 (9), 4185–4194.

Van Zanten, H.H.E., Van Ittersum, M.K., De Boer, I.J.M., 2019. The role of farm animals in a circular food system. Global Change Biology 21, 18–22.

Wilkinson, J.M., 2011. Re-defining efficiency of feed use by livestock. Animal 5, 1014–1022. doi:https://doi.org/10.1017/S175173111100005X.

Willett, W., Rockström, J., Loken, B., Springmann, M., Lang, T., Vermeulen, S., et al., 2019. Food in the Anthropocene: the EAT–Lancet Commission on healthy diets from sustainable food systems. The Lancet 393 (10170), 447–492.

Wilkinson, J., Lee, M., 2018. Use of human-edible animal feeds by ruminant livestock. Animal 12 (8), 1735–1743.

# Foodshed analysis and carrying capacity estimation

*Christian J. Peters*

**USDA, Agricultural Research Service, Food Systems Research Unit, Burlington, VT, United States.**

## 6.1 Introduction

Have you ever wondered about the potential of a place to feed its people? Perhaps you asked yourself where your food comes from. Maybe you debated about the type of diet that feeds the most people. It could be you pondered population growth and how many people the world can feed. If any of these are true, then you have already, perchance unwittingly, begun to think in terms of foodsheds and carrying capacity.

The methods covered in this chapter address two related issues. Foodshed analyses inquire about the past, present, or future sources of food for a particular place. Carrying capacity studies examine the ability of a specific geographic area to feed people. These two issues are so closely linked that they can be hard to tell apart. Nonetheless, the nuance is important. The first issue focuses on food consumption – how do we feed the people in this place – whereas the second issue emphasizes food production – how many people can this place feed? As you will see in this chapter, there is merit in thinking about food from both directions.

Before diving into the details, a word of caution is warranted. The study of foodsheds and carrying capacity are emerging areas of research and neither the terminology nor the methods are standardized. Thus, the goals of this chapter are to explain the conceptual foundations of each approach, to distill the essential elements of foodshed and carrying capacity models, and to show what each approach can teach us about food systems. Giving deference to the older term, we will begin with carrying capacity.

## 6.2 History and conceptual foundations

### 6.2.1 Carrying capacity of human populations

The term "carrying capacity" has a complicated history. Among biologists and demographers, the concept gets traced to Thomas Malthus's famous work *An Essay on the Principle of Population* in which Malthus claimed that populations naturally grow at an exponential rate but must ultimately be limited by the available food supply (Seidl and Tisdell, 1999). Malthus' principle was expressed mathematically by Francois Verhulst in the 1830s and Raymond Pearl in the 1920s using a model of population growth now known as the Verhulst-Pearl logistic Eq., which states that a population grows at a rate *r* that slows as the population approaches a certain size *K* at which point the population plateaus (Seidl and Tisdell, 1999). However, as Sayre (2008) has pointed out, the term "carrying capacity" was not used to refer to this population limit until the mid-20th century. Rather, the term "carrying capacity" originally applied to measurements of a ship's cargo storage capacity and morphed over time to be used in the context of rangeland management, the ability of land to support livestock, and then to the capacity of ecosystems to support a certain size population, eventually being applied to humans (Sayre, 2008). The term now commonly refers to Earth's capacity to support people (see, for example, Cohen, 1995 and Butler, 2004).

Taken at face value, this idea seems reasonable. No population can grow forever without reaching some limit. Quantifying this limit, however, is where problems arise. Cohen (1995) summarized the results of 65 global carrying capacity estimates conducted between 1679 and 1994 and identified six distinct methodologies whose results span three orders of magnitude, from one billion to one trillion persons. Cohen (1997) argues that differences in methodology explain why different studies reach different conclusions and examining a particularly high estimate proves illustrative. Franck, von Bloh, Müller, Bondeau, and Sakschewski (2011) follow the logic of a 1967 study that estimated Earth's carrying capacity at one trillion people based on the rate of terrestrial photosynthesis. Using a global vegetation model to calculate net primary productivity, Franck et al. (2011) trim the number down to 282 billion people. Bear in mind, that this estimate assumes humanity's food needs are met by feeding all people a *vegetarian* diet using *all* productive land at *maximum* potential caloric output. As the authors demonstrate, sparing land for forests and accounting for increased urban land use decreases the estimate further. In essence, the assumptions behind the carrying capacity analysis matter.

For the purposes of this chapter, I will follow the definition offered by Susiarjo, Sreenath, and Vali (2006), "Carrying capacity is the maximum population size of a given species that an area can support without reducing its ability to support the same population of the species in the future." This definition has the advantage of being applicable at multiple spatial scales and being sensitive to change over time. At the risk of stating the obvious, be aware that the definition of 'support' lies in the mind of the modeler, particularly when applying carrying capacity to people. In other words, determining how much a system can supply and what people need requires making assumptions about the technology and resources used and the quality of life the system will provide. Carrying capacity is a constantly moving target.

Before launching into models of carrying capacity, a few related concepts must be introduced. First, the concept of an "ecological footprint" refers to the equivalent area of land required to both supply natural resources and manage wastes from human economies, implying that carrying capacity is a limit to the 'load' people place on the environment rather than the number of people the environment can support (Rees, 1996). Similar to this idea, the concept of "planetary boundaries" refers to limits within which humanity's activities must operate in order to avoid irreparably damaging the natural systems and services upon which we depend, such as a stable climate (Rockström et al., 2009). Finally, "planetary overload" refers to a condition in which human carrying capacity has been exceeded (Butler, 2016).

Within food systems research, carrying capacity takes a Malthusian bent, referring to the ability to meet human food needs. Nevertheless, carrying capacity need not be a loaded term. Rather, it can be interpreted as a measure of the productive output of an agricultural landscape counted not in the biomass harvested as crops or livestock products but instead as the number of persons fed (e.g., Cassidy, West, Gerber, and Foley, 2013; Peters, Fick, and Wilkins, 2003). Seen in this light, carrying capacity is a property of an ecosystem rather than of a species, measuring the potential of the land to feed people given a set of assumptions about the amount people eat, the composition of the diet, and the productivity of available agricultural technology.

## 6.2.2 The concept of a "foodshed"

The origin of the term "foodshed" can be traced to a little known and out-of-print book entitled *How Great Cities Are Fed* (Hedden, 1929). Written in the 1920s, the book describes the food systems of major American cities, primarily New York City, as they existed in the early 20th century. The book owes its existence to research done in the face of a possible nationwide railroad strike to understand how the disruption might impact New York's food supply. No strike occurred, and crisis was averted. Yet the book illuminated an eerie truth about the food system, namely, no one fully understood it.

Hedden used the term foodshed in the context of economic geography, trying to explain why particular foods are sourced from certain places and not others. He describes the concept as similar to a watershed but distinct in that "the barriers which guide and control movements of foodstuffs are more often economic than physical" (Hedden, 1929 p.17). He elaborates with a range of examples, showing how distance, tariffs, freight rates, and seasonality influence where a city sources its food. In this context, the term foodshed can be defined as the geographic area that supplies an area with food.

The term "foodshed" dropped into obscurity to be resurrected decades later in the context of local and alternative food systems. In the seminal paper, "Coming into the foodshed," Kloppenburg, Hendrickson, and Stevenson (1996) offer an additional interpretation of the foodshed concept. Rather than describing the flow of food, a foodshed can also be viewed as a place for taking action, a specific geography in which new food systems can be organized and created. This second definition has resonated with the local and alternative food movements and should be viewed as related to but distinct from Hedden's original idea (Peters, Bills, Wilkins, and Fick, 2009).

Since Kloppenburg et al. (1996), a growing body of literature on foodshed assessment has emerged. Reviews of this literature trace the start of foodshed analysis to the early 2000s with most papers and reports published since about 2007 (Freedgood, Pierce-Quiñonez, and Meter, 2011; Horst and Gaolach, 2015; Schreiber, Hickey, Metson, Robinson, and MacDonald, 2021).

This work focuses primarily on the capacity for self-supply of food at different spatial scales and on flows of food at the sub-national scale (Schreiber et al., 2021). From this body of work, it appears that both Hedden's definition and Kloppenburg's definition of foodshed are still active, with some work focusing on existing foodsheds and other work examining potential foodsheds. While the definitions appear to be stable, the methods have evolved.

Early work on foodshed assessment in North America focused on understanding the capacity to localize food supplies across a range of scales, from an individual city to a multi-state region (Horst at Gaolach, 2015). Over time, the range of scales considered has expanded to include studies conducted at the national or global scale that consider the foodsheds of many cities (or population centers) simultaneously (Schreiber et al., 2021). In addition, while the methods used are not standardized, foodshed assessments tend to examine either capacity for providing food within a given geographic area or on measuring the flows of food (Schreiber et al., 2021). This chapter will focus on the study of capacity, distinguishing between two methods used in this research, the use of net-balance approaches to estimate self-sufficiency and spatial analysis to estimate the size of potential foodsheds.

### 6.2.3 Relationship between the two ideas

Foodsheds and carrying capacity are related but distinct concepts. Think of them as two sides of the same coin. Carrying capacity measures the ability of an ecosystem, most likely an agro-ecosystem, to feed people. Foodsheds demarcate the geographic area from which food is, or could be, supplied. Bear in mind, however, that this terminology is not consistent in the literature, so when in doubt, refer to how the authors define and use these terms.

## 6.3 Methodology

The methods used to estimate carrying capacity and to assess foodsheds continue to evolve. This chapter covers three types of approaches commonly encountered in the literature: (1) net balance approaches, (2) carrying capacity models, and (3) spatial optimization. Each is discussed in a general way and elaborated upon with case study examples.

### 6.3.1 Net balance approaches

One common approach to understanding the capacity of a place to meet its food needs is a net-balance or self-sufficiency analysis. A net-balance is like a budget. You tabulate inflows and outflows to determine if a system has a surplus or a deficit. Many fields outside of business and economics use budgets to understand dynamics in systems. For example, in agricultural and environmental sciences, nutrient budgets help to determine the amount of fertilizer needed to support crop growth or to identify areas in which excess application might lead to nutrient runoff and associated pollution. Within the realm of food systems, net-balance approaches are used to determine if a place is a net-importer or exporter of food. Such calculations are commonly performed at the national scale, measuring the balance of trade in terms of dollars or tons of product. In the context of understanding local or regional food production capacity, these studies are also done at the sub-national scale.

Estimates of self-reliance compare production to consumption, generating a self-sufficiency ratio (SSR) or percentage (Eq. (6.1)). The SSR is reported as the ratio of production (P) to consumption (C) in a particular place (i) for a given food (j) at a specific point in time (k). Values >1.0 indicate the region is a net producer, and values <1.0 indicate that a region is a net consumer. However, one cannot infer much about the flow of food from an SSR number. A net-producing region may still import the same product from other regions, particularly if the product is perishable and produced only seasonally. Moreover, self-reliance assessments consider the balance of food produced and consumed but not necessarily the source of the inputs used (such as fertilizer) to grow those foods.

$$SSR_{ijk} = P_{ijk}/C_{ijk} \qquad (6.1)$$

As hypothetical examples, consider three products: lettuce, apples, and fluid milk. Lettuce is highly perishable, so providing a year-round supply requires year-round production. Apples, on the other hand, can be stored for up to a year under controlled atmosphere conditions, so a year-round supply is feasible even with a single production season. Milk, like lettuce, has a limited shelf life, and is typically produced year-round in U.S. systems, regardless of location. In each case, interpretation of the SSR value requires additional information about the production systems and the consumption patterns. Before delving into interpretation, however, it is worth examining how SSR calculations are performed in greater detail. The simplicity of Eq. (6.1) belies the complexity involved.

### 6.3.1.1 Sample calculation – fluid milk SSR

To illustrate how net-balance approaches work, consider the case of fluid milk in New England. Six states comprise this region of the United States: Connecticut, Maine, Massachusetts, New Hampshire, Rhode Island, and Vermont. Fluid milk is produced throughout the US, though historically, production has concentrated in the humid, temperate climates within the conterminous U.S. State-level estimates of production are available (USDA National Agricultural Statistics Service, 2021) but regional consumption must be estimated from national per capita estimates of consumption (USDA Economic Research Service, 2021) and state-level estimates of population (U.S. Census Bureau, 2016). Data needed to calculate the SSR are described in Table 6.1.

Pause a moment to reflect on the data used in this example. The product, fluid milk, is produced in bulk on farms and is only rarely consumed directly. Milk sold or served for human consumption is pasteurized and packaged before being ingested, but the example relies on milk production at the farm gate. In other words, it is a proxy for dairy products. Following this logic, per capita consumption is represented in the aggregate using the *fluid milk equivalent* of all dairy products included in the U.S. food supply, the raw material needed to make the assortment of whole and reduced-fat milk, cream, butter, yogurt, cheeses, and ice cream. An advantage to this approach is that consumption inherently includes product lost along the supply chain due to processing losses, spoilage, plate waste or other reasons. However, the consumption estimate derived by multiplying New England's population by U.S. per capita consumption may not reflect regional consumption patterns.

**TABLE 6.1**  Metadata for a self-sufficiency ratio calculation. Estimating potential regional self-reliance of the New England states to supply fluid milk.

| Descriptor | Production | Per Capita consumption | Population |
|---|---|---|---|
| Data product | Annual Survey Data | Food Supply Data | Intercensal Estimates of Population |
| Data source | USDA National Agricultural Statistics Service | USDA Economic Research Service | U.S. Census Bureau |
| Data item | Milk production | Per capita availability, all dairy products | Population on July 1 |
| Description | Fluid milk production from dairy cows | Total consumption of dairy products on a milk equivalent basis | Annual estimates of population made between decennial censuses |
| Units | pounds | pounds per capita | persons |
| Years | 1991–2010 | 1991–2010 | 1991–2010 |
| Geography | New England, by state | National, US | New England, by state |

Following Eq. (6.1), the self-reliance of New England for supplying its own fluid milk in the year 2010 can be calculated as follows:

$$P_{ijk} = 4,027,500,000 \text{lbs}$$
$$C_{ijk} = 603.2 \text{lbs fluid milk per capita} \times 14,457,499 \text{persons}$$
$$8,721,141,793 \text{lbs}$$
$$SSR_{ijk} = 4,027,500,000/8,721,141,793$$
$$= 0.46181$$
$$= 46.2\% \text{ (rounded to three significant digits)}$$

Based on this example, the SSR of New England for fluid milk in 2010 was 0.462. In other words, production of fluid milk in 2010 was equivalent to 46.2 percent of the region's total fluid milk consumption.

### 6.3.1.2 *Value of the net-balance approach*

The preceding example demonstrates some of the limitations of available data to assess self-sufficiency (Freedgood et al., 2011; Horst and Gaolach, 2015; Schreiber et al., 2021). Analyses tend to rely on the same data sources, using *national* per-capita estimates of food consumption and *state-level* population data to estimate regional consumption, missing any regional differences in food preferences. Likewise, production is estimated either at the farm-scale or the first stage of food processing, since publicly available data do not estimate production further "downstream" in the food supply. Imports and exports are carefully tracked, but data collection at the subnational scale permit only a limited understanding of food flows. Thus, self-sufficiency analyses are truly mass-balance assessments. They do not indicate how much a geographic area actually supplies its own food.

**TABLE 6.2**  Summary results for a study of food self-reliance of the Northeast U.S. using a net-balance approach. Adapted from Griffin et al. (2015).

| Food group | Weight basis | Production (10³ Mg) | Consumption (10³ Mg) | Self-sufficiency ratio |
|---|---|---|---|---|
| Grains | weight of milled product | 1,150 | 14,627 | 0.08 |
| Pulses | weight of dry beans, peas, or lentils | 15 | 212 | 0.07 |
| Vegetables | fresh weight | 2,953 | 11,837 | 0.25 |
| Fruits | fresh weight | 1,389 | 7,622 | 0.18 |
| Dairy | fluid milk equivalents | 13,043 | 17,079 | 0.76 |
| Eggs | eggs in shell | 676 | 946 | 0.71 |
| Meat | carcass weight | 2,411 | 11,496 | 0.21 |
| Fish and shellfish | edible weight | 395 | 1,360 | 0.29 |
| Oils | product weight | 1,396 | 14,398 | 0.10 |
| Sweeteners | dry weight | 290 | 3,752 | 0.08 |

Despite these deficiencies, net-balance approaches provide valuable insight into the capacity of a geographic area to supply food. To illustrate, consider an analysis by Griffin, Conrad, Peters, Ridberg, and Tyler (2015), which includes a thorough explanation of the calculations and decisions regarding data sources involved in conducting a self-reliance analysis. Griffin et al. (2015) rely on agricultural censuses and surveys and on fisheries landings combined with data on food processing conversions to estimate the quantity of primary food commodities produced in the 12-states of the Northeast U.S. (Connecticut, Delaware, Maine, Maryland, Massachusetts, New Hampshire, New Jersey, New York, Pennsylvania, Rhode Island, Vermont, and West Virginia). As in the fluid milk example, Griffin et al. (2015) use per capita food availability as a proxy for consumption and multiply national per capita estimates times the Northeast population to calculate regional demand. The approach is simple, but the results are illuminating.

Summarized by food group, the results clearly show that the Northeast U.S. is a net food importer (Table 6.2). Dairy products have the highest self-sufficiency ratio, but still fall below 1.0. Indeed, the region has a higher self-sufficiency for animal-sourced foods than plant-sourced foods, despite the fact that the region grows a diversity of crops. The analysis considered more than 100 food commodities tracked in the food supply, so the averages hide a few cases where production of an individual crop, like cabbage, exceeds regional consumption. On the other hand, subsequent research showed that the region is also a net feed importer, meaning that the Northeast's self-sufficiency in producing livestock products should be adjusted downward (Conrad, Tichenor, Peters, and Griffin, 2016). Overall, the net balance approach makes a convincing case that this populous region of the United States is a strong net food importer.

## 6.3.2 Carrying capacity modeling

### 6.3.2.1 *Comparison of different approaches*

Carrying capacity can be determined by any factor that influences the rates of reproduction or mortality of a species. However, carrying capacity analysis often focuses on the ability of natural resources to feed people. Within this vein, some studies explore highly theoretical limits to the ability of Earth to feed humanity under scenarios in which all terrestrial photosynthesis is appropriated for food production (Binder, Holdahl, Trihn, and Smith, 2020; Franck et al., 2011). Other studies examine historical examples of societies reaching the carrying capacity of their agricultural systems in the rare instances where sufficient data are available (Fanta, Šálek, Zouhar, and Sklenika, 2018; Lee, 2014). Here we consider those studies that estimate the number of people a geographic area can feed from its agricultural resources. Such studies do not predict future populations, but rather assess the sufficiency of resources to meet existing demands or to understand how per capita consumption and production potential influence how many people a geographic area can feed.

No standard approach exists for conducting such analyses, but the panoply of methods ultimately answer the same three questions: (1) how much food does a person need, (2) how much food can the local resources produce, and (3) how are production and consumption interrelated? To illustrate the point, Table 6.3 compares two carrying capacity analyses conducted at different spatial scales using different methods with distinct motivations. Despite the differences, focusing on how each study answers the three core questions highlights some of the common features between these studies. Any paper on carrying capacity analysis can be better understood by dissecting the methods down to these questions.

On the consumption side, the amount of biological productivity required to support a person depends on average caloric needs and the make-up of the average diet. Both studies consider dietary composition, but in different ways. Sakschewski, von Bloh, Huber, Müller, and Bondeau (2014) estimates food need based on trends in consumption observed in the food supply, exploring scenarios of how food consumption will change as economies grow. Food is represented in a simplified fashion, as calories from plant and animal sources. In contrast, Peters et al. (2016) estimate food need for ideal diets according to national recommendations, exploring complete diets with different levels of meat, dairy, and egg consumption. While the approaches for estimating consumption differ, both studies consider an important common variable - the proportion of the diet from plant-based versus animal-sourced foods.

On the production side, the studies take quite different approaches. Sakschewski et al. (2014) use a global vegetation model to measure how plant productivity varies over the surface of the Earth and correlate plant productivity with food production potential. This technique models productivity based on key drivers of photosynthesis, like temperature and precipitation, enabling the authors to consider how climate change and $CO_2$ fertilization will affect future food supplies. Peters et al. (2016) estimate productivity based on mean crop yields for a ten-year period, holding yields fixed for all scenarios. Sakschewski et al. (2014) consider availability of irrigation water, whereas Peters et al. (2016) consider how variation in land suitability for cultivated cropping would affect food production.

Not surprisingly, each study uses a somewhat different approach to integrate production and consumption estimates. Sakschewski et al. (2014) estimate total global production of plant

**TABLE 6.3** Summary of core questions answered and methods employed in two carrying capacity studies.

| Aspect of study | Key details | Sakschewski et al. (2014) | Peters et al. (2016) |
|---|---|---|---|
| General description | Motivation | To understand the capacity of agriculture to meet future food demand in light of climate change and dietary shifts | To understand how dietary choices influence the number of people that can be fed from available agricultural land |
| | Spatial scale | Global, disaggregated by country | National, conterminous US |
| | Time period | 1990–2100 | 2001–10 |
| Consumption: How much food does a person need? | Representation of food | Total calories required to supply food from both plant and animal sources | Complete diets, including all food groups used in national dietary recommendations |
| | Estimate of need | Based on calories in the food supply | Based on dietary recommendations |
| | Variation explored | Change in food demand with economic growth | Range of animal-sourced protein, from current consumption levels to vegan diet |
| | Scenarios | Demand held constant at year 2000 or based on projected future GDP | Ten diets: two based on actual consumption and eight healthy diets with different amounts of meat |
| Production: How much food can resources produce? | Resources considered | agricultural land and surface water (for irrigation) | agricultural land |
| | Availability of resources | land use patterns equal to year 2000, irrigation constrained by surface water supply | land use patterns equal to year 2007 |
| | Productivity per unit land | Global vegetation model of plant productivity based on factors that influence photosynthesis | Based on actual crop yields and estimated grazing potential |
| | Variation explored | climate change, $CO_2$ fertilization | Proportion of cropland used for cultivated crops |
| Integration: How are production and consumption inter-rated? | Conceptual basis | Carrying capacity is limited by total production of calories from agricultural biomass | Liebig's Law: Carrying capacity determined by the most limiting land resource |
| | Method | Carrying capacity equals total yield (Y) of plant calories divided by total global calorie demand (W) per person (P) | Population supported by the most limited pool of agricultural land: cultivated cropland, total cropland, or all cropland and grazing land |
| | Final units | number of people fed | number of people fed |

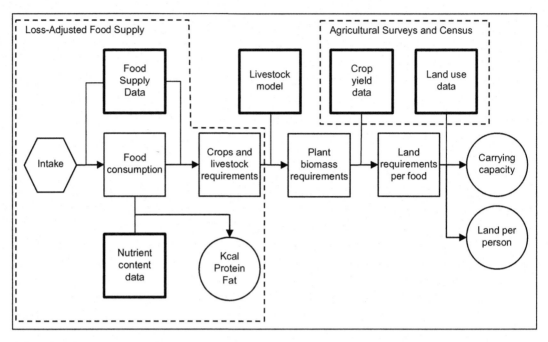

**FIGURE 6.1** Flow diagram of the U.S. carrying capacity model. Boxes indicate data sources (thick lines) or intermediate calculations performed within the model (thin lines). The hexagon indicates food intake which is treated as a variable input in the model. Circles indicate key outcome variables calculated by the model.

calories from agricultural land and calculate carrying capacity based on the plant calories needed per person to supply a complete diet. Peters et al. (2016) use an approach similar to Liebig's Law of the minimum, estimating carrying capacity for complete diets based on the land required per capita to supply food and availability of the limiting pool of land. These approaches are different paths to the same outcome. Both studies calculate the number of people that could be fed, one at the global scale and the other at the scale of the United States.

In reading literature on carrying capacity, do not expect to compare estimates across papers unless you are ready to parse the methodological differences yourself. Rather, it is better to discern what the study shows across the range of scenarios examined and look for common patterns across papers.

### 6.3.2.2 An illustrative example – U.S. carrying capacity analysis

Looking inside the "black box" of a model makes it easier to show the major steps involved and to understand its strengths and weaknesses. In doing so, we will also explore model design and the sensitivity of results to model assumptions. Here we will delve deeper into one of the studies discussed in Table 6.3, the carrying capacity analysis of the 48 conterminous United States (Peters et al., 2016).

The U.S. carrying capacity model estimates per capita land requirements of *complete diets* and the number of persons who could be fed from the agricultural land of the conterminous U.S. (Fig. 6.1). To achieve this outcome, the model leverages publicly available data from the U.S. Department of Agriculture, in particular the Loss-Adjusted Food Supply of the Food Availability (Per Capita) Data System (USDA Economic Research Service, 2021) and production data from annual agricultural surveys and the Census of Agriculture. The food availability data estimate intake of food across more than 100 primary food commodities by multiplying food supply estimates by conversion factors that account for losses and waste that occur between the point of production (e.g., the farm gate or first stage of processing) and the point of consumption. Working in reverse, the US carrying capacity model allows the user to input the estimated intake of food groups which are decomposed into constituent food commodities and converted into equivalent amounts of crop and livestock commodities. Livestock feed requirements from a separate study were used to determine the plant biomass (feed crops, forages, and grazed material) needed to produce the animal-based foods in the diet. Land requirements for each constituent food, feed, or forage crop were determined by dividing by crop yields then aggregated into land requirements per person. Per capita land requirements were divided into the area of available land to determine carrying capacity.

A key aspect of the model's design is to define diet using the same food groups as the Dietary Guidelines for Americans. Food group servings get converted into foods using the set of commodities already tracked for the country's food supply data. While the food commodities are akin to raw ingredients, such as flour, rather than foods people eat, like bread, the approach has the advantage of relying on durable data sets. The Dietary Guidelines for Americans are updated every five years and the food supply data are produced annually. Following this approach, the model can calculate food needed to supply a wide range of diets and covert those diets into a set of agricultural commodities.

At this point, the model design hit a snag. Facing a dearth of publicly available data on livestock feed conversion factors and ration composition, these numbers had to be estimated separately. The livestock model integrates information on livestock nutrition, animal performance, and ration balancing to estimate the feed needs of complete systems for six major livestock classes found in the US: beef cattle, dairy cattle, broilers, laying hens, turkeys, and swine (Peters, Picardy, Darrouzet-Nardi, and Griffin, 2014). From that point forward, the U.S. carrying capacity model again leverages available data, using crop yield estimates from annual surveys and the Census of Agriculture and land use data collected through the Census of Agriculture and compiled by USDA in the Major Land Uses data set. Reliance on regularly collected data ensures that the model can be updated over time.

Once the data are assembled in the spreadsheet model, calculating the land requirements of each component food is simple algebra. Calculating per capita land requirements and carrying capacity is not as straightforward. The model considers three interactions between production and consumption. First, certain crops, such as soybean, get processed into multiple products each of which support production of different foods. The oil portion of the soybean gets extracted and sold as vegetable oil or used as an ingredient in a wide range of processed foods. The residual left after extracting the oil, called soybean meal, is rich in protein and valuable as animal feed. The model accounts for the multi-use nature of such crops to accurately assess the total amount of crop needed to support a complete diet. A second interaction regards the relationship between beef and dairy breeds of cattle. Culled cattle from dairy farms enter the food supply and male calves born to dairy cows typically get raised for beef or veal. Thus, the system used to produce beef depends on both the amount of beef consumed and the amounts of dairy products consumed. Third, and finally, the model differentiates between different types of land to account for suitability of land for different purposes. Cropland could be used

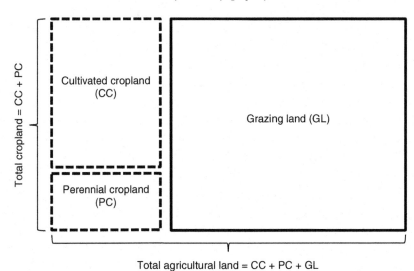

**FIGURE 6.2** Pools of agricultural land considered in the U.S. carrying capacity model. Classes of arable land shown in dashed lines and non-arable land shown in solid lines.

for grazing, but the reverse is not true, meaning that sparing land from livestock production does not always equate to increased capacity for plant-sourced foods. While not exhaustive, these three interactions illustrate the inter-relationships between production and consumption that make the model look more like a system and less like a simple balance sheet.

Carrying capacity is calculated based on the limiting pool of land. As shown in Fig. 6.2, the model differentiates between arable land (total cropland) and non-arable land (grazing land), and arable land is further divided into cultivated cropping and perennial forages. Which pool limits carrying capacity depends on how the amount of land required to produce a given diet compares to the amount of land available. Specifically, carrying capacity depends on how the ratio of required land use per capita compares to the ratio of land availability. For example, say the diet land requirements are 50 percent cropland and 50 percent grazing land. If only 30 percent of total agricultural land is cropland, then cropland determines the carrying capacity. Following this logic, the model estimates the number of people that could be fed per year (person-years of food) by the limiting pool of land dividing the area of land in the pool by the area required per capita.

These interactions enable the U.S. carrying capacity model to explore the potential ecological synergies of livestock systems in supplying complete, nutritionally adequate diets. Animal scientists have long contended that livestock can convert inedible byproducts and forages into edible food, but the extent to which this actually expands the food system is debatable, since intensified livestock systems often rely on feed grains to achieve high productivity. The model captures the dual-use nature of certain crops and sets limits on the amount of land that can be used as cultivated cropland. As it turns out, the model results are highly sensitive to the limit on cultivated cropland use.

Fig. 6.3 shows the dynamics of this relationship, comparing eight diets that comply with the Dietary Guidelines for Americans and are, by that measure, nutritionally equivalent.

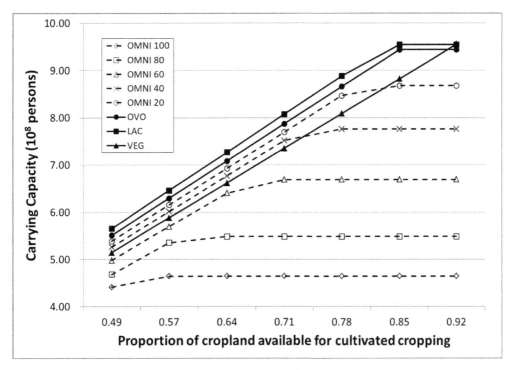

**FIGURE 6.3** Carrying capacity of U.S. agricultural land as a function of diet and proportion of cropland available for cultivated cropping from Peters et al. (2016). Creative Commons BY 4.0 International.

Increasing the percentage of cropland that can be used for cultivated crops increases carrying capacity up to the point where the dietary demands for perennial forage crops equal the land restricted to those crops. After that point, a diet cannot take advantage of the extra "tillable" land and carrying capacity for that diet plateaus. Producing edible food from land restricted to perennial forages enables some omnivore diet scenarios to reach higher carrying capacities than the vegan diet at the current proportion of cultivated to perennial forage cropland (70:30). However, vegan diets outperform even the lacto-vegetarian and ovo-lacto vegetarian scenarios at very high levels of cultivation (near 100 percent). Moreover, the absolute carrying capacities achieved at high levels of cultivation are greater than those at lower levels of cultivation by *hundreds of millions of persons*. Why does this variation matter? It matters because the results are so strongly influenced by this particular assumption – percentage of land in cultivation. When identified and explained, such findings can expand our understanding of carrying capacity, but beware the unexamined assumption!

## 6.3.3 Spatial optimization of foodsheds

### 6.3.3.1 *Spatial analysis of foodsheds*

Spatial analyses of foodsheds estimate the geographic extent of a population's actual or potential food supply. Extent is measured in units of distance, either as the maximum distance within a local foodshed or the weighted average within which food can be sourced. These analyses leverage existing spatial data, such as land cover, municipal boundaries, population, and soils, to map the spatial distribution of production and consumption. Given the limits of publicly available data, foodshed analyses typically assess the geographic extent within which the raw agricultural commodities could be supplied rather than examining the vast number of processed food products and prepared foods that exist in the marketplace. The exact data sources vary from study to study, but the purpose remains the same – to calculate the spatial extent of one or more foodsheds.

### 6.3.3.2 *The limits of available data*

No one has yet mapped the spatial distribution of food consumption completely and in a real-time manner. Imagine, for a moment, where each bite of food consumed by every person in a particular place is actually eaten. People eat at home, at work or school, in restaurants and fast-food establishments, and everywhere else food is served. The idea is dizzying. Yet food consumption need not be understood at such a granular level to be meaningful for foodshed analysis. Rather, the spatial scale at which food is consumed is so small relative to the large geographic areas required to produce the crops and livestock as to be irrelevant for the purposes of estimating the extent of the food supply. For this reason, foodshed analyses estimate the consumption needs of large bodies of people, such as cities or metropolitan areas. Aggregate food needs can be calculated simply, relying on estimates of population and per capita consumption.

Spatial estimation of food production could be equally overwhelming if one tried to capture the thousands of forms in which food is prepared. From fruits and vegetables eaten raw to ready-to-eat packaged products, production takes place on farms, in processing facilities, in commercial kitchens, in homes, and anywhere else food is cooked, baked, or otherwise prepared. Capturing this resolution would be valuable for understanding the flow of foods through food supplies, but publicly available data do not readily permit such an analysis. Accordingly, foodshed analyses estimate capacity for food across gridded surfaces (such as land cover) or geographic units (such as counties) for which production is already estimated.

Once consumption and production have been estimated spatially, the final step of a foodshed analysis is to determine the extent within which food needs can be supplied. One common way of approaching this problem is to use optimization techniques to solve this mathematical problem. Optimization distributes available food production capacity to meet food needs of consuming centers according to an optimization function or goal. Often, the model minimizes the food distance traveled to supply food to all consumption centers, an approach used across scales (see, for example, Galzki, Mulla, and Peters, 2015; Hu, Wang, Arendt, and Boeckenstedt, 2011; Peters et al., 2009; Zumkher and Campbell, 2015; and Kinnunen et al., 2020). However, the approach need not focus on food distance directly. For example, Peters, Bills, Lembo, Wilkins, and Fick (2012) examined which foods might be prioritized for local consumption based on maximizing returns to land from agricultural production. Optimization results can be used to calculate the sourcing distance for the entire study area and for each consumption center within it.

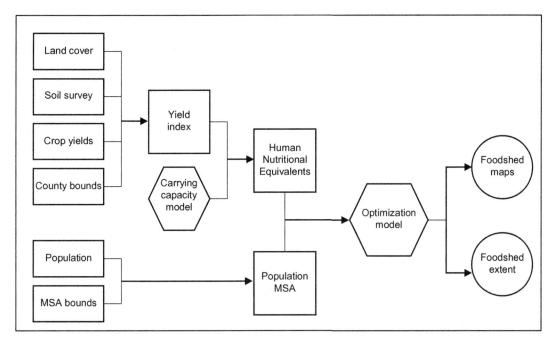

**FIGURE 6.4** Data flow diagram of the spatial model used by Kurtz et al. (2020) to map foodsheds for the conterminous US.

### 6.3.3.3 Sample approach – U.S. foodshed analysis

As with carrying capacity models, considering a specific foodshed analysis helps to demonstrate the essential steps in the approach. Spatial analysis of foodsheds requires the integration of data in a geographic information system, use of computational approaches to determine how food is distributed through the model system, and estimation of the extent of the foodsheds. To this end, we will explore a study by Kurtz, Woodbury, Ahmed, and Peters (2020) on the biophysical capacity to localize food supplies for metropolitan areas in the 48 states of the conterminous U.S. This study estimated the minimum distance within which metropolitan areas could supply food needs from U.S. agricultural land.

A simplified diagram of the method is shown in Fig. 6.4. Reading the diagram from left to right, the approach begins with the assembling of publicly available spatial data relevant to estimating food production and food consumption, such as land cover, the soil survey, and the boundaries of metropolitan areas. Spatial data on the productive capacity of land was used to create an index that was combined with a carrying capacity model to estimate the number of person's worth of food (human nutritional equivalents) that can be produced within each county in the study area. Likewise, data on population and metropolitan areas was used to estimate the spatial distribution of food need. These intermediate calculations provided input to an optimization model that determined the minimum distance within which all 378 metropolitan areas could be fed, and output from the model was used to create foodshed maps and estimates of foodshed extent. While these methods apply specifically to the analysis U.S.

foodsheds, the diagram shows a process common to other analyses, namely the integration of spatial data into a multi-step procedure for estimating foodsheds.

The analysis by Kurtz et al. (2020) models the food system as a world of producers and consumers, represented by counties (which supply food) and metropolitan areas (which demand food). Their approach then "localizes" the food supply using optimization, a family of computational approaches used to estimate a desired outcome, in this case, determining the minimum possible distance in which food can be supplied. How does this work? An optimization problem has three main components: an optimization function, a set of variables, and a set of constraints. The function represents the outcome you want to optimize, in this case, the distance food travels. The variables represent the possible outcomes to consider, in this case, the amount of food supplied from each county to each metropolitan area. The constraints represent conditions that must be met by the solution, for example, all metropolitan areas must have their food needs fully satisfied. Optimization problems use computer algorithms, called solvers, to find the solution that minimizes (or maximizes) the response function while satisfying all the constraints.

Kurtz et al. (2020) use optimization to determine the minimum total distance within which all people residing in the 378 metropolitan areas could be fed from U.S. agricultural land. You can think of this as the total "food miles" traveled in the model food system. The extent of the foodshed of each individual center is then measured in a normalized fashion using weighted average source distance, or WASD (Eq. (6.2)). For any given metropolitan area, the WASD calculation averages distance (D) between producer (i) and consumer (j) weighted by the amount of food (F), in this case the number of human nutritional equivalents s, shipped from each producer to each consumer.

$$\text{WASD} = \sum \left( D_{ij} \times F_{ij} \right) / \sum F_{ij} \qquad (6.2)$$

### 6.3.3.4 Summarizing results of foodshed analyses

The advantage to optimization is that it enables one to consider the food needs of multiple population centers simultaneously, such that the model simulates competition between foodsheds. However, this generates an enormous amount of data. Average values, such as weighted average source distance, describe the central tendency, but visualization is important for conveying the range of results in an accessible way.

Cumulative distribution functions are one commonly used method. The example in Fig. 6.5 shows the percentage of the population that could be fed from agricultural land as the weighted average food distance of the foodshed increases (Hu et al., 2011). This Fig. aggregates data for the hundreds of cities and counties across the eight-state study area. The message is clear. All population centers could theoretically source the entirety of their food needs from agricultural land located within 70 miles or less of the city (or county) center.

Maps are another commonly used tool for visualizing the spatial extent of foodsheds. These maps take many forms (Fig. 6.6). Foodshed maps demarcate the area covered by a foodshed, such as the potential local foodsheds shown for cities in southeastern Minnesota (panel A). Heat maps show how foodshed extent varies depending on where a city is located within the study area, showing cities with relatively large and relatively small potential local foodsheds (panel B). Flow maps, less commonly seen in work published to date, show the preferential

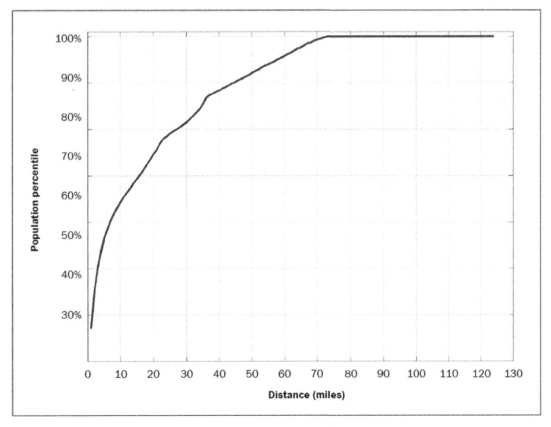

**FIGURE 6.5** Cumulative distribution function showing distance within which food needs could be satisfied for cities in the Midwestern U.S. according to a foodshed analysis by Hu et al. (2011). Creative Commons BY 4.0 International.

paths taken to supply consumers with food (Panel C). While not exhaustive, these examples give a sense of the variety of ways data is conveyed in spatial analyses of foodsheds.

## 6.4 Results and implications

### 6.4.1 What does a regional self-reliance study teach you?

Self-sufficiency ratios (SSR) or self-reliance ratios are the typical output of a regional self-reliance style study. These estimates are intuitively easy to grasp. They show the balance between production and consumption, with values *greater than or equal to* 100 percent signaling theoretical capacity for self-sufficiency and values *less than* 100 percent signaling varying degrees of import dependence. These estimates provide a general sense of how much food

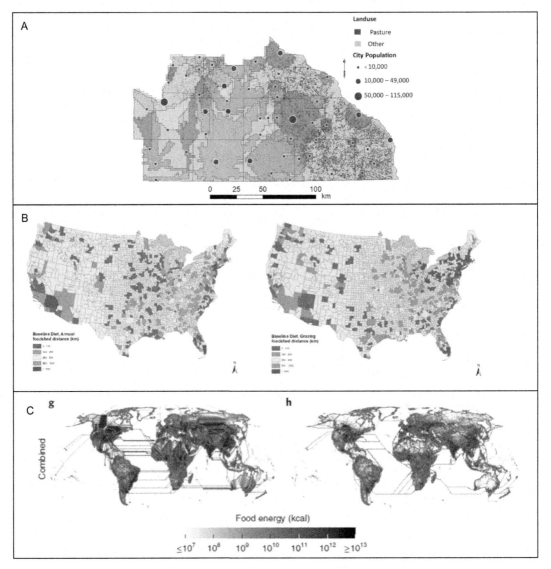

**FIGURE 6.6** Sample forms of foodshed maps. **Panel A** shows the extent of potential local foodsheds for foods derived from pastureland for cities and towns in southeastern Minnesota (Galzki, Mulla, & Meier, 2017). Creative Commons BY 4.0 International. **Panel B** shows heat maps of the weighted average source distance of foodsheds for foods derived from cultivated cropland and grazing land for 378 metropolitan areas in the U.S. Reprinted (adapted) with permission from Kurtz et al. (2020). Copyright 2020, American Chemical Society. **Panel C** shows flows of cereals according to shortest distance versus shortest time optimization for a study of global foodsheds. Reprinted by permission from Springer Nature (Kinnunen et al., 2020).

could come from within the local or regional area, and for this reason, calculation of self-sufficiency ratios from a net-balance analysis is part of some foodshed analyses. Likewise, carrying capacity analyses often express results in proportion to the size of the population, another form of SSR. However, SSR estimates tell you nothing about the actual flows of food. How much stays in the study region, and how much moves out?

Collection of related data can help SSR estimates to tell a more complete story (Peters, Gómez, and Griffin, 2021). For example, the USDA Agricultural Marketing Service reports information on weekly and monthly shipments of fresh produce from major supply points, such as key production regions (for domestically produced fruits and vegetables) and major ports (for imports). Such data provide insight into the role a region plays seasonally in supplying perishable food commodities. Likewise, case studies of supply chains can shed light on the importance of regionally produced products for meeting regional food demand, such as the share of product sourced from regional farms. This triangulation of information can help to interpret the degree to which a region really relies on its own production.

Even when one leverages supporting data, it is impossible to draw any normative conclusions from SSR estimates. How does one judge whether a particular value is too high or too low? Yet, this does not mean SSR estimates are useless. On the contrary, they provide benchmarks for understanding a locale's existing, or historical, capacity for self-supplying its food. Determining whether a particular SSR value is too high or too low depends on context. Consider a region with a low SSR for fresh vegetables, for sake of example say the SSR is 5 percent. If people in this region have ample access to vegetables imported from outside the region, and if the region has little agricultural land for producing vegetables, an SSR of 5 percent may be entirely appropriate. In contrast, if access to vegetables is limited or if people desire access to certain varieties of vegetables not available in local markets, and resources exist to produce vegetables locally, then perhaps SSR is too low. Rather than serving as ends in themselves, perhaps the greatest value to SSR estimates is to serve as a starting point for further conversation.

## 6.4.2 The meaning of carrying capacity estimates

Carrying capacity studies typically estimate the *maximum* population that a geographic area can feed given some set of starting conditions. The meaning is clear, but the practical implications are not. As with SSR estimates, how low is too low? Does a geographic area need to stay below its carrying capacity, or can it rely on supplemental capacity in the form of food imports? Moreover, maximum population may not be the most meaningful indicator. Rather, countries might be constrained by the carrying capacity in the poorest year.

To have meaning, carrying capacity estimates, like all results from models, must be put into context. What are the starting assumptions and input data? The amounts and types of food needed depend on the quantity of energy needed to support physical activity, the quality of the diet relative to nutritional recommendations, the cultural preferences of the diet (particularly for animal products), and the losses and waste that occurs in the supply chain. Capacity to produce food depends on crop yields, livestock feed needs, availability of irrigation water, and the areas of arable and grazing lands. These factors vary not only from place to place but also over time. They vary across different views of what constitutes a sufficient diet or a sustainable food system. Rather than a source of weakness or uncertainty, differences in approach invite conversations about the feasibility of starting assumptions or the reliability of input data and how the results would change given different assumptions.

Two important questions should be resolved by such conversations. First, how sure are we in our estimates? Statistically speaking, one needs to know the uncertainty in the estimates introduced by factors for which variability can be quantified, such as annual variations in crop yield. Second, what are we aiming for? Assumptions about the need for food, for example, implicitly involve making decisions about what kind of eating pattern is acceptable. These implicit assumptions reveal what we value in the food system.

Once we understand the factors that influence the carrying capacity estimates, we must revisit the question, "What do the estimates mean?" One may be comforted by a carrying capacity *greater* than the current population, but today's buffer in carrying capacity may be consumed by tomorrow's population growth. On the other hand, one might be alarmed by a carrying capacity estimate *smaller* than the resident population, yet many countries rely on net imports of food. In other words, what is a desirable number? It depends. At the global scale, a carrying capacity equal to or greater than the current population is certainly desirable. However, more is not necessarily better. For example, a high carrying capacity might be achieved by feeding everyone a subsistence diet with input-intensive agriculture. Like any other number, carrying capacity must be interpreted in context.

### 6.4.3 The allure and limits of foodshed analysis

Foodshed analysis seems to promise a satisfying answer to the question, "Where does my food come from?" Tempting as that question may be, the field has not yet reached the point where it is an easy question to answer. Nonetheless, foodshed analysis provides two kinds of information useful to understanding the geographic scale of food supplies.

Mathematically, the spatial extent of actual or potential foodsheds is estimated by the weighted average source distance (WASD) for supplying food to a given location. A poor correlation exists between food miles and environmental impact, meaning that WASD conveys no meaningful information about the optimal scale of food distribution for better sustainability. Nonetheless, the measure is quite descriptive indicating that extent to which population centers must rely on distant sources for food.

Since the approach is inherently geographical, foodshed analyses typically use maps to display information about the actual or potential foodsheds. Beyond providing a visualization of the estimates of WASD, maps convey spatial patterns in foodsheds, making it easier to see where potential for food system localization is greatest. As with SSR, WASD, and carrying capacity, foodshed maps do not necessarily tell you how food is actually sourced, so at this juncture, foodshed analysis is most meaningful for understanding theoretical potential for local and regional food.

### 6.4.4 Key lessons

Foodshed analysis, carrying capacity estimation, and regional self-reliance studies share a common goal, namely, to quantify the capacity of geographic areas to supply human food needs. These types of analysis are notoriously difficult to interpret in a normative fashion. While food supply data provide useful information for understanding national food security, the tools described herein do not provide clear guidance on how the food system should be changed for better sustainability. Rather, the estimates obtained must be interpreted in the context of additional information.

Despite these challenges, these tools have value. Two particular uses stand out:

1. Gaining perspective. All three tools can be used to set benchmarks. The tools described in this chapter address concepts that are not well or widely understood. For example, while many people are excited by the prospect of supporting local farmers and food producers, how many deeply understand the potential for supplying food in such a fashion?
2. Synthesizing information. Conducting such analyses inevitably requires the researchers to determine how to synthesize information, and much can be learned about the food system in the process. For example, one might come to appreciate new insights gained from exploring how dietary preferences influence carrying capacity or the ability to source food from within a given region.

We would be wise to value these models for how they help us to think about the productive capacity of food systems, even if they have limited predictive power.

## Acknowledgements

This work was supported in part by the U.S. Department of Agriculture, Agricultural Research Service.

## References

Binder, S., Holdahl, E., Trihn, L., Smith, J.H., 2020. Humanity's fundamental environmental limits. Human Ecology 48, 235–244.

Butler, C.D., 2004. Human carrying capacity and human health. Plos Medicine 1 (3), 192–194.

Butler, C.D., 2016. Planetary overload, limits to growth and health. Current environmental health reports. 3, 360–369.

Cassidy, E.S., West, P.C., Gerber, J.S., Foley, J.A., 2013. Redefining agricultural yields: from tonnes to people nourished per hectare. Environmental Research Letters 8, 1–8. Available at Web site http://iopscience.iop.org/1748-9326/8/3/034015/pdf/1748-9326_8_3_034015.pdf.

Cohen, J.E., 1995. Population growth and Earth's human carrying capacity. Science 269, 341–346.

Cohen, J.E., 1997. Population, economics, environment, and culture: an introduction to human carrying capacity. Journal of Applied Ecology 34 (6), 1325–1333.

Conrad, Z., Tichenor, N., Peters, C., Griffin, T., 2016. Regional self-reliance for livestock feed, meat, dairy, and eggs in the Northeast U.S. Renewable Agriculture and Food Systems 32 (2), 145–156. https://doi.org/10.1017/S1742170516000089.

Fanta, V., Šálek, M., Zouhar, J., Sklenika, P., 2018. Equilibrium dynamics of European pre-industrial populations: the evidence of carrying capacity in human agricultural societies. Proc. R. Soc. B 285, 20172500. http://dx.doi.org/10.1098/rspb.2017.2500.

Freedgood, J., Pierce-Quiñonez, M., Meter, K.A., 2011. Emerging assessment tools to inform food systems planning. Journal of Agriculture Food Systems and Community Development 2 (1), 83–104.

Franck, S., von Bloh, W., Müller, C., Bondeau, A., Sakschewski, B., 2011. Harvesting the sun: new estimations of the maximum population of planet Earth. Ecological Modeling 222, 2019–2026.

Galzki, J., Mulla, D., Peters, C., 2015. Mapping the potential of local food capacity in Southeastern Minnesota. Renewable Agriculture and Food Systems 30 (4), 364–372.

Galzki, J.C., Mulla, D.J., Meier, E., 2017. Mapping potential foodsheds using regionalized consumer expenditure data for Southeastern Minnesota. Journal of Agriculture, Food Systems, and Community Development 7 (3), 181–196. http://dx.doi.org/10.5304/jafscd.2017.073.013.

Griffin, T., Conrad, Z., Peters, C., Ridberg, R., Tyler, E.P., 2015. Regional self-reliance of the Northeast Food System. Renewable Agriculture and Food Systems 30 (4), 349–363.

Hedden, W.P., 1929. How Great Cities Are Fed. D.C. Heath and Company, New York.

Horst, M., Gaolach, B., 2015. The Potential of Local Food Systems in North America: a Review of Foodshed Analyses. http://cambridge.org/core/about 30 (5), 399–407.

Hu, G., Wang, L., Arendt, S., Boeckenstedt, R., 2011. An optimization approach to assessing the self-sustainability potential of food demand in the Midwestern United States. Journal of Agriculture, Food Systems, and Community Development 2 (1), 195–207. http://dx.doi.org/10.5304/jafscd.2011.021.004.

Kinnunen, P., Guillaume, J.H.A., Taka, M., D'Odorico, P., Siebert, S., Puma, M.J., et al., 2020. Local food crop production can fulfill demand for less than one-third of the population. Nature Food 1, 229–237. https://doi.org/10.1038/s43016-020-0060-7.

Kloppenburg Jr, J., Hendrickson, J., Stevenson, G.W., 1996. Coming in to the foodshed. Agriculture and Human Values 13 (3), 33–42.

Kurtz, J.E., Woodbury, P.B., Ahmed, Z.U., Peters, C.J., 2020. Mapping U.S. food system localization potential: the impact of diet on foodsheds. Environmental Science & Technology 54 (10), 12434–12446. https://doi.org/10.1021/acs.est.9b07582.

Lee, H.F., 2014. Climate-induced agricultural shrinkage and overpopulation in late imperial China. Climate Research 59 (3), 229–242.

Peters, C.J., Fick, G.W., Wilkins, J.L., 2003. Cultivating better nutrition: can the Food Pyramid help translate dietary recommendations into agricultural goals? Agronomy Journal 95, 1424–1431.

Peters, C.J., Bills, N.L., Wilkins, J.L., Fick, G.W., 2009. Foodshed analysis and its relevance to sustainability. Renewable Agriculture and Food Systems 24 (1), 1–7.

Peters, C.J., Bills, N.L., Lembo, A.J., Wilkins, J.L., Fick, G.W., 2012. Mapping potential foodsheds in New York State by food group: an approach for prioritizing which foods to grow locally. Renewable Agriculture and Food Systems 27 (2), 125–137.

Peters, C.J., Picardy, J.A., Darrouzet-Nardi, A., Griffin, T.G., 2014. Feed conversions, ration compositions, and land use efficiencies of major livestock products in U.S. agricultural systems. Agricultural Systems 130, 35–43.

Peters, C.J., Picardy, J., Darrouzet-Nardi, A.F., Wilkins, J.W., Griffin, T.S., Fick, G.W., 2016. Carrying capacity of U.S. agricultural land: ten diet scenarios. Elementa: Science of the Anthropocene 4, 000116. http://doi.org/10.12952/journal.elementa.000116.

Peters, C.J., Gómez, M.I., Griffin, T.S., 2021. Roles of Regional Production in a Global Food System. Renewable Agriculture and Food Systems 36, 432–442. https://doi.org/10.1017/S1742170519000401.

Rees, W.E., 1996. Revisiting carrying capacity: area-based indicators of sustainability. Population and Environment 17 (3), 195–215.

Rockström, J., Steffen, W., Noone, K., Persson, Å., Chapin III, F.S., Lambin, E., …, Foley, J., 2009. Planetary boundaries: exploring the safe operating space for humanity. Ecology and Society 14 (2), 32. http://www.ecologyandsociety.org/vol14/iss2/art32/.

Sakschewski, B., von Bloh, W., Huber, V., Müller, C., Bondeau, A., 2014. Feeding 10 billion people under climate change: how large is theproduction gap of current agricultural systems? Ecological Modelling 288, 103–111.

Sayre, N., 2008. The genesis, history, and limits of carrying capacity. Annals of the Association of American Geographers 98 (1), 120–134.

Seidl, I., Tisdell, C.A., 1999. Carrying capacity reconsidered: from Malthus' population theory to cultural carrying capacity. Ecological Economics 31, 395–408.

Schreiber, K., Hickey, G.M., Metson, G.S., Robinson, B.E., MacDonald, G.K., 2021. Quantifying the foodshed: a systematic review of urban food flow and local food self-sufficiency research. Environmental Research Letters 16, 023003.

Susiarjo, G., Sreenath, S.N., Vali, A.M., 2006. Optimum supportable global population: water accounting and dietary considerations. Environment, Development, and Sustainability 8, 313–349.

U.S. Census Bureau. (2016). National Intercensal Tables: 2000-2010. https://www.census.gov/data/Tables/time-series/demo/popest/intercensal-2000-2010-national.html.

USDA Economic Research Service (ERS). (2021). Food Availability (Per Capita) Data System [on-line]. https://www.ers.usda.gov/data-products/food-availability-per-capita-data-system/.

USDA National Agricultural Statistics Service. (2021). Quick Stats Database. https://quickstats.nass.usda.gov/.

Zumkehr, A., Campbell, J.E., 2015. The Potential for Local Croplands to Meet US Food Demand. Frontiers in Ecology and the Environment 13 (5), 244−248.

# 7

# Market and supply chain models for analysis of food systems

*Charles F. Nicholson*[a,b] *and Miguel I. Gómez*[b]

[a] School of Integrative Plant Science, Cornell University, Ithaca, NY, USA [b] Charles H. Dyson School of Applied Economics and Management, Cornell University, Ithaca, NY, USA

## 7.1 Introduction

Food systems embody many interactions among diverse actors, resulting in social, environmental, health and economic outcomes. As noted in other chapters of this book, the economic incentives are among the key drivers of outcomes and the evolution of food systems. This is particularly true in market economies and for agricultural commodities, but the economic performance of food systems has global relevance. Thus, the analysis of markets for food—especially from a systems perspective—provides insights about how to improve outcomes from food systems. In addition, the organizational structures that link different actors in a food system can often be usefully considered from the perspective of supply chains. As a result, economic and supply chain analyses of food systems issues can complement other analytical perspectives and support informed decision making to improve desired outcomes.

There are many different types and methods of economic analysis, but most focus on outcomes associated with company or market performance metrics, such as costs, prices, quantities supplied or consumed, profitability and efficiency in the use of resources. Some types of economic analysis also allow a focus on the important question of the *distribution* of costs, profits and consumption among different food system actors, that is, who might benefit or be worse off through patterns of food system evolution. These outcomes are important to the many different types of businesses (formal or informal) that are key actors in food systems and to what we might think of as the 'end-users' of food, the individuals, households and communities that are food consumers. Economic outcomes are also key drivers of food system evolution, providing incentives for production, marketing, distribution and innovation. Public or private interventions in food systems to improve outcomes often operate through mechanisms that attempt to modify these economic incentives and the responses of consumers and businesses. Economic analyses therefore often play a key role in the design and implementation of food system interventions.

*Food Systems Modelling.*
DOI: https://doi.org/10.1016/B978-0-12-822112-9.00001-1

Economic and supply chain analyses are not only about economic outcomes: they can also incorporate and complement other indicators of food system performance. One example to be discussed subsequently (Nicholson et al., 2015) is an economic analysis of how localizing food systems can affect greenhouse gas emissions from transportation as well as costs and prices. Another example (Nicholson and Monterrosa, 2021) indicates how various supply chain interventions are likely to affect household consumption of fruits and vegetables, with important implications for food security and health outcomes.

Given the plethora of different economic and supply chain models and their diverse applications, a comprehensive review is not appropriate for present purposes. This chapter emphasizes three commonly-used types of economic and supply chain models: spatial optimization models, partial equilibrium models, and dynamic supply chain models based on participatory group model building and system dynamics. For each, we will provide an overview of the mathematical structure of the model, discuss the kinds of food system issues for which these models are most appropriate, indicate data needs and suggest priority areas for future development and application. We also illustrate a small number of environmental and food security metrics that can be included in analyses based on these methods.

## 7.2 Spatial optimization models (Transportation and transshipment models)

Food system economic problems often involve decisions on how to bring food and agricultural products from source locations (e.g., farms) to demand locations (e.g., grocery stores). In these types of problems, the spatial location of businesses that participate in the supply chain (e.g., producers, transporters, manufacturers, wholesalers, retailers) typically plays a critical role in identifying cost-minimizing (or profit-maximizing) decisions regarding the flow of products from supply to demand locations. Moreover, the spatial configuration of the food supply chain has implications for non-economic impacts such as the use of scarce natural resources including land and water, greenhouse gas emissions, and even dietary intakes of the population.

A common challenge for food supply chains is how to locate processing facilities and choose transportation routes to minimize the costs of delivering product to customers. Low costs are particularly important for agricultural commodities, which tend to be homogeneous and do not receive price premiums from buyers. *Transportation problems* involve choosing the least cost way to transport products from supplier locations to customer locations when the amounts at supply and demand locations are fixed. Transportation costs tend to be large for agricultural commodities because they are bulky or perishable. A *transshipment problem* involves choosing the least-cost way to move product through a supply chain when there are additional locations (called *transshipment nodes*) between producer and consumer locations. For example, supermarket chains usually ship products from grower/packer/shippers to distribution centers rather than directly to individual supermarket stores. Transshipment nodes can also be processing centers: beef cattle are produced on the farm and then is delivered to slaughter and processing plants.

For both transportation and transshipment problems we can develop *optimization models* to assess which of many possible combinations result in the lowest cost. *Linear programming* (LP) is the most common type of optimization model used for these problems. LP determines

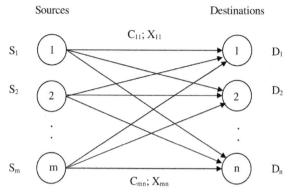

Sources        Destinations     **FIGURE 7.1** Structure of the basic transportation model.

the optimal value of a linear *objective function* subject to a set of linear constraints (Hiller and Lieberman, 2021). For food systems, the objective function is typically the sum of costs for moving product through the supply chain—a value that we want to minimize. This section describes how to define transportation and transshipment models and the type of information they provide food system decision makers. These models also provide information relevant for understanding other market and supply chain models, discussed in subsequent sections. We also discuss applications of these models to examine economic and environmental impacts of the configuration of supply chains in the context of food systems.

## 7.2.1 The transportation model

The assumptions of the basic transportation model (Fig. 7.1) are:

- One agricultural commodity is produced and sold in different geographic locations;
- There are m supply locations (nodes) where the commodity is produced with supply at each node denoted as $S_i$ for $i=1, 2, 3…m$;
- There are n demand locations (nodes) where the commodity is consumed, with demand at each node denoted as $D_j$ for $j=1, 2, 3, n$;
- Supply $S_i$ and demand $D_j$ at each location are known and constant;
- The objective function is to minimize total transportation costs from supply to demand nodes. The only costs are transportation costs, (i.e., the product has already been produced or processed);
- The cost of transportation per unit from each supply node i to each demand node j is known and constant (i.e., the cost does not vary with volume);
- The amount of total supply is greater than or equal to the amount of total demand.

Note that in this case, any supply node i can sell to any demand node j. The model determines the movement (flow) of product (denoted as the variables $X_{ij}$) from supply node i to demand node j based on the allowable network connections that minimizes the total costs of transportation. The specific components are as follows:

- $C_{ij}$ = the unit transportation costs from supply i to demand j (e.g., $C_{13}$ is the unit transportation cost from supply location 1 to demand location 3);

- $X_{ij}$ = the amount of product shipped from supply origin i to demand destination j (e.g., $X_{32}$ is amount of commodity X that is transported from supply location 3 to demand location 2);
- $S_i$ = supply (fixed farm production capacity) at supply node i; and
- $D_j$ = demand (fixed consumption level) at demand node j.

The *objective function*, Z, is the sum of the cost of each of the possible product flows, $X_{ij}$, times the unit cost of transportation $C_{ij}$ from supply node *i* to demand node *j*:

$$Z = \sum_{i=1}^{m} \sum_{j=1}^{n} C_{ij} \cdot X_{ij}$$

This linear objective function is minimized subject to physical constraints on amounts of product:

$$\sum_{i=1}^{n} X_{ij} \geq D_j$$

which indicates that the sum of the amounts shipped from supply nodes i must be greater than or equal to the quantity demanded at demand node j. For example, if demand destination 1 consumes 200 units of the product, then this constraint requires that at least 200 units be shipped to destination 1 from the supply origins. In addition, the constraint

$$\sum_{j=1}^{J} X_{ij} \leq S_i$$

This constraint indicates that the amount shipped from any supply origin i to the demand destinations cannot exceed the available supply from supply origin i. For example, if supply origin 2 has 100 units of product to distribute, then this constraint requires that the shipment out of origin 1 to the demand destinations not exceed 100 units. In addition, as a mathematical (and practical) condition, we constrain the values of the $X_{ij}$ to be non-negative:

$$X_{ij} \geq 0, \; for \; all \; i, j$$

This LP model determines amounts that should be shipped from each supply location i to each demand location j to minimize the overall transport costs given the location-specific constraints on supply and demand.

As a simple example, consider a large grower-packer-shipper that produces fresh blueberries in three different locations in the U.S. The company sells its blueberries to three different demand destinations in the US, implying a network with three supply nodes (=1,2,3) and three demand nodes (*j*=1,2,3) in which any supply node i can ship to any demand node j. The costs of shipping one unit of product from i to j is both known and constant. There are nine possible shipments in the network, three for each supply node to the three demand locations. The model will determine the amounts of product that should be shipped from each supply node i to each demand node j to minimize costs. The information about supply and demand is given by:

Sources    Destinations

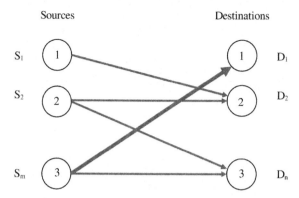

**FIGURE 7.1A** Solution to the transportation problem using the network structure from Fig. 7.1 (width of the arrow indicates the relative size of the flow).

| Node | $S_i$ | $D_j$ |
|------|-------|-------|
| 1 | 100 | 200 |
| 2 | 200 | 200 |
| 3 | 300 | 200 |

In this example, demand is equal at each of the j nodes, but the supply is different at each of the i nodes.

In this case $X_{ij}$ indicates the amount of blueberries shipped from packing plant i to demand node j. The parameter $S_i$ represents the weekly production capacity of each packing plant (e.g., $S_1 = 100$) and the parameter $D_j$ represents the weekly demand in each destination (e.g., $D_1 = 200$). In addition, a matrix of unit costs ($C_{ij}$) indicates shipping costs ($/100 boxes) from packing plant I to demand location j:

| Origin Node | (To) Demand Node | | |
|-------------|----------------|-------|-------|
| | $D_1$ | $D_2$ | $D_3$ |
| $S_1$ | 40 | 47 | 80 |
| $S_2$ | 72 | 36 | 58 |
| $S_3$ | 24 | 61 | 71 |

In this case (see Excel spreadsheet example), the lowest-cost solution formulated with Excel's Solver is $X_{12} = 100$, $X_{22} = 100$, $X_{23} = 100$, $X_{31} = 200$, $X_{33} = 100$, and all other $X_{ij} = 0$. It is sometimes useful to represent these values graphically with a modification of the structure from Fig. 7.1 as in Fig. 7.1a. This supply chain configuration yields the minimum transportation costs of $26,000. This assignment of shipments from i to j is consistent with the constraints: no more is shipped than is available at each supply node i and enough is shipped to each demand node j to meet the demand at that location. This model thus provides guidance about how to keep transportation costs as low as possible. It can also be used to assess how

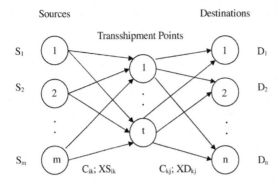

FIGURE 7.2 Structure of the basic transshipment model.

to minimize transportation costs if the basic assumptions are changed, such as if quantities supplied or demanded or costs change over time. The model can also be used to evaluate whether investments in new supply locations would lower costs (i.e., by adding additional supply nodes with estimated costs to demand nodes) or if the costs associated with serving new demand locations are warranted. The transportation model is thus a very flexible tool for the assessment of food supply networks.

### 7.2.2 The transshipment model

The *transshipment model* is an extension of the *transportation model* in which intermediate "transshipment points" are added between the supply and demand locations. A common example of an intermediate node is a processing location located between farms and retail stores—a simple supply chain. More generally for transshipment models, shipments of product are permitted between any pair of nodes, meaning that shipments can occur between supply and demand locations, between supply and transshipment locations, or between transshipment and demand locations (Fig. 7.2). Some transshipment models also include shipments between the intermediate nodes (as in Nicholson et al. 2015), but this will not be a focus of our discussion here.

We consider a simple example in which all supply must be shipped to an intermediate node before it is shipped to the final demand destination. This is common in supply chains for agricultural commodities, where additional activities associated with product marketing (sorting, grading, cleaning or packaging) need to occur prior to delivery to a retail store. We ignore the costs associated with these additional activities, but they are relatively straightforward to include. The model now includes three types of nodes: m for supply, t for transshipment and n demand (Fig. 7.2). The basic assumptions for the transportation model also apply in this case. We define $XS_{ik}$ for shipments from supply node i to transshipment node t and $XD_{kj}$ for shipments from transshipment node k to demand node j. Determining least-cost shipments using LP requires a revised objective function with two types of shipments:

$$Z = \sum_{i=1}^{m}\sum_{k=1}^{t} C_{ik} \cdot XS_{ik} + \sum_{k=1}^{t}\sum_{j=1}^{n} C_{kj} \cdot XD_{kj}$$

where the $C_{ik}$ and the $C_{kj}$ are the (constant) per-unit cost of shipping from supply node i to transshipment node k and from the transshipment node k to demand node j, respectively. Similar to the transportation model, we need constraints on amounts shipped to demand nodes meeting the required quantities Dj:

$$\sum_{k=1}^{t} XD_{jk} \geq D_j$$

and constraints limiting the amount shipped from supply nodes to the amount available, Si:

$$\sum_{k=1}^{t} XS_{ik} \leq S_i$$

In addition, we must ensure that product shipped from each transshipment node to demand nodes is less than the amount of product shipped into that transshipment node from the supply points:

$$\sum_{j=1}^{n} XD_{jk} \leq \sum_{i=1}^{m} XS_{ik}$$

To illustrate this model, consider adding two transshipment nodes to the previous transportation model. In this example, assume that the amounts $S_i$ and $D_j$ are as in the previous transportation model. Then, we need additional information about the $C_{ik}$ and $C_{kj}$, the unit costs of shipment to and from transshipment points:

|  | (To) Transshipment Node | |
| --- | --- | --- |
| Origin Node | $K_1$ | $K_2$ |
| $S_1$ | 10 | 12 |
| $S_2$ | 12 | 10 |
| $S_3$ | 15 | 12 |

and

|  | (To) Demand Node | | |
| --- | --- | --- | --- |
| Transshipment Node | $D_1$ | $D_2$ | $D_3$ |
| $K_1$ | 20 | 22 | 40 |
| $K_2$ | 22 | 20 | 32 |

Solving the LP problem (see Excel spreadsheet example) indicates the shipments that result in the lowest transportation cost: $XS_{11} = 100$, $XS_{21} = 100$, $XS_{22} = 100$, $XS_{32} = 300$, $XD_{11} = 200$, $XD_{22} = 200$, $XD_{23} = 200$. Two hundred units of product are transported to transshipment node 1 (from supply nodes 1 and 2) and 400 units of product are transported to transshipment node 2 (from supply nodes 2 and 3). Transshipment node 1 then supplies demand node 1 and transshipment node 2 supplies demand nodes 2 and 3. The transportation cost is $21,200 (which is not directly comparable to transportation model results given different assumptions about the unit costs).

### 7.2.3 Examples and extensions

Modifications allow the transshipment model to represent more realistic supply chains. First, the volumes that can be "processed" at a transshipment node may be limited by available capacity and this could be represented by an additional set of constraints. Second, product transformation (e.g., from a raw material to a finished product) might be done at "transshipment" nodes (which would then be thought of as "processing locations" and generally imply additional costs). This may also allow the assessment of multiple products simultaneously—rather than just a single commodity. Third, the costs of transporting product may not be constant with volume; often larger shipments will have lower per-unit costs. Finally, it may be appropriate to allow amount supplied and demanded at each location to vary. Each of these is relatively straightforward to include in a transshipment model but make the model formulation more complex. We discuss many potential modifications in the next section.

Atallah et al. (2014) developed a transportation model to assess economic and environmental impacts of larger broccoli production-distribution networks in the eastern U.S. that would reduce the shipments of broccoli produced in California. This analyzes a regionalized alternative to the current system in response to changing consumer preferences. Consumers are increasingly aware of the social and environmental impacts of food systems sometimes resulting in shifts to public policy. In response to consumer concerns, growers and retailers want to examine alternative ways to produce and distribute food such as regional and local food systems (while controlling costs). Atallah et al. used a production and transportation model to determine optimal locations for expanded broccoli production in the northeastern U.S. and the associated changes in supply chain costs, consumer prices and the weighted average source distance (WASD) travelled. Their analysis indicated it was possible to increase seasonal broccoli production by 30 percent without an increase in the overall supply chain costs—higher production costs were offset by lower transportation costs. Distances traveled by broccoli to meet demand in the east coast could be reduced, with lower greenhouse gas emissions.

Another empirical example of a large-scale transshipment model is Nicholson et al. (2015). This model of the US dairy industry had more than 200 supply nodes, 400 demand nodes and 600 processing nodes. The model included one raw material (farm milk) that could be transformed into more than 20 dairy products. Farm milk was required to be processed into a product before shipment to a demand node, but the model also allowed product to be shipped from one processing node to another. This model was used to assess the impacts on costs and transportation emissions from reconfiguring the supply chain to minimize the distance traveled by beverage milk in the Northeastern US, consistent with efforts to "localize" food systems. Localizing was implemented in the model by limiting the locations from which fluid milk could be supplied and processed. This study found that localizing beverage milk resulted in both *increased* total costs and emissions from transportation for the Northeast region and for the US, in part because localizing beverage milk had the impact of increasing distances traveled for other dairy products.

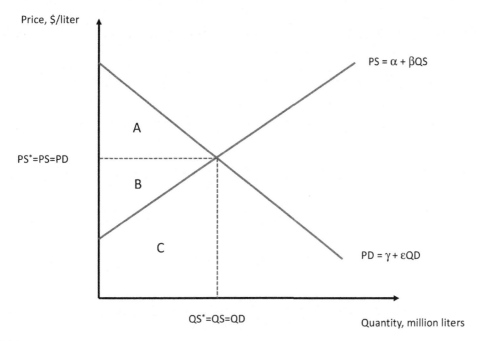

FIGURE 7.3 Supply and demand curves for soft drinks.

## 7.3 Partial equilibrium models

### 7.3.1 Partial equilibrium model characteristics and metrics

Economists often assess food system interventions with partial equilibrium (PE) models of markets—models that can be conceptual or empirical. These models extend those described previously because they include supply and demand of products. In PE models the amounts supplied and demanded will depend on various factors, particularly the price received by a supplier or paid by a buyer. The outcomes associated with a *market equilibrium* are of interest, such as the quantities produced (sold), the quantities demanded (purchased), prices, revenues to sellers and expenditures by buyers. This kind of analysis typically employs *comparative statics*, where the focus is on *changes in the equilibrium conditions* with a change in market conditions or due to an intervention, without consideration of dynamics or the process of adjustment. PE models tend to focus on one or a limited number of specific markets and not the overall conditions for an entire economy typically modeled with general equilibrium models that consider all markets for goods and services simultaneously.

A conceptual example is the market for soft drinks (Fig. 7.3). This shows price on the vertical axis and quantity on the horizontal axis. A *supply curve* shows the response of the quantity of soft drinks produced to price. Based on economic concepts of profit maximization, a higher price means that more soft drinks will be produced. A *demand curve* shows the response of buyers to price. Based on economic concepts of (buyer) utility maximization, as the price increases the quantity purchased will decrease. Here, both the supply and demand curves are lines, but the relationships can also be nonlinear. A market equilibrium exists when the same price $P^*$ is paid by the buyer[1] and received by the seller, and where the quantities produced and sold both are equal, $QD^*=QS^*$. The diagram shows revenues received by sellers (the areas $B+C$, equal to the price times the quantity) and the expenditures by buyers (also $B+C$).

Economists also consider *consumer surplus* and *producer surplus* as measures of (economic) welfare. Consumer surplus is the triangular area A. This area represents a "surplus" in the sense that all consumers in the market pay the price $P^*$, but some consumers would be willing to pay a higher price; the demand curve shows how much would be demanded at prices higher than $P^*$. The difference between what the consumers actually pay ($P^*$) and what they would be willing to pay, summed up for all of values of prices higher than $P^*$ is a type of economic benefit to consumers. Similarly, many producers would be willing to produce at prices lower than $P^*$ (as shown by the supply curve at prices lower than $P^*$), but they actually receive the price $P^*$. The area B sums up the differences between what producers were paid and what they would have been willing to accept and is the producer surplus, a type of economic benefit to producers. Important economic impacts of food system interventions include changes in prices ($P^*$), quantities ($Q^*$), supplier revenues, buyer expenditures, producer surplus and consumer surplus.

### 7.3.2 Example of analysis with a PE model: a tax on sugary soft drinks

Let's examine a food system intervention using this conceptual model. One proposed response to rising obesity rates and diabetes incidence in higher- and middle-income countries is taxes on foods considered to be unhealthy. For example, some cities have implemented taxes on sugared soft drinks in an effort to reduce their consumption (e.g., Fletcher et al. 2010; Redondo et al. 2018). If the tax is T (measured in \$/unit) and is charged to producers of soft drinks, a new with-tax supply curve ($S_{Tax}$) lies above the previous supply curve. At any quantity produced, costs for soft drink producers are higher because they pay the tax on each unit. A new equilibrium occurs where a new quantity purchased (consumed) of $QS^{**}$, which is lower than the previous $QS^*$ (Fig. 7.4).

The higher price paid by the buyers, $P_{Buyer}$ reduces consumption of sugary soft drinks—the intended effect.[2] Producers receive the price $P_{Buyer}$ but have paid the tax, so the *net price* they receive is now $P_{Seller}$ ($= P_{Buyer}-T$). This graphical analysis indicates that a tax will decrease producer net revenues (to $P_{Seller}$ times $QD^*$). Consumer expenditures can either increase or decrease—this depends on the nature of changes in price and quantities—and the government entity imposing the tax receives the amount $P_{Buyer}-P_{Seller}$ times $QD^{**}$ (or T times $QD^{**}$).

---

[1] It is common in economic analyses to refer to the equilibrium values of P and Q with notation like $P^*$ and $Q^*$, where the * indicates that this is an *equilibrium* value, not just *any* value of P or Q.

[2] In this simple example, we assume that customers have no alternative sources of soft drinks and cannot avoid the tax. Often it is cities that have imposed this kind of tax, which means that consumers often can purchase product in other nearby cities that do not tax and this has limited the effectiveness of soft drink taxes as a public health measure (Cattaneo et al., 2021). This analysis also does not represent the potential to switch to artificially-sweetened soft drinks, which can have their own negative public health effects. Additional analyses would be needed to evaluate the effectiveness of a specific tax policy, but our analysis illustrates basic effects and trade-offs.

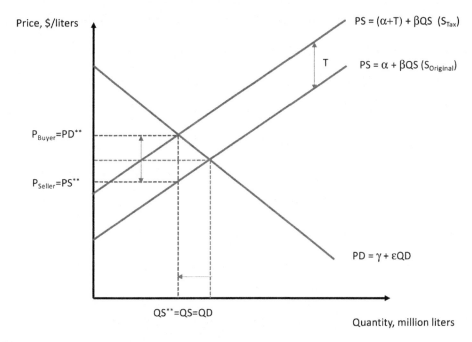

**FIGURE 7.4** Supply and demand curves for a tax on sugared soft drinks.

Consumer surplus and producer surplus are both smaller with the tax, and the government now receives as revenue some of the value previously comprising consumer and producer surplus. However, the total of producer surplus, consumer surplus and government receipts is also smaller with the tax. In Fig. 7.3, the producer and consumer surplus were the combined area of $A+B$, that is, the area of a triangle under the demand curve but above the supply curve for Q values to the left of QD*. With the tax (Fig. 7.4) the area under the demand curve and above the supply curve but to the left of QD** is clearly smaller than the previous $A+B$ because it no longer includes any of the triangular area to the right of QD**. This smaller value of the combined consumer and producer surplus and government receipts is sometimes referred to as the "welfare loss" (or "deadweight loss") effect of a tax because it is a loss of economic value that no longer accrues to consumers, producers or government. This is only a "welfare loss" in the narrow sense of the lost economic value in this one market and does not consider the changes in demand for complementary products or the potential health benefits that accrue to consumers due to less consumption of sugary soft drinks.

In the case of a single product with linear supply and demand curves, it is usually possible to solve for the market equilibrium using algebra. For example, with supply and demand curves specified algebraically as:

$$PS = \alpha + \beta \cdot QS \ (Supply\,equation)$$

$$PD = \gamma - \varepsilon \cdot QD \ (Demand\,equation)$$

we can use the condition that in equilibrium PS=PD and QS=QD to solve for the value of QS:

$$QS^* = \frac{\gamma - \alpha}{\beta + \varepsilon}$$

With the value of QS*, we can calculate the value of PS* and the other outcomes. A unit tax is like adding T to the supply Eq., which would modify the equilibrium to

$$QS^{**} = \frac{\gamma - \alpha - T}{\beta + \varepsilon}$$

which indicates that QS** < QS*, in agreement with our previous graphical result.

### 7.3.3 PE model optimization formulation

In most empirical settings it is preferable to develop tools that allow for the analysis of market equilibrium that also include transportation costs between buyer and seller and/or the costs of transforming raw materials into a final product. It is also useful to have a framework that can allow for the integration (or interaction) of more than one market in different locations, as in the two countries who might import from or export to each other. Optimization models extending those from the previous section can be used to incorporate these elements of food systems. The Eqs. for a general specification with multiple supply and demand locations would include the objective function to be maximized:

$$\sum_{k=1}^{K} \gamma_k \cdot QD_k - 0.5 \cdot \varepsilon_k \cdot QD_k^2 - \sum_{i=1}^{I} \alpha_i \cdot QS_i + 0.5 \cdot \beta_i \cdot QS_i^2 - \sum_{j=1}^{J}\sum_{k=1}^{K} DC_{jk} \cdot XFP_{jk}$$

$$- \sum_{j=1}^{J} PC_j \cdot QP_j - \sum_{I=1}^{I}\sum_{j=1}^{J} AC_{ij} \cdot XRM_{ij}$$

where

QD and QS are as previously defined, but now for multiple demand locations (k) and supply locations (i) that can differ;

$XFP_{jk}$ is the amount of product shipped from a processing location (j) to a demand location (k)

$QP_j$ is the amount of product processed at a location (j);

$XRM_{ij}$ is the amount of a farm product shipped from production location (i) to processing location (j);

$DC_{jk}$ is the cost per unit to transport product from processing location (j) to demand location (k);

$PC_j$ is the cost per unit to process (or package) the product at location (j);

$AC_{ij}$ is the cost per unit to transport product from a farm location (i) to a processing location (j)

This objective function represents the area under the demand curve to Q less the area under the supply curve to $Q^3$ less the other costs associated with moving or transforming product from its production location to a consumption location. In addition, we require that physical constraints are met, such as:

$$\sum_{j=1}^{J} XRM_{ij} \le QS_i$$

The total amount of product shipped from production location i to processing location j must be less than or equal to the amount produced (supplied) at location i;

$$QD_k \le \sum_{j=1}^{J} XFP_{jk}$$

The total amount of product shipped from processing locations j to demand locations k must be greater than the amount consumed (demanded) at location k;

$$QP_j = \theta_j \cdot \sum_{i=1}^{I} XRM_{ij}$$

The amount processed at a location j is equal to the sum of product transported to j from all i locations, multiplied by a transformation factor that converts arriving (raw) product into processed product.

$$\sum_{k=1}^{K} XFP_{jk} \le QP_j$$

The total amount transported from processing location j to demand locations k must be less than or equal to the amount of product processed at j.

This objective function and the constraints solve for market equilibrium for a single market (location) or multiple markets (multiple production, processing and demand locations). Models with multiple market locations facilitate the assessment of the impacts of trade policy on different countries or regions. This model has a *nonlinear* objective function (because of squared variables like $QD^2$) and therefore requires a nonlinear programming (NLP) approach to optimization rather than a LP approach. Policies like subsidies, support programs, quotas or taxes can be included as additional costs or restrictions.

### 7.3.4 Empirical examples of PE models

We can use this framework to analyze a tax on sugared soft drinks if we assume one production location, one demand location, and no additional costs for transportation or processing. With values $\alpha=5$, $\gamma=15$, $\beta=2$ and $\varepsilon=2$ and a tax $T=2$, the tax has the impact of

---

[3] The first two components of the objective function are derived from calculus. For example, the integral of a function represents the area under a curve, and in this case the initial function is $PD = \gamma - \varepsilon \bullet QD$. The integral of this function is $\gamma\,QD - 0.5 \bullet \varepsilon \bullet QD^2$. The derivative of this integral function equals the original function $PD = \gamma - \varepsilon \bullet QD$.

TABLE 7.1    Comparative statics analysis of a tax on sugary soft drinks.

| Outcome | Initial Market Equilibrium | With Tax Market Equilibrium | Difference |
|---|---|---|---|
| QS | 2.5 | 2.0 | −0.5 |
| QD | 2.5 | 2.0 | −0.5 |
| PS | 10.0 | 11.0 | +1.0 |
| PD | 10.0 | 9.0 | −1.0 |
| Consumer expenditures | 25.0 | 22.0 | −3.0 |
| Producer revenues | 25.0 | 18.0[a] | −7.0 |
| Government revenues | 0.0 | 4.0 | +4.0 |
| Consumer + Producer Surplus | 12.5 | 8.0 | −4.5 |

[a] In this case, revenues net of the tax paid to the government.

TABLE 7.2    Comparative statics analysis of an effective behavior change communication program on sugary soft drinks.

| Outcome | Initial Market Equilibrium | With Tax Market Equilibrium | Difference |
|---|---|---|---|
| QS | 2.5 | 2.0 | −0.5 |
| QD | 2.5 | 2.0 | −0.5 |
| PS | 10.0 | 9.0 | −1.0 |
| PD | 10.0 | 9.0 | −1.0 |
| Consumer expenditures | 25.0 | 18.0 | −7.0 |
| Producer revenues | 25.0 | 18.0 | −7.0 |
| Government revenues | 0.0 | 0.0 | 0.0 |
| Consumer + Producer Surplus | 12.5 | 8.0 | −4.5 |

reducing soft drink consumption by 20 percent (from 2.5 to 2.0; Table 7.1). Consumer prices increase (but by less than the amount of the tax) and producer prices decrease (but also by less than the amount of the tax). Consumer expenditures and producer revenues both decrease, and the government gains revenue. The sum of consumer surplus and producer surplus is decreased, consistent with our previous conceptual analysis of "welfare".

We can also use this framework to analyze the impacts of "behavior change communication" programs (Kennedy et al., 2018) on soft drink consumption. Effective programs would have the effect of reducing the demand for sugary soft drinks, which is equivalent to a "shift down" in the demand curve. Changing $\gamma$ from 15 to 13 (a "shift down") would result in the same reduction in soft drink consumption (Table 7.2) but with different impacts on other

TABLE 7.3 Assumptions for the analysis of international trade with a partial equilibrium model.

| Assumption (Parameter) | Country A (Low Cost, Low Demand) | Country B (High Cost, High Demand) |
|---|---|---|
| Supply intercept, $\alpha$ | 5 | 10 |
| Supply slope, $\beta$ | 2 | 2 |
| Demand intercept, $\gamma$ | 15 | 20 |
| Demand slope, $\varepsilon$ | 2 | 2 |
| Assembly costs within country, AC | 1 | 1 |
| Assembly costs to other country, AC | 6 | 6 |
| Processing costs, PC | 1 | 1 |
| Transformation coefficient, | 1 | 1 |
| Distribution cost within country, DC | 1 | 1 |
| Distribution cost to other country, DC | 2 | 2 |

NOTE: 'Low cost' is indicated by a smaller supply intercept value in Country A than Country B and 'low demand' is indicated by a smaller demand intercept value in Country A than Country B. This will usually imply that the equilibrium price in the absence of trade will be lower in Country A than Country B, which provides economic incentives for the two countries to trade.

outcomes: prices decrease for both consumers and producers, expenditures and revenues decrease, and government receives no revenue.

As a final empirical example, consider analysis of trade between two countries. In an initial situation (Table 7.3), there are no barriers to trade, and although both countries produce and deliver a food product to consumers in their own countries, one country has lower production costs and less demand and is thus able to export to the other country (which has higher production costs and higher demand).

Without trade barriers, Country A produces more of the product than Country B and exports more than 70 percent of its production (2.0/2.75) to Country B (Table 7.4). As a result, imports from Country A provide more than 70 percent (2.0/2.75) of the product consumed in Country B. Producers in Country B might observe the large quantity of imports from Country A (and their much higher revenues—a measure of the size of the industry) and propose to the government of Country B that they should receive protection from imports (or, "foreign competition") in the form of a tariff on imports from Country A.

A per-unit tariff adds a constant value to the distribution costs (so that DC from A to B is now the previous DC + IT, where IT means import tariff). With a tariff of 5 per unit to product imported from Country A, the new distribution cost from Country A to Country B rises from 2 to 7.[4] This distribution cost is now large enough that it is no longer profitable for Country A to export to Country B, and all imports from A are eliminated. Both Country A

---

[4] Note that we also need relatively high costs for transporting raw product "assembled" from production to processing, or this could be a way to circumvent a tariff on a "finished" product that is "distributed". In many cases, a country might levy a tariff cost that affects imports of both farm and finished products.

TABLE 7.4    Analysis of international trade with a partial equilibrium model.

| Outcome | Initial Market Equilibrium with Trade | | Market Equilibrium with Tariffs on Imports from Country A | | Difference with the Import Tariff | |
|---|---|---|---|---|---|---|
| | Country A | Country B | Country A | Country B | Country A | Country B |
| QS | 2.75 | 0.75 | 1.75 | 1.75 | −1.00 | +1.00 |
| QD | 0.75 | 2.75 | 1.75 | 1.75 | +1.00 | −1.00 |
| PS | 10.50 | 11.50 | 8.50 | 13.50 | −2.00 | +2.00 |
| PD | 13.50 | 14.50 | 11.50 | 16.50 | −2.00 | +2.00 |
| Imports from Country A | – | 2.00 | – | 0.00 | – | −2.00 |
| Consumer expenditures | 10.12 | 39.87 | 20.12 | 28.87 | +10.00 | −11.00 |
| Producer revenues | 28.87 | 8.62 | 14.87 | 23.62 | −14.00 | +15.00 |
| Government revenues | 0.00 | 0.00 | 0.00 | 0.00 | 0.00 | 0.00 |
| Consumer + Producer Surplus in both countries | 16.25 | | 12.25 | | −4.00 | |

and Country B are now "self-sufficient" in the product—all of their consumption comes from production in their own country. The tariff reduces production in Country A (by more than one-third) and increases it in Country B (by more than 100 percent). Consumption increases in Country A (due to lower consumer prices) and decreases in Country B (due to higher consumer prices). Producer revenues in Country A are decreased (by nearly 50 percent) and increased (more than 100 percent) in Country B. In this case, the tariff prevents any trade from occurring, so the government does not collect any revenues. If the goal of implementing the tariff was to support producers in Country B, we would probably consider the tariff policy a success. However, the combined value of consumer and producer surplus in these two countries is reduced by nearly 25 percent. This reduction in economic value often is a reason why economists favor "free trade"—trade barriers reduce combined producer and consumer surplus.

Although this is a simple example, it illustrates key points from analysis of trade in food products:

- A policy of "self-sufficiency" generally will negatively affect consumers in the importing country and producers in the exporting country;
- A policy of "self-sufficiency" generally will benefit consumers in the exporting country and producers in the country implementing the tariff;
- The policy actions of Country B also affect stakeholders in Country A, which could give Country A incentives or justification to 'retaliate' through changes to trade policies on products imported from Country B.

### 7.3.5 Data needs and model calibration

The information required to implement empirical partial equilibrium models typically comprises the nature of supply and demand responses and the costs for transportation and processing in the markets to be analyzed. For modeling trade, it may be necessary to include information on trade (imports and exports) and tariffs. Often the required information is available from previous studies or can be estimated with secondary data. Empirical models are also calibrated to approximate observed outcomes through modifications to the supply curves, demand curves and costs.

### 7.3.6 Examples and extensions for food systems

A vast number of studies use PE models of markets and trade, based on an initial formulation by Takayama and Judge (1964). We describe here only a few relevant examples. Wailes et al. (2015) used a global model of rice markets (Arkansas Global Rice Model) to assess regional and national approaches to improving food security for rice consumption in West Africa. These authors examined the strategies and consequences of pursuing self-sufficiency in rice. They found that the elimination of rice imports (i.e., 'self-sufficiency' in West Africa) would reduce overall global rice prices (negatively affecting other rice-producing countries) and increase total costs at the global level. Further, if self-sufficiency made domestic rice uncompetitive with imported rice, this would result in a significant price discount for domestic rice, reducing benefits to producers and consumers. Dorosh et al. (2016) combines the analysis of price and quantity data over time with the development of a PE model of multiple grain markets to assess food security impacts of alternative policies in South Sudan. Their analyses indicated that if total imports of cereals are reduced by one-third, domestic prices of cereals could rise by 45 percent or more. They concluded that government policy should maintain incentives for private sector imports to avoid destabilizing market supplies, domestic prices, and ultimately, food consumption of the poor.

Other extensions include "agriculture sector models" (Hazel and Norton, 1986) in which the supply is modeled based on the allocation of land area and yields. The models described above assume a "perfectly competitive market" (with a large number of producers and consumers, none of whom have market power to affect outcomes). These assumptions are not appropriate for all markets, although they may apply to agricultural commodities. Modifications to the framework above can account for situations of monopoly (single seller), oligopoly (a few large sellers) or their counterparts on the buying side, monopsony and oligopsony (Harker, 1986). Durond-Murat and Wailes (2010) describe an alternative approach to SPE models that solves a system of Eqs. derived from economic theory.

## 7.4 Dynamic supply chain models

The activities of the multiple actors in food systems often can be usefully viewed through the lens of food supply chains. A supply chain is "a set of structures and processes a group of organizations uses to deliver an output to a customer" (Sterman, 2000). This definition emphasizes interactions given both the 'structures and processes' and the involvement of multiple organizations, and can be applied to both complex and simpler supply chains (such as an individual farmer selling product at a farmers' market). The supply chain perspective is useful because it explicitly incorporates behaviors and interactions among food system participants to a greater extent than do transportation, transshipment or spatial price equilibrium models. Many supply chain analyses examine how outcomes change over time. This can be particularly useful because many food supply chains demonstrate considerable variation in prices, costs, supplies and profits over time. For example, both dairy product markets (Nicholson and Stephenson, 2015a) and beef markets (Rosen et al. 1994) show evidence of short-term price volatility and longer-term cycles of prices, supplies and profitability. We use tools and concepts from the system dynamics (SD) modeling approach to represent the food supply chain. SD focuses on stocks, flows and feedback processes that create outcomes varying over time (Sterman, 2000). SD models can incorporate multiple supply chain actors, their behaviors and interactions. It is also possible to develop SD models of supply chains through participatory processes with stakeholders (e.g., Vennix, 1996).

### 7.4.1 Model structures

Supply chain models using SD methods often have at their core a "stock management structure" (SMS). A stock is any quantity that can be quantified at a given time, including a physical quantity such as the amount of grain in inventories or less tangible elements such as expectations of future demand. For food supply chains, it is common for supply chain actors (managers) to base the actions (decisions) on observed values of a few key stocks that they attempt to maintain within an acceptable range.

The SMS represents the information tracking and decision-making structure required to maintain the stock as close as possible to a desired level (Fig. 7.5); the goal is to maintain the value of the *Stock* equal to the *Desired Stock*. Actors in food supply chains often want to maintain inventories of their product at a desired level, for example. The decision maker observes the value of the *Loss Rate*, which can be thought of as shipments to the next stage of the supply chain. Based on the rate at which product flows out of the *Stock* based on the *Loss Rate*, the manager also forms expectations about the future values of the *Loss Rate*, which are then used as one component of the *Desired Acquisition Rate*. If the pattern of orders from customers is changing over time, it makes sense for a manager to update expectations about future demand, but probably not immediately in response to every change in demand, so managers often based expectations on 'smoothed' values of demand such as a moving average. The *Desired Acquisition Rate* also includes component called *Adjustment for Stock*. This variable reflects the idea that changes in demand will affect the *Loss Rate* and thus the value of the *Stock* and reductions in stock must also be replaced to maintain the stock at a desired level.

The stock to be managed is shown as a box. Quantities coming into the stock (inflows) and leaving the stock (outflows) are double-line arrows and a valve symbol. Blue single-line arrows show the linkages (causality) among variables; a variable being pointed to by an arrow is caused by the variable preceding it. *Polarity* of the linkage between variables is shown by the + or - associated with each arrow. A + indicates that the 'cause' variable and the 'effect' variable would change in the same direction. The diagram also illustrates that there is 'feedback' in this system—through the arrows that begin with the stock and link to loss rate, expected loss rate, desired acquisition rate and back to the stock.

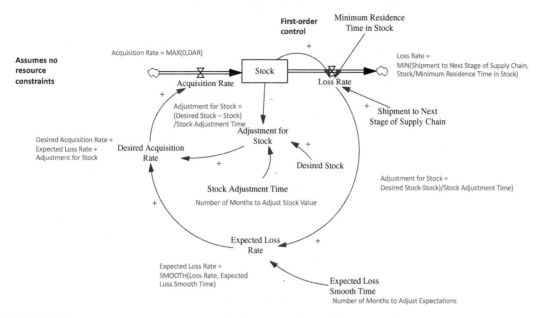

**FIGURE 7.5** Elements of the stock management structure for a supply chain actor.

The SMS is a core element of how many supply chains are managed, but a more realistic supply chain would include multiple actors, where prices and ordering decisions to suppliers (upstream in the supply chain) are influenced by profitability experienced by the each of the supply chain participants. Another common feature is a "supply line" that includes the amount of product that is either currently in production but not yet finished (like growing crops), or for which a customer has placed an order but the product has not yet arrived. Thus, a more appropriate SMS also includes a stock for the supply line (Fig. 7.6).

### 7.4.2 Empirical example: vegetable supply chains in Kenya

A more realistic supply chain includes multiple interacting SMS components, such as a vegetable supply chain with vegetable producers (farmers), wholesalers, retailers and consumers (Fig. 7.7). This structure shows the linkages among the different supply chain actors, where each type of actor is represented by multiple companies or operating units (e.g., multiple farms, wholesalers, retailers and consumers). The stock to be managed by farmers, wholesalers and retailers is their inventory of a vegetable product. The stock is managed by each of farmers, wholesalers and retailers by monitoring inventory coverage—the number of weeks of product they have in inventory at current rates of sale to their customers. Stock managers also track profits and adjust their ordering decisions to suppliers. If inventories are high relative to current sales, they reduce prices to their customers and place fewer orders with their suppliers. If inventories are low, they raise prices to their customers and place more orders with their suppliers. For farmers, profits determine whether to increase or decrease the acres planted (and therefore, production of the vegetable crop). Consumers are defined as the end users, and

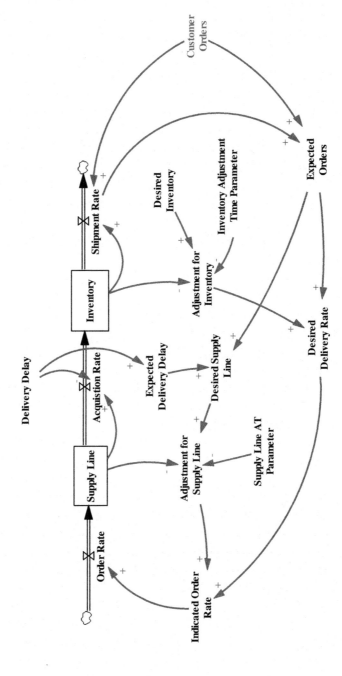

FIGURE 7.6   Stock management structure for a supply chain actor with a supply line.

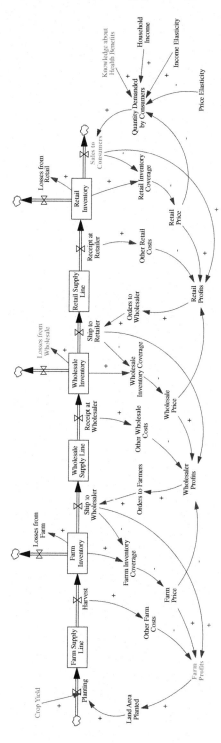

**FIGURE 7.7** Conceptual SD model of a vegetable supply chain with farmers, wholesalers, retailers and consumers.

the quantity they demand of vegetables is determined by prices, incomes, and knowledge of the healthfulness of vegetables.

This model also modifies each the stocks (inventories) to include losses—perishability of vegetables can be a high proportion of the product provided to the next stage of the supply chain. Another addition to this structure is the inclusion of other costs of operating each of the businesses in the supply chain, because these also determine profits. We assume unit costs apply to product that arrives at each stage of the supply chain (whether sold or part of 'losses') and revenues are received only for product actually provided to the next stage of the supply chain or customers. This supply chain comprises multiple interacting stocks, flows and feedbacks, with physical flows of product moving through the stocks and the values of the stock influencing profit and decisions about how much to produce or order.

To illustrate application to analysis of proposed interventions, suppose that the goal is to increase the consumption of vegetables. Many studies (e.g., Micha et al., 2015) have indicated that vegetable consumption levels are below 'optimal' for human health and also below the recommendations of the World Health Organization (WHO, 2004). There are multiple potential intervention points (Fig. 7.7), and often we want to assess which intervention will have the largest impact and(or) is the most cost-effective with resources available. Here, we focus on three interventions: increasing farm yields by 5 percent, reducing wholesale losses by 25 percent and increasing education of consumers about the value of healthy eating. We will focus on the impacts each has on vegetable consumption, but also on profits for farmers. We compare outcomes to a 'benchmark' model in 'dynamic equilibrium' where flows of product move through the supply chain, but all the elements are balanced to keep inventory levels, prices and consumption constant. We assume that intervention can be implemented instantaneously and without cost, a best-case scenario. Combining SD methods with implementation science (Bauer et al. 2015) can assess more realistic processes that required time and funding for implementation.

The simulation model is programmed in Vensim® software (https://vensim.com) and is simulated for 100 weeks to allow for sufficient time after an intervention for system adjustments to occur. We use a computation interval (the 'time step' for the model of 0.125 weeks) and assume that each of the interventions is implemented at 10 weeks. We will examine the pattern of behaviors for the key variables for each of the interventions and its average value during weeks 10 to 100.

### 7.4.3 Simulation results for three interventions

The simulation analyses indicate that each of the proposed interventions will increase vegetable consumption, albeit by relatively small amounts of at most 1 percent (Fig. 7.8 and Table 7.5). However, they also indicate that for two of the interventions, the profits earned by farmers will decrease (Fig. 7.9 and Table 7.5), sometimes by as much as 30 percent.

An increase in yield will increase the amount produced per acre and increase the *Farm Supply Line* and *Harvest* quantity (per the relationships in Fig. 7.7). If the costs per kg of production are the same (and yield increases could either increase or decrease costs per kg), this will increase *Other Farm Costs. Farm Inventory* will also increase, signaling that prices should be decreased to maintain the desired inventory coverage. Increased costs and lower prices per kg

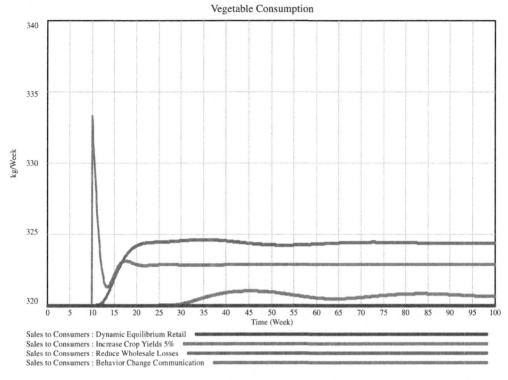

**FIGURE 7.8** Simulated impacts of supply chain interventions on vegetables consumed.

**TABLE 7.5** Simulated results of three interventions to increase vegetable consumption, mean values from week 10 to 100.

| Outcome | Dynamic Equilibrium | Increase Crop Yields 5 percent | Reduce Wholesale Losses 25 percent | Behavior Change Communication |
|---|---|---|---|---|
| Sales to Consumers, MT/week | 320.0 | 320.5 | 324.6 | 322.9 |
| Percentage change in Sales to Consumer | – | +0.2 percent | +1.3 percent | +0.9 percent |
| Farm profits, $000/week | 1,875 | 1,297 | 1,457 | 1,931 |
| Percentage change in Farm Profits | – | −30.8 percent | −22.3 percent | +3.0 percent |

Note: Sales to Consumers and Farm Profits are shown in bold orange in Fig. 7.7.

will result in an initial decrease in *Farm Profits*. However, the lower *Farm Price* will also increase *Wholesaler Profits*, which increases *Orders to Farmers* and the value of *Ship to Wholesaler*. This will offset the increase in *Farm Inventories*, increase *Wholesaler Inventory*, lower *Wholesale Price* and increase orders from *Retailers*. Ultimately, this process will result in lower *Retail Price* and (small) increased *Quantity Demanded by Consumers*. The magnitude of the changes depends on

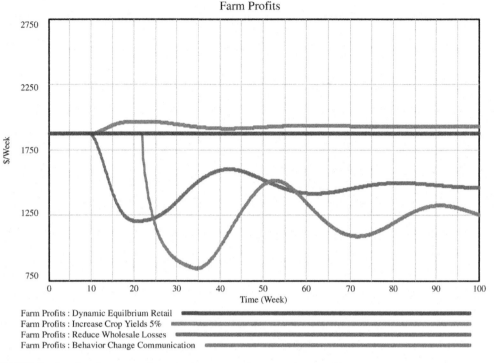

FIGURE 7.9 Simulated impacts of supply chain interventions on farm profits.

the specific values of the parameters but it is quite possible for higher yields to reduce farmer profits but benefit consumers and other actors in the supply chain. Simões et al. (2019) noted a similar effect of increasing productivity in the dairy supply chain of Brazil.

Many efforts have been made to reduce 'post-harvest' losses (Sheahan and Barrett, 2017; Cattaneoet al., 2021) often under the assumption that this will substantially increase available supplies of food products. In this case, a reduction in *Losses from Wholesale* will increase *Wholesale Inventory* and decrease *Wholesale Price,* and it is sensible to reduce the amount ordered from farmers because with lower losses, less product is needed to meet the quantity that retailers will demand. As with the previous example, the specific outcomes depend on the values of the parameters, but it is possible to achieve only modest increases in consumption even with substantial reductions in losses. The supply chain adjusts to the new requirements for inventory and product available for consumption. This intervention also reduces *Wholesaler Profits,* while increasing *Retail Profits,* which implies a re-distribution of earnings in the supply chain.

Many organizations have committed resources to programs (such as behavior change communication) to increase consumer demand for healthy foods. The existing evidence (e.g., Rekhy and McConchie, 2014) suggests that it is difficult to modify consumer purchase preferences. Increased demand for healthy foods may also have to work against the offsetting increases in price. A 50 percent increase in the knowledge of the healthfulness of fruits and vegetables results in a short initial increase in consumption with increased knowledge, but much of this effect is quickly offset through adjustments in *Retail Price* due to lower *Retail Inventory*. The effects of larger sales and higher prices are transmitted through the rest of the supply chain, with similar effects on wholesalers and farmers. Of the three intervention options, this is the only one that increases both consumption (by less than 1 percent) and Farm Profits (by 3 percent).

These examples suggest three summary points:

- Food supply chain often are sufficiently complex so that both the direction and magnitude of responses to interventions are difficult to predict with intuition alone or development of only conceptual models or diagrams;
- Values of the parameters matter to the outcomes, especially those for the responsiveness of supply chain actors;
- Interventions can worsen some outcomes and create instability in food supply chains as shown by the lower and oscillating patterns in farm profits (Fig. 7.9).

### 7.4.4 Data needs, sensitivity analysis and participatory stakeholder modeling

The data needed to develop even a simple supply chain model such as the one described in this chapter can be challenging, although many descriptive studies of food supply chains exist (Gómez and Ricketts, 2013; de la Peña et al., 2018; Ridoutt et al. 2019). Data on prices at different points in the supply chain and quantities tend to be more readily available than information on product losses, other costs or the responsiveness of supply chain actors to changes in inventory levels or profitability. Information on consumer responsiveness to prices and income are often available from previous econometric studies, although in some cases these may be dated and use aggregated product categories. In most cases, some data will be missing or will need to be adapted from other studies.

There are two principal approaches to dealing with limited data in these dynamic supply chain models. *Sensitivity analysis* can evaluate the impacts of uncertainty in model parameters by selecting a large number of different values within a specified range and evaluating their impact on model outcomes. When outcomes are quite sensitive to different assumptions about parameter values, this places limits on the certainty to which interventions can be assessed, but also highlights priority information needs and focus for future data collection efforts. Another approach is to involve supply chain stakeholders themselves in the process of developing the model and defining the parameters. *Group Model Building* (Vennix, 1996) involves iterative meetings with stakeholders (farmers, wholesalers, retailers, consumers, and policy makers) to define the outcomes that interventions should improve, map system interactions, and develop values of unknown parameters. GMB can improve the quality of supply chain models, improve stakeholder understanding of the sources and solutions to problems, and enhance consensus for future action (Rouwette and Franco, 2015).

### 7.4.5 Examples of SD modeling of food supply chains

SD models of supply chains are very common, but applications to food systems are fewer but growing. Nicholson and Stephenson (2015) developed an SD-based model of the US dairy supply chain with multiple products and analyzed impacts of government support policies on farm incomes and government costs. Kopainsky et al. (2017) showed that GMB can be adapted to groups at the community level with low or no formal educational background through work with farmers in Zambia. Following the recommendation of Nordhagen (2020), there is a small but growing body of literature using participatory SD modeling methods to assess the impacts of supply chain (sometimes, 'value chain') interventions on food security and nutrition outcomes (e.g., Cooper et al., 2020 and Nicholson and Monterrosa, 2021). The flexibility of SD models and the importance in many systems of stock-flow-feedback dynamics supports their expanded use for environmental and social issues, and especially to address global problems of food and nutrition security under the growing challenges of climate change.

## 7.5 Concluding comments

Market and supply chain models can complement other types of food system analyses when economic outcomes (costs, prices, incomes) are of interest, and can be extended to include other impacts, such as those on the environment (e.g., Nicholson et al., 2015) and nutritional impacts (e.g., Nicholson and Monterrosa, 2021). Each of the models discussed in this chapter derives from core economic or supply chain management principles that will often be important to consider when assessing food system interventions. Spatial optimization models are often used to provide input about where production and processing activities should be located to result in the lowest cost, which can benefit both food supply chain companies and consumers. The examples from Atallah et al. (2014) and Nicholson et al. (2015) illustrate how these models can be extended to examine the cost and environmental implications of localizing or regionalizing food systems. Spatial price equilibrium models are widely used for the analysis of agricultural and trade policies and provide a framework to assess the impacts of policy on a variety of food system actors (such as farmers or consumers). The empirical examples from Wailes et al. (2015) and Doresh et al. (2016) show how these models can be used to assess proposed policies for country-level food self-sufficiency. Global models based on this framework can be used to assess the impacts of diets on health and climate change (e.g., (Springmann et al., 2018; Mason D'Croz, 2019). Dynamic models of supply chains based on SD can be useful for a variety of objectives, including understanding the tradeoffs among food system actors and dynamics resulting from interventions. Their flexibility and processes to engage directly with stakeholders imply they may be particularly useful when applied to the analysis of 'nutrition-sensitive value chains' (de la Peña and Garrett, 2018) that currently constitute a principal approach to improve nutritional outcomes in low-income country settings. The development of empirical information required for implementation of these models can range from relatively straightforward (unit costs of shipment in a transportation model) to more challenging (consumer responses to price changes or price setting in response to inventory holdings). However, even when information specific to a location or issue is not entirely available, use of market and supply chain models for sensitivity analysis, to assess how different assumption do or don't modify relevant outcomes) can be useful.

## References

Atallah, S.S., Gómez, M.I., Björkman, T., 2014. Localization effects for a fresh vegetable product supply chain: broccoli in the eastern United States. Food Policy 49, 151–159. https://doi.org/10.1016/j.foodpol.2014.07.005.

Bauer, M.S., Damschroder, L., Hagedorn, H., Smith, J., Kilbourne, M.A., 2015. An introduction to implementation science for the non-specialist. BMC Psychology 3, 32. https://doi.org/10.1186/s40359-015-0089-9.

Cattaneo, A., Sánchez, M.V., Torero, M., Vos, R., 2021. Reducing food loss and waste: five challenges for policy and research. Food Policy 98, 101974. https://doi.org/10.1016/j.foodpol.2020.101974.

Cooper, G.S., Rich, K.M., Choudhury, D.K., Sadek, S., Shankar, B., Kadiyala, S., et al., 2020. Using Participatory System Dynamics Approaches to Evaluate the Nutritional Sensitivity of a Producer-Facing Agricultural Intervention in India and Bangladesh. In: Paper presented at the 3RD Asia Pacific System Dynamics Conference. Brisbane, Australia, 2nd-4th February 2020.

de la Peña, I., Garrett, J.L., 2018. Nutrition-sensitive Value Chains: A Guide for Project Design, I and II. International Fund for Agricultural Development (IFAD), Rome Volumes.

Dorosh, P.A., Rashid, S., van Asselt, J., 2016. Enhancing food security in South Sudan: the role of markets and regional trade. Agricultural Economics 47, 697–707.

Durand-Morat, A., Wailes, E., 2010. RICEFLOW: A Multi-Region, Multi-Product, Spatial Partial Equilibirium Model of the World Rice Economy. Department of Agricultural Economics and Agribusiness University of Arkansas, Fayetteville Division of Agriculture July. [Staff Paper 03 2010].

Fletcher, J.M., Frisvold, D., Tefft, N., 2010. Can Soft Drink Taxes Reduce Population Weight? Contemporary economic policy 28, 23–35. https://doi.org/10.1111/j.1465-7287.2009.00182.x.

Gómez, M.I., Ricketts, K.D., 2013. Food value chain transformation in developing countries: selected hypotheses on nutritional implications. ESA Working Paper No. 13-05. Agricultural Economics Development Division. FAO 2013.

Harker, P.T., 1986. Alternative Models of Spatial Competition. Operations Research 34, 410–425.

Hazell, P.B.R., Norton, R.D., 1986. Mathematical Programming for Economic Analysis in Agriculture. Macmillan Publishing Co., New York.

Hiller, F.S., Lieberman, G.J., 2021. Introduction to Operations Research, 11th edition McGraw-Hill, New York.

Kennedy, E., Stickland, J., Kershaw, M., et al., 2018. Impact of Social and Behavior Change Communication in Nutrition Sensitive Interventions on Selected Indicators of Nutritional Status. Journal of Human Nutrition 2, 24–33. http://doi.org/10.36959/487/279.

Kopainsky, B., Hager, G.M., Herrera, H., Nyanga, P.H., 2017. Transforming food systems at local levels: using participatory system dynamics in an interactive manner to refine small-scale farmers' mental models. Ecological Modelling 362, 101–110. http://doi.org/10.1016/j.ecolmodel.2017.08.010.

Mason-D'Croz, D., Bogard, J.R., Sulser, T.B., Cenacchi, N., Dunston, S., Herrero, M. et al.. Gaps between fruit and vegetable production, demand, and recommended consumption at global and national levels: an integrated modelling study. www.thelancet.com/planetary-health Vol 3 July 2019.

Micha, R., Khatibzadeh, S., Shi, P., et al., 2015. Global, regional and national consumption of major food groups in 1990 and 2010: a systematic analysis including 266 country-specific nutrition surveys worldwide. BMJ Open 5, e008705. 2015 http://doi.org/10.1136/bmjopen-2015008705 .

Nicholson, C.F., He, X., Gómez, M.I., Gao, H.O., Hill, E., 2015. Environmental and Economic Analysis of Regionalizing Fluid Milk Supply Chains in the Northeastern U.S. Environmental Science & Technology 49, 12005–12014. http://doi.org/10.1021/acs.est.5b02892.

Nicholson, C.F., Stephenson, M.W., 2015a. Price Cycles in the U.S. Dairy Supply Chain and their Management Implications. Agribusiness: An International Journal 31, 507–520. http://doi.org/10.1002/agr.21416.

Nicholson, C.F., Stephenson, M.W., 2015b. Dynamic Market Impacts of the Dairy Margin Protection Program. Journal of Agribusiness 32, 165–192.

Nicholson, C.F., Monterrosa, E., 2021. Participatory Value Chain Modeling to Enhance Fruit and Vegetable Consumption Nairobi, Kenya. In: Global Alliance for Improved Nutrition Working Paper.

Nordhagen, S., 2020. Food supply chains and child and adolescent diets: a review. Global Food Security 26, 100415.

Redondo, M., Hernández-Aguado, I., Lumbreras, B., 2018. The impact of the tax on sweetened beverages: a systematic review. American Journal of Clinical Nutrition 108, 548–563. 2018 https://doi.org/10.1093/ajcn/nqy135 .

Rekhy, R., McConchie, R., 2014. Promoting consumption of fruit and vegetables for better health. Have campaigns delivered on the goals? Appetite 79, 113–123.

Ridoutt, B., Bogard, J.R., Dizyee, K., Lim-Camacho, L., Kumar, S., 2019. Value chains and diet quality: a review of impact pathways and intervention strategies. Agriculture 9.

Rosen, S., Murphy, K., Scheinkman, J., 1994. Cattle Cycles. Journal of Political Economy 102 (3), 468–492.

Rouwette, E. and Franco, L.A.. 2015. Messy problems: practical interventions for working through complexity, uncertainty and conflict.

Sheahan, M., Barrett, C.B., 2017. Review: food loss and waste in Sub-Saharan Africa. Food Policy 70, 1–12. http://dx.doi.org/10.1016/j.foodpol.2017.03.012.

Simões, A.R.P., Nicholson, C.F., Novakovic, A.M., Protil, R.M., 2019. Dynamic impacts of farm-level technology adoption on the Brazilian dairy supply chain. International Food and Agribusiness Management Review 23 (1). http://doi.org/10.22434/IFAMR2019.0033.

Springmann, M., Mason-D'Croz, D., Robinson, S., Wiebe, K., Godfray, H.C.J., Rayner, M., et al., 2018. Health-motivated taxes on red and processed meat: a modelling study on optimal tax levels and associated health impacts. Plos One 13 (11), e0204139.

Sterman, J.D., 2000. Business Dynamics: Systems Thinking and Modeling for a Complex World. Irwin/McGraw-Hill, Boston.

Takayama, T., Judge, G., 1964. Equilibrium among Spatially Separated Markets: a Reformulation. Econometrica 32, 510–524. http://doi.org/10.2307/1910175.

Vennix, J.A.M., 1996. Group Model Building: Facilitating Team Learning Using System Dynamics. John Wiley and Sons, Chichester, UK.

Wailes, E.J., Durand-Morat, A., Diagne, M., 2015. Regional and National Rice Development Strategies for Food Security in West Africa. Food Security in an Uncertain World, pp. 255–268.

World Health Organization (WHO), 2004. Joint FAO/WHO Workshop on Fruit and Vegetables for Health. In: Report of a Joint FAO/WHO Workshop, 1-3 September 2004. Kobe, Japan.

# 8

# Using input-output analysis to estimate the economic impacts of food system initiatives

*Becca Jablonski[a], Jeffrey K. O'Hara[b],*
*Allison Bauman[a], Todd M. Schmit[c] and*
*Dawn D. Thilmany[a]*

[a] Colorado State University, Department of Agricultural and Resource Economics, Fort Collins, CO, United States [b] U.S. Department of Agriculture, Agricultural Marketing Service, Washington, DC, United States [c] Cornell University, Dyson School of Applied Economics and Management, Ithaca, NY, United States

## 8.1 Introduction

As policymakers invest resources in food systems programming, they increasingly want to understand the resulting economic impacts (O'Hara and Pirog, 2013; Rahe, Van Dis, and Gwin, 2019). Economic impact assessment models document the relationship between one sector and the rest of the economy within its region. In this chapter, we focus on the local foods sector, although the economic concepts, models, and methodological refinements we discuss can be applied to any sector of the economy.

Specifically, economic impact studies provide estimates for the following types of questions:

- How does supporting an increase in sales by one sector in a local economy (e.g., food manufacturing) impact the production of other goods and services in the economy. In other words, to increase sales, additional inputs and labor are required. To the degree those are local, additional impacts accrue.
- How will exogenous shocks, through private or public investment, support industries of interest and, through them, other backward-linked industries? What will be the impact

on employment or labor income in the region? In what sectors and/or in what household income categories will changes occur?

In this chapter, we discuss how input-output (I-O) models can be used to estimate these impacts. In the first section, we explain how economic impact can result from local and regional food systems by describing export-oriented food production systems versus import substitution "local food" strategies, and subsequently describe the methods for conducting economic impact assessments and some techniques to more accurately capture the economic impacts of exogenous shocks to the food system. This includes comparing I-O models with competitive general equilibrium models. We also discuss how the characteristics of local food systems may result in higher economic impacts than commodity-oriented production. We further provide specific steps that researchers can take to augment traditional data sources to increase the accuracy of model outputs. In the second section, we provide an example of I-O modeling from a recently completed assessment of nutrition incentive programs. This example could guide those interested in undertaking their own I-O studies. Finally, we discuss future research needs. We propose integrating economic impact assessments with methodological approaches that incorporate other types of impacts resulting from food systems efforts – including environmental – in order to understand broader effects and potential trade-offs.

### 8.1.1 An overview of economic impact methods

Money moves throughout the economy in a circular fashion. Households consume goods and services from firms and institutions. Households make purchases with money that they do not save, which they have earned by offering labor to firms to produce goods and services. In addition to goods purchased for consumption, the value of goods and services in an economy includes those for investment, purchase by governments, and net exports to other economies.

Net exports are the difference between exports and imports. An "export-based" approach to economic development is premised on increasing the production of goods that are exported outside of the region. The payment for these goods brings money into the region. In general, economic development strategies are typically premised around attracting large employers that produce goods for export, instead of promoting entrepreneurship (Goetz, Partridge, and Stephens, 2018). Despite the emphasis on export-based approaches to economic development, reducing imports also increases a region's economy by reducing the money leaving the economy through "leakage." However, import substitution has received less emphasis as an economic development strategy from economists. This is because it can be less expensive for a region to purchase some of its goods outside of its region than to produce them internally. Indeed, one cannot generalize about whether local food programs increase societal welfare, since it depends, in part, on assumptions that must be made about why consumers buy local foods and the extent to which agricultural markets are uncompetitive. See Winfree and Watson (2017 and 2021) for more details.

Changes in either the supply of or demand for a good will lead to changes in its price. A fundamental issue in economic models pertains to how the model will incorporate price adjustments that could arise from an external demand or supply shift. Economic models that reflect price changes are called competitive general equilibrium (CGE) models. In these

**TABLE 8.1** Comparison of CGE and I-O models.

| Topic | CGE models | I-O models |
|---|---|---|
| Prices | Price adjustments stipulated from Eqs. of consumer preferences and firm behavior, which are derived from economic theory | Prices do not adjust |
| Model input limitations | Requires compiling and standardizing economic statistics across countries with different data collection procedures | For the US, parameter values typically reflect "average" expenditure purchasing patterns for the sectors |
| Geographic scale | Typically used to examine trade flows across broader regions (e.g., a few countries or groups of countries) | Typically used for smaller geographic regions, like a single state or county |
| Sector scale | Typically used to examine a select number of aggregated sectors due to data and computational challenges | Can accommodate more sectors and there are no restrictions on adjusting the model parameters for the sector being examined |
| Accessibility to researchers | Some CGE models not available to outside researchers; others require extensive training | IMPLAN sells off-the-shelf software that is straightforward to use |

models, a change in demand or supply to one sector in one region will impact all sectors in all regions, under the premise that the initial shift will subsequently impact the supply of and/or demand for goods elsewhere. After one sector in one region is shocked, these models determine new equilibrium prices so that demand equals supply across all markets. However, the incorporation of price adjustments (and whether they are needed or not) depends on the level of change in supply or demand. So, depending on what one is modeling and the economy that one is analyzing, price adjustments could be of little consequence to the results. The use of CGE models in the context of local foods is atypical, although there are exceptions (Cantrell et al., 2006).

Local food economic impact studies more typically use I-O models (Appendix 8.1). Unlike CGE models, prices in I-O models do not change in response to shocks that a researcher might impose. Instead, I-O models are premised on a linear relationship between the final demand for a sector and its output. Additionally, I-O models do not have supply constraints.

Before describing the Eqs. within I-O models, we provide several reasons why I-O modeling has been more pervasive in food systems analysis than CGE modeling (also see Table 8.1):

- I-O models can accommodate more sector-specific or region-specific detail than CGE models because they are computationally easier to solve. This practically implies that CGE models may only include a select number of regions and sectors. For instance, the regions in a CGE model may be a group of a few countries with perhaps a few regions within those countries. By contrast, I-O models can accommodate many sectors and can be undertaken within narrower geographic areas, like a state or county. This can allow the researcher to include more details in the model about the sector they are interested in examining.

- In some contexts, the assumptions within I-O models may be more appropriate for food systems studies than CGE models. A key limitation of I-O models is that prices do not change. In CGE models, prices are an endogenous variable that is determined through Eqs. for consumer preferences and firm behavior. However, this approach also has limitations. First, the functional forms of the Eqs. in CGE models require behavioral assumptions and parameter values. This implies that the accuracy of CGE model output depends on the plausibility of the parameter values that are used, but obtaining defensible parameter values can be challenging. Second, and related, is that while the mathematics for I-O calculations are simple, CGE models can be "black boxes" that make it difficult to discern how a model is being solved. The assumption that prices do not change may be reasonable for smaller shocks.
- While we discuss the data limitations with I-O models in detail in this chapter, data quality is also a concern with CGE models. Parameterizing CGE models may require compiling and standardizing economic statistics across countries that do not collect data in the same format or at a consistent level of quality.
- I-O models are more accessible to researchers than CGE models. I-O modeling is undertaken more frequently, in part, because a company called IMPLAN (www.implan.com) sells off-the-shelf databases and software that users can operate in a straightforward fashion. The U.S. Forest Service first developed IMPLAN as a linear programming model in 1976, although IMPLAN is now an independent corporation. CGE models are less accessible. For instance, the Global Trade Analysis Project (GTAP) at Purdue University's Department of Agricultural Economics has developed a CGE model that researchers can purchase to use directly. However, users typically need to take a one-week short course in order to begin using the model in its most basic form. Other equilibrium models of agricultural markets are completely inaccessible to external researchers.

### 8.1.1.1 How do i-o models work?

I-O models consist of linear Eqs. that parameterize how the expansion of one sector affects the economy in a specific region. Structurally, I-O models are premised on the notion that the economic output of one sector either provides inputs for another sector or is consumed by end users. Eq. (8.1) formalizes this concept:

$$x = Ax + d \tag{8.1}$$

Suppose there are $n$ sectors in the economy. In (8.1), $d$ is an $n$x1 vector that represents the final demand by consumers for the goods that each sector produces. The $n$x1 vector $x$ represents the total output for each sector. $A$ is an $n$x$n$ matrix. For a given row $i$ within the matrix $A$, the coefficients correspond to the number of units that every sector needs to produce one unit of sector $i$. In summary, Eq. (8.1) implies that the total output for a given sector ($x$) equals the sum of the product's use as an intermediate input ($Ax$) and the product's final demand ($d$).

In order to solve for $x$, Eq. (8.1) can be rewritten as:

$$x = (I - A)^{-1}d \tag{8.2}$$

In (8.2), $I$ is an *nxn* identity matrix. The elements of an identity matrix consist entirely of zeros, except for the diagonal elements, which have values of one. In Eq. (8.2), a sector receives an external shock to demand ($d$) that is designed by the researcher (i.e., the shock is "exogenous"), which then leads to changes in output ($x$).

Following Schmit and Boisvert (2014), we provide an example of I-O transactions in Table 8.2. For simplicity, the Table only presents three industrial sectors: agriculture, manufacturing, and retail trade. In an I-O Table, the values of the cells are implicitly associated with a specific time interval, which is typically one year.

The rows represent how the output from one sector is allocated to other sectors. The shaded portion of the Table represents sales of intermediate inputs between industries. For instance, the first row represents sales of inputs from agricultural producers to other agricultural producers and other sectors of the economy. The row elements for the "final demand" are the contribution that producing industries provide to end-use markets. These end-use markets include personal consumption by households, government purchases, capital investments, and exported goods.

The matrix columns represent the input values needed from sectors across the economy for one sector to produce an output. The shaded cells in the first column are the intermediate inputs that agricultural producers purchase both from agricultural producers and other sectors. The final "value-added" rows are the input purchases of labor and non-capital inputs that firms make. This includes employee compensation, proprietor income, property income (such as rent, interest, corporate profits, and depreciation), and indirect business taxes (Schmit and Boisvert, 2014). The row representing "imports" corresponds to the labor and intermediate inputs that the sector purchases for production from outside of the economy.

### 8.1.1.2 I-O model outputs and key metrics

The economic impacts from shocks in I-O models are commonly reported as "multipliers." An expansion of sales by a sector leads to a "direct" economic impact equal to the size of the expansion. These sales will also lead to a) "indirect" effects arising from increased sales by input suppliers to the sector and b) "induced" effects associated with an increase in local spending from employees and proprietors in both directly and indirectly affected sectors. This increase in spending arises from increases in employee and proprietor income. Thus, a sector's economic "multiplier" is the sum of its direct, induced, and indirect effects divided by its direct effects. We present this relationship in Eq. (8.3) and, also, in Fig. 8.1.

$$\text{multiplier} = \frac{\text{direct effects } + \text{ indirect effects } + \text{ induced effects}}{\text{direct effects}} \tag{8.3}$$

Researchers use I-O models to estimate multipliers for the following relevant economic development metrics: employment (i.e., the annualized number of full-time or part-time jobs), labor income, gross regional product, and output. Output equals the sales value of all goods and services produced in a region. Labor income, which represents the wages/salaries of workers and proprietor income, is a subset of gross regional product because it also encompasses property income, investment income, and indirect tax payments. While these four metrics are common, multipliers can also be estimated for other metrics.

**TABLE 8.2**   An example of an I-O transaction Table for a regional economy.

| Industry sector | Intermediate demand | | | Final demand | | | |
|---|---|---|---|---|---|---|---|
| | Agriculture | Manufacturing | Retail trade | Households | Government | Investment | Exports |
| Agriculture | | | | | | | |
| Manufacturing | | | | | | | |
| Retail trade | | | | | | | |
| Value added component | | | | | | | |
| Employee compensation | | | | | | | |
| Proprietor income | | | | | | | |
| Other property income | | | | | | | |
| Indirect business taxes | | | | | | | |
| Imports | | | | | | | |

Note: The shaded portion of the Table contains inter-industry linkages. For example, the cells in the first row contain the sales of agricultural goods to the agricultural sector itself, and to the other "downstream" sectors of the economy. In addition to these sales to other industries that use agricultural goods as intermediated inputs, the elements in any row for the remaining columns in the Table (labeled Final Demand) record deliveries from a product section (or industry) to final markets for such things as personal consumption (households), government purchases of goods, capital investment and goods exported to other markets in other regions of the country or abroad. Each column of the shaded portion of the Table records intermediated inputs required by a particular industry to produce its output. The sector columns and rows (shaded area) generally make up the largest area in the inter-industry transactions Table. In the 2020 IMPLAN data, the full inter-industry transactions Table includes 546 individual sectors. The final row, imports, accounts for both intermediated inputs and labor purchased by the sector from regions outside the local economy. Final demand (i.e., households, government, investment, and exports) does not purchase value added components (Schmit and Boisvert 2014).

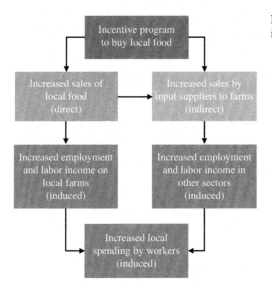

**FIGURE 8.1** Flow chart example of direct, indirect, and induced effects.

### 8.1.1.3 *Best practices in undertaking I-O studies*

I-O scenarios are most accurate when modeling smaller economic shocks to one sector. This is because I-O models assume that a) prices in an economy do not change, b) firms exhibit constant returns to scale, and c) there are no resource constraints for inputs. From the perspective of economists, perhaps the most limiting feature of I-O models is their lack of insight into the broader welfare implications of economic shocks (O'Hara and Pirog, 2013). For instance, converting from fossil fuel energy production to renewable energy production may have large environmental benefits for society. However, the net change in energy sector jobs from such a shift may be minimal, since renewable energy jobs would increase while fossil fuel jobs decrease (Partridge, 2013). We will return to this issue in the final section of the chapter.

There are two recommendations to improve the accuracy of economic impact assessments. First, under certain contexts that we describe below, opportunity costs should be considered (e.g., Hughes et al., 2008; Jablonski et al., 2016). Opportunity costs represent the value of the forgone market activity that would have occurred, in our context, in the absence of a local food project.

For instance, Hughes et al. (2008) studied the economic impacts of farmers markets in West Virginia. They assumed that an increase in farmers market sales would result in a reduction in expenditures at grocery stores. They found that expanding farmers market sales could have a positive impact on farms, as well as service sectors that support farmers. At the same time, there could be negative impacts to the food retail sector, like grocery stores, from displaced sales. There could also be negative impacts on sectors that support food retailers, like truck transportation, wholesale trade, and air transportation. Overall, Hughes et al. (2008) found that the increase in economic activity from farmers market sales exceeded the reduction in economic activity from reduced grocery store sales.

Technically, Hughes et al. (2008) and Thilmany et al. (2021), the latter of which we use to develop the case study later in the chapter, are "contribution analyses" instead of economic impact analyses. The key distinction between the two types of studies is how opportunity costs are considered. We use the phrase "economic impact study" throughout the chapter for simplicity and accessibility, although we provide further details in the side box.

---

**SIDEBOX: Contribution analysis vs. economic impact analysis**

*Both economic impact and contribution studies use the same methodology. Following Watson, Wilson, and Thilmany (2007), the difference between the two types of studies is the assumption of the source of funding for a hypothetical shock. Economic impact assessments estimate how a region's economy is impacted by an increase in expenditures that is exogenous, or external, to the region. This exogenous shock implies that spending within the region is not displaced. A contribution analysis examines shocks that can occur from within a region and potentially displace other expenditures within it.*

*For example, suppose we wanted to estimate the impacts of a food incentive program on a state's economy. Further suppose that the program received 50 percent of its funding from its state's government and 50 percent from the federal government. For simplicity, we assume that all of the federal government funding is external to the state. We would consider the influence that the federal government's support for the program would have on the state's economy as an "economic impact." Considering opportunity costs from displaced expenditures is not required, since no expenditures within the state were displaced.*

*To estimate the influence of the state's funding from state sources, we would need to consider the opportunity costs of those funds from within the state. For example, state support for an incentive program may displace funds to build a new school, prison, or sports stadium. In Hughes et al. (2008), farmers market sales occurred at the expense of purchases from grocery stores in the state.*

---

In another example, Swenson (2010) studied the economic impacts of farms in the upper Midwest growing and selling fruits and vegetables to residents in the region's metropolitan areas. However, there is not necessarily spare land available for this production in the upper Midwest. Thus, Swenson (2010) assumed that acreage needed for local fruit and vegetable production would displace acreage currently being used for conventional commodity crops. This implies that the net benefits equal the difference between the economic impacts of local fruit and vegetable production and the impacts currently provided by commodity crop production.

Bear in mind that an increase in local food purchases may decrease purchases of nonlocal food, meaning that region-specific economic impact estimates cannot be aggregated. For instance, suppose southern California schools started buying lettuce from southern California farmers instead of from northern California farmers, and vice versa. There would be positive localized economic impacts to the growers in each respective region from these shifts. More specifically, a shift by southern California schools to purchase local lettuce would benefit southern California farms, and a similar pattern would occur in northern California. However, there would be no statewide economic impacts in California from these simultaneous changes since they are effectively canceling each other out. From a statewide perspective, southern

California farms are increasing sales to southern California schools but are also losing sales to northern California schools, and vice versa. Expenditures are being reallocated within the state instead of increasing in magnitude. This implies that the net impacts may be minimal relative to gross impacts once economic "winners and losers" are identified and taken into consideration.

For the second recommendation to improve the accuracy of economic impact assessments, we suggest careful consideration of modification of the IMPLAN data. IMPLANI-O estimates increase in accuracy when the user modifies IMPLANs proxy "production function" parameters with study-specific data that more accurately reflects how a sector operates in a region. A production function in IMPLAN represents the breakdown in expenditures across all sectors that one sector undertakes to produce goods and services.

IMPLANs default parameter values represent aggregated industry conditions by construction. However, the spending patterns of firms within industries can vary considerably from aggregate conditions. For this reason, IMPLANs default parameters have been shown to underestimate the impact of local food projects in I-O calculations. This underestimation is partially because, as a percentage of total expenditure, local food operations hire more local labor and buy more inputs locally compared to conventional farms (e.g., Jablonski and Schmit, 2015; Schmit, Jablonski, and Mansury, 2016; Jablonski, Bauman, and Thilmany McFadden, 2020).

Where is data available for researchers to modify IMPLANs default parameter values? One approach is to undertake surveys of local farms to ask about their expenditure patterns (e.g., Jablonski and Schmit, 2015). However, collecting primary data through surveys can be expensive and time-consuming. Another possibility is to use secondary data. The USDAs Agricultural Resource Management Survey (ARMS) is its primary source of information on the financial condition of U.S. farm businesses. It is a national dataset that researchers could use to customize IMPLAN data to reflect local food sector expenditures. Since 2008, ARMS data includes questions about farm sales through local food channels and provides a sufficiently large sample of producers participating in these markets (Low and Vogel, 2011). Accordingly, these data can be used to understand expenditure patterns and modify agricultural sector data in I-O models to account for differential economic flows that ARMS data reveal among producers operating in local and regional food markets. However, although ARMS data can help with gross spending coefficients, it does not contain information on the proportion of farm spending that is local.

## 8.1.2 Case study: nutrition incentive programs (including gusnip, formerly called FINI)

To illustrate an economic impact assessment of a local food project using the methodological approaches and best practices described above, we provide an example of a recently completed study on nutrition incentive programs (Thilmany et al., 2021). Nutrition incentive programs support projects to increase the purchase of fruits and vegetables among low-income consumers, most often those participating in the Supplemental Nutrition Assistance Program (SNAP), by providing incentives at the point of purchase. A relatively new USDA program, the Gus Schumacher Nutrition Incentive Program (GusNIP, formerly the Food Insecurity

Nutrition Incentive Program), is one example from a portfolio of programs nationwide that aims to increase fruit and vegetable purchasing among SNAP consumers by providing incentives that stretch their food dollars (USDANIFA, n.d). These funded projects are likely to have a positive impact across a diverse set of stakeholders through improvements in food security; positive economic contributions to local farmers and grocery stores; and, more broadly, through spillovers to local economies.

Although the GusNIP program does not require its grant recipients to operate nutrition incentive programs in a manner that leverages grant dollars to support local producers, many of them do. The community groups operating these programs often have missions that align with supporting local farmers as a vital aspect of more regionally focused food systems. They choose to add local components, but there has been no previous evidence of the effectiveness of this strategy. Accordingly, there is interest in exploring the potential economic impact within communities of marrying nutrition incentives with procurement programs that support increased purchases from local farmers and ranchers.

In response to these interests, SPUR and the Fair Food Network, in collaboration with partnering nonprofit organizations, governmental agencies, and Colorado State University, came together to frame and estimate the potential impacts of a variety of expansion scenarios for healthy food incentives at grocery stores and farmers markets across various geographies. In addition to allowing individual programs across the U.S. to assess the economic contributions of their programs to their state economies, the nonprofit organizations leading this work felt that the compilation of these analyses is an important element of building the policy case for expanding incentives to a larger share of SNAP participants and markets across the US.

### 8.1.3 Data

This case study explores how healthy food incentives may differentially affect several sets of stakeholders in the food supply chain. It was commissioned, in part, to provide data to educate policymakers at the state and national level, with the hope that policymakers might consider increasing the amount of funding available for these programs. Given that incentives can be provided to support either fresh fruit and vegetable sales that are direct to consumer or that move through grocery stores, we estimate different scenarios to account for various supply chain scenarios. To do this, we:

1) Customize our I-O model using farm financial data from USDA ARMS on direct/local farm expenditures, to refine estimates based on national averages so they better reflect the sector when local marketing linkages are stronger.
2) Use incentive redemption data (i.e., nutrition incentive money spent by low-income households on food) from SPUR, the Fair Food Network, and other currently operating incentive programs along with USDA Food and Nutrition Service (FNS) SNAP redemption data to estimate total incentives spent by market channel.
3) Refine incentive redemption data in the I-O model by considering SNAP user spending pattern data from the USDA Economic Research Service (ERS) to allocate what sectors in the economy capture the new dollars that flow to SNAP households.

### 8.1.3.1 *Incentive redemption and snap reimbursement data*

To estimate the economic impacts of food incentive programs in the nine participating states (California, Colorado, Hawaii, Iowa, Michigan, New York, North Carolina, Texas, and Washington), we use data from currently operating incentive programs. Note that we conduct each state's analysis independently, and thus results are provided at the state level. Additionally, herein we present results from only one state, Colorado.

These data were collected through a survey and individual meetings with teams from each incentive program in our study. Numerous nonprofits and government agencies run these programs using a variety of program designs that are marketed with different names. Some of the programs partner with just a few markets or stores, while others offer incentives through hundreds of retail locations. Due to the unique nature of each program, conversations with each team provided the research team with valuable context for conducting the analysis. Collected data included:

- incentives redeemed at each of the market outlets where they operate;
- SNAP reimbursements at those same market outlets; and
- details on program design (including if they required grocery store purchases to be local).

### 8.1.3.2 *Farm financial data*

We used ARMS data to delineate expenditure patterns for farms with direct-to-consumer sales (i.e., farmers market, CSA, farm stand, and other direct). We then used these data to create a local food sector in IMPLAN. Due to a small sample size, we used data from producers across the U.S. and customized each state's IMPLAN model. Using this newly created sector, we were able to more accurately model the economic contribution of incentives spent at farmers markets, CSAs and farm stands (for more details on how to modify a sector's expenditure patterns, also called a production function, see Schmit and Jablonski, 2017 and Thilmany McFadden et al., 2016).

### 8.1.3.3 *SNAP user spending pattern data*

To evaluate potential estimated contributions if current programs were scaled to reach locations across the state, we used SNAP reimbursement data from the USDAFNS, estimated targets for expansion (e.g., incentives available at 90 percent of eligible grocery stores and 100 percent of farmers markets), and gathered an average incentive redemption to SNAP reimbursement ratio from current incentive programs (calculated as total incentives redeemed by market channel divided by SNAP redemptions at the same market outlet).

Beyond the customization completed for the farm sectors, which were of primary interest, we used the default data from IMPLAN to estimate the broader economic contributions of programs. Since there was a keen interest in whether local components to some programs had been effective in their intention to support economies, there was a focus on comparing results of a program with a local purchasing requirement for grocery store purchases to those without local purchasing requirements (modeled by changing the regional purchase coefficient, see *Additional Model Customization* section for more details). The results provided a customized IMPLAN multiplier and economic contributions for a broad set of economic outcomes for each of the participating states where healthy food incentive initiatives are currently being undertaken.

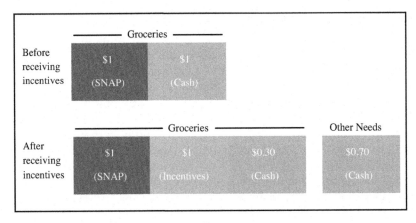

**FIGURE 8.2**  What happens to consumer spending when consumers have more money because of incentives? Amount spent by source of funds.

### 8.1.3.4 Opportunity cost

To avoid overestimating contributions, an important component to consider is the displacement of spending in the economy as a result of the incentive program, also referred to as opportunity cost. When a person receives incentives, their purchasing power increases (referred to below as "new money"). While incentives can only be used to purchase fresh fruits and vegetables, participants are not prohibited from using other (preexisting) funds they had previously intended to use for food purchases on nonfood items. When a person receives incentives, their total household spending increases (see Fig. 8.2). A couple of the participating program partners did ask participants questions getting at this idea, but the detailed level of data necessary to allocate displaced spending across the economy was not available in the primary data. Fortunately, a recent study by the USDA ERS on the economic impacts of SNAP (Canning and Stacy, 2019) provided evidence that when a person receives additional SNAP benefits and their total household spending increases, they put just 30 percent of their increased total household spending toward "food at home" purchases (i.e., groceries).[1] The remaining 70 percent of their increased total household spending goes toward "other expenses", including prepared food, transportation, housing, utilities, clothes, and many of the items that households spend money on in the economy.[2] None of the additional benefits are saved. Using these data, we were able to better estimate where the "shock" might occur due to the nutrition incentive.

### 8.1.3.5 Additional model refinements and customizations

In IMPLAN, when modeling a shock to a retail or wholesale sector, the shock must be margined. For both retail and wholesale sectors in IMPLAN, the cost of goods sold (also

---

[1] While incentive dollars may be spent differently than monies through the broader SNAP program, this study provided the most reliable estimate that we found to be relevant and credible.

[2] As is often the case, correctly matching IMPLAN sectors to spending in different sectors of the economy is not trivial. In some cases that were not clearly aligned with a specific sector, we made educated decisions on how to allocate those monies.

Producer value = $50     Transportation and wholesale costs = $25     Retail markup = $25

Total revenue/purchase price = $100
Marginal revenue/retail producer price = $25

FIGURE 8.3 Tracing a sale of $100 at a grocery store through the economy.

referred to as COGS) is not included at the wholesale or retail level. For example, if we want to model a produce purchase from a grocery store, the portion of the "shock" that represents the retail markup will be allocated to grocery stores, while the remainder of the "shock" will be allocated to the wholesale, transportation, and fruit and vegetable farming sectors (backward-linked supply chain sectors). Margins allow us to allocate sales to the correct industries by tracing consumer expenditures through retail, wholesale, and distribution sectors back to the industries that manufactured the products that are ultimately sold (see Fig. 8.3); this is called the value chain. Margining is an important concept for us to model accurately. When we want to model the contribution of an additional $100 spending at grocery stores, we do not give full credit and increase spending at grocery stores by $100. Instead, we go backwards through the value chain in the economy and allocate that $100 to various sectors along the way, from the retail store to transportation services to the last place a meaningful transformation of the product was made, which in the case of incentive program spending at grocery stores is at the farmgate. Note that all incentives spent at farmers markets, CSAs, and farm stands go directly to the newly created local food sector in IMPLAN and are not margined (see Schmitt and Boisvert, 2014 for more information).

Because participants spend their increased household spending from incentives in different sectors throughout the economy, many of which are retail, margining can present some challenges when the sector that produced a good is not clear. To allocate spending throughout the value chain for each retail (i.e., margined) sector of the economy where we assume an increase in spending, we increase spending in the sector in which the last meaningful transformation occurred and use IMPLANs default sector margins to allocate spending along the supply chain. Determining the sector where the last meaningful transformation was made is not always clear. For example, we see an increase in sales at some retail sectors, but there are multiple choices from the associated producing sector. We used the detailed spending categories provided in the USDA ERS SNAP report and made our best guesses as to the appropriate IMPLAN sectors. IMPLAN provides margins for the majority of commodities down to the 4-digit North American Industry Classification System. To use these margins, the producing/manufacturing sector is the sector and IMPLAN provides the margined sector and the margin value for all sectors along the value chain. These margins add up to 100 percent.

Once consumer spending changes are fully margined, the next challenge is determining the percentage of each purchase that was made locally (i.e., within the state). We use the state-level regional purchase coefficient (RPC), the proportion of total demand for a commodity by all users in the region that is supplied by producers located in the region, as an estimate for the percent purchased locally. IMPLAN uses a national data to estimate trade flows between regions, which produces the estimated RPC for each commodity. The implication is that all users purchasing a commodity buy exactly the same amount of local versus nonlocal product. We assume that all retail purchases are 100 percent local and use the RPCs already populated in IMPLAN for the remaining sectors along the value chain as a best guess estimate for the percent purchased locally.

For the incentive programs that require the purchase of local produce from grocery stores, we modify the RPC for all users that buy products from the fruit and vegetable farming sectors and the proportion of the produce purchased by wholesale trade and truck transportation. To estimate current contributions of incentive programs, we increase the RPC for all users that buy products from local fruit and vegetable farms, as well as the proportion of fruits and vegetables purchased by wholesale trade, and the proportion of fruits and vegetables purchased by truck transportation by the percentage of produce required to be purchased locally by grocery stores. To estimate current contributions of existing programs, we increase the RPC by the current local purchasing requirement of the program. To estimate contributions of incentive programs, we assume a local purchasing requirement of 20 percent and increase the existing RPCs by 20 percent. By increasing the RPC, we are assuming that as a result of the incentive program's local purchasing requirement, participating stores purchase 20 percent more local produce than they did in the absence of the program and the total output of the farming sector is increasing as a result of a change in demand. The changes we model are less than 1 percent of total industry output, and thus it is reasonable to assume there will not be any resulting structural changes in the industry when increasing production to meet this new demand.

### 8.1.3.6 Results for nourish colorado's[3] 2018 double up food bucks program

Here we present the results from one of the participating project partners: Nourish Colorado's 2018 Double Up Food Bucks program. Nourish's program operates in farmers markets, farm stands, CSAs, small-scale retailers, and large grocery stores. In 2018, all incentive dollars were required to be spent on Colorado-grown produce.[4]

In 2018, Nourish provided $167,911 in incentives, of which $137,909 went directly to farmers through sales at farmers markets, farm stands, and CSAs, and $30,002 to food retailers, where customers were required to purchase only Colorado-grown produce (Table 8.3). Using the data and methods described above, we found that the nutrition incentive program contributed

---

[3] Nourish Colorado was previously known as LiveWell Colorado.

[4] For more information about the program, visit https://doubleupcolorado.org/.

**TABLE 8.3** Incentives, economic contribution, and implied output multiplier for Nourish Colorado's 2018 incentive program.

| | | Economic contribution | | | |
|---|---|---|---|---|---|
| | Incentives ($) | Output ($) | Employment (jobs) | Labor income ($) | Output multiplier |
| Incentives spent on farm-direct sales to local F&V farmers | 137,909 | 402,964 | 1.9 | 86,669 | 2.9 |
| Incentives spent on F&V at retail food sales | 30,002 | 49,473 | 0.4 | 17,986 | 1.6 |
| All incentives | 167,911 | 452,438 | 2.3 | 104,655 | 2.7 |

$452,000 to Colorado's economy, including 2.3 jobs[5] and $105,000 in labor income.[6] The implied output multiplier[7] (output contribution/incentives) is 2.9 for farm-direct sales, 1.6 for incentives spent at food retailers, and, weighting these effects, an overall multiplier of 2.7. This means that for every dollar spent of incentive dollars at farm-direct outlets, the contribution to the economy was 2.9 dollars – 1 dollar of the incentive plus and 1.9 in additional economic activity. For every dollar spent at food retailers, the contribution to the economy was 1.6 dollars.

We also estimate the economic contributions if this incentive program were scaled up, reaching a larger share of retail food outlets (and thus, SNAP participants) across the state of Colorado. We present an upper bound (scenario A), assuming high SNAP participation (based on FY2013 participation data) and high market penetration (reaching 90 percent of eligible grocery stores, 25 percent of eligible corner stores, and 100 percent of eligible farmers markets). We also present a lower bound (scenario B), assuming low SNAP participation (based on FY2019 participation data) and low market penetration (reaching 60 percent of eligible grocery stores, 10 percent of eligible corner stores, and 80 percent of eligible farmers markets). We compare results for a program in which there is no local purchasing requirement for grocery stores to one in which grocery stores are required to purchase at least 20 percent of their produce from Colorado farmers. Because a different allocation of spending occurs across market outlets in the expansion scenarios than it did in Nourish Colorado's 2018 program, we cannot simply assume a linear expansion of economic contributions due to the differences in multipliers between the different market outlets. Additionally, the local purchasing requirement for grocery stores differs between the current program and the expanded program.

[5] Employment is a job that can either be a full time or a part-time job, and a person can hold more than one job, so the job count and number of persons employed is not necessarily the same.

[6] Labor income, which includes employee compensation and proprietor income, is included in output, but it is often broken out due to interest from policymakers.

[7] The most commonly used multiplier is a Type SAM multiplier (if not mentioned, this is likely the type of multiplier being presented). Type SAM multiplier = total effect/direct effect, where total effect = direct effect + indirect effect + induced effect, and effect references either impact or contribution.

**TABLE 8.4**  Potential economic contributions across the state's economy if incentive programs were scaled statewide in Colorado for Scenario A, assuming high SNAP participation rates and high market penetration.

| | Incentives ($) | No local component for retail food sales | | | Local component for retail food sales | | |
| --- | --- | --- | --- | --- | --- | --- | --- |
| | | Output ($) | Employment (jobs) | Labor income ($) | Output ($) | Employment (jobs) | Labor income ($) |
| Incentives spent on farm-direct sales to local F&V farmers | 336,814 | 984,157 | 5 | 211,671 | 984,157 | 5 | 211,671 |
| Incentives spent on retail food sales to all F&V farmers | 15,077,212 | 21,306,757 | 172 | 7,961,339 | 37,352,393 | 174 | 8,074,028 |
| All incentives | 15,414,026 | 22,290,913 | 176 | 8,173,010 | 38,336,550 | 179 | 8,285,699 |

Notes: High market penetration assumes the program will reach 90 percent of eligible grocery stores, 25 percent of eligible corner stores, and 100 percent of eligible farmers markets.

**TABLE 8.5**  Potential economic contributions across the state's economy if incentive programs were scaled statewide in Colorado for Scenario B, assuming low SNAP participation rates and low market penetration.

| | Incentives ($) | No local component | | | Local component | | |
| --- | --- | --- | --- | --- | --- | --- | --- |
| | | Output ($) | Employment (jobs) | Labor income ($) | Output ($) | Employment (jobs) | Labor income ($) |
| Incentives spent on farm-direct sales to local F&V farmers | 208,522 | 609,292 | 3 | 131,046 | 609,292 | 3 | 131,046 |
| Incentives spent on retail food sales to all F&V farmers | 7,706,678 | 11,013,053 | 89 | 4,115,468 | 19,186,612 | 90 | 4,147,979 |
| All incentives | 7,915,200 | 11,622,345 | 92 | 4,246,514 | 19,795,904 | 92 | 4,279,025 |

Notes: Low market penetration assumes the program will reach 60 percent of eligible grocery stores, 10 percent of eligible corner stores, and 80 percent of eligible farmers markets.

If Colorado's incentive program were scaled statewide, the upper bound for total incentives spent in the state is $15.4M (Table 8.4). This would result in an estimated economic contribution to the state's economy (for a program without a local purchasing requirement) of $22.3M, 176 jobs, and $8.2M in labor income. If the program required local purchasing, the estimated upper bound for the economic contribution would increase to $38.3M, 179 jobs, and $8.3M in labor income. The lower bound for total incentives spent in the state is $7.9 million (Table 8.5). This

**TABLE 8.6** Employment contributions for the farm and grocery sectors for Nourish Colorado's 2018 incentive program.

| | Incentives ($) | Employment (jobs) | Labor income ($) |
|---|---|---|---|
| All ag sectors (not just F&V farmers) | 167,911 | 0.50 | 19,645 |
| Fruit and vegetable farmers | | 0.40 | 17,605 |
| Retail food sector | | 0.05 | 1,676 |

would result in an estimated economic contribution to the state's economy (for a program without a local purchasing requirement) of $11.6M, 92 jobs, and $4.2M in labor income. If the program had a local purchasing requirement, the estimated lower bound for the economic contribution would increase to $19.8M, 92 jobs, and $4.3M in labor income.

In looking at employment contributions to the agricultural sector, we see that Nourish Colorado's 2018 incentive program created $20,000 in labor income in the agricultural sector, with the majority coming from fruit and vegetable farmers (Table 8.6). Due to the complexities of employment in the agricultural industry, we assert that labor income is a better measure of employment contribution than jobs for the agricultural sector. Employment contributions of Nourish's program to the retail sector were 0.05 jobs and $1.7K in labor income.

### 8.1.3.7 Conclusion and key takeaways from gusnip case study

Regional economists have long been asked to explore how policy changes or program investments will affect the economy, and they use numbers in educational or political campaigns to inform policy decisions or garner support. As expected, programs that emerge from local and regional food system innovations, such as the Nourish Colorado incentives program, are of particular interest because they have such a focus on local versus nonlocal spending. Consider some of what the contribution analysis of this program illustrates:

- Because Nourish framed its programs in partnership with a variety of different markets and stores (from farmers markets to traditional food retailers), it offers up a particularly rich case study to explore how incentives flow through the economies of the surrounding communities, and how those flows may vary by types of markets.
- One of the richest aspects of such analysis in the context of food systems is the "linked" tracking of benefits through communities to households, retailers, farmers, and broader economies, highlighting how important strong linkages, social capital, and economic structure are in the determination of how much "gain" is captured by the local economy.
- Although based on assumptions, scenarios related to different pathways of program expansion can provide plausible estimates. For example, we found that every $1 invested in a healthy food incentive program generates up to $3 in economic activity.
- IMPLAN is a powerful tool, but the analysis conducted on the incentive program is only relevant because it was informed by those who framed, operationalized, and evaluated how the program is impacting recipients, the markets where they shop, and the procurement choices made by the markets where direct sales from farmers is not possible.

### 8.1.4 Integrating economic impact assessment methods with other food system modeling efforts

As explained above, I-O models provide important information regarding interindustry linkages in a local economy, and thus can help to examine the equilibrium effects of a "shock," including the distribution of the economic effects throughout the local economy. However, in modeling food systems, the goals of policymakers, economic developers, and other stakeholders are often broader than just economic. Further, there can be important trade-offs associated with initiatives – from environmental or social perspectives – that one would want to capture in order to more fully consider the impacts of a particular program or policy.

For this reason, IMPLAN has started to integrate some environmental data sets into its data library, working with the U.S. Environmental Protection Agency (EPA) to bring its emissions data to the IMPLAN industry sectors. By so doing, IMPLAN and the EPA provide users with the ability to see environmental impacts tied to economic impacts (IMPLAN, 2020). The EPAs environmentally extended I-O model data includes a full-economy U.S. life cycle model at a resolution of approximately 400 goods and services (Ingwersen, 2018).

Acknowledging the importance of integrating environmental information with I-O data, Britain and the European Union maintain satellites to their national I-O accounts, which cover activities linked to the economy but are not part of the core economic accounts. One of the focuses of their satellite accounts is natural capital (i.e., biodiversity, including ecosystems that provide essential goods and services). With the addition of satellite accounts for natural capital, these governments can measure the changes in the stock of natural capital at a variety of scales and integrate the value of ecosystem services into accounting and reporting systems at both the Union and national levels (European Commission, 2019).

Although the integration of natural capital accounts into I-O models is an important step towards taking a more systems approach to evaluating the impacts of food systems efforts, there are many potential important impacts that have still not been captured. A recent report by the United Nations Environment's TEEBAgriFood (2018) addresses the core theoretical issues and controversies underpinning the evaluation of the nexus between the agrifood sector, biodiversity and ecosystem services, and externalities, including human health impacts from agriculture on a global scale. It argues the need for a "systems thinking" approach, drawing out issues related to health, nutrition, equity, and livelihoods. Moreover, it discusses, in some detail, how economic impact assessments need to be augmented with other approaches. However, it does not provide specific guidance on how to do this in practice.

Johnson et al. (2014) provide the most detailed theoretical framework from which to draw, laying out a complex I-O model that incorporates satellite accounts across other types of capital, including social, physical, human, natural, cultural, and political capital. They essentially take concepts from welfare economics and consider theoretically how to incorporate them into an I-O framework. However, as with the TEEB (2018) report, their work lacks detail on application.

The reason that these existing publications fall short in how to put these frameworks into practice is that data needs are significant – and challenging! As an example, TEEBAgriFood (2018) acknowledges that both positive and negative externalities need to be incorporated. However, how does one disentangle positive and negative social capital using available data? How does one incorporate the distribution of costs and benefits, for example, benefits going to more powerful rather than marginalized groups, and more costs being placed on marginalized groups? Accordingly, although integrating this broader array of tangible and intangible assets as satellite accounts into systems of national accounts is a recommended and important next step, there are many challenging, unanswered questions.

## 8.2 Conclusion

There has been substantial progress over the last 10 years in considering best practices in conducting economic impact assessments, thanks, in part, to the USDAs Toolkit (Thilmany McFadden et al., 2016) and the proliferation of case studies it spawned (Appendix 8.1). Further, new data sources from the USDA allow for much more robust consideration of the local food sector in economic impact assessments by facilitating modifications to existing IMPLAN farm sector data. As a result, projects such as the Nutrition Incentive Program case study described above can be conducted in more accurate ways, with additional, and less invasive, data collection from producers.

**APPENDIX 8.1**   Summary of selected food system I-O studies

| Study | Region / Market channel |
| --- | --- |
| Bauman and McFadden (2017) | CO farm to school, wine sectors |
| Christensen et al. (2019) | GA/MN farm to school |
| Conner et al. (2008) | MIF&V farms |
| Duval, Bickel, and Frisvold (2019) | AZ farm to school |
| Henneberry, Whitacre, and Augustini (2009) | OK farmers markets |
| Hughes et al. (2008) | WV farmers markets |
| Hughes and Isengildina-Massa (2015) | SC farmers markets |
| Jablonski, Schmit, and Kay (2016) | NY food hub |
| Miller et al. (2015) | MI local foods |
| Mon and Holland (2006) | WA organic apple farms |
| O'Hara and Parsons (2013) | MN/VT organic dairy farms |
| Pesch and Tuck (2019) | MN direct-to-consumer vegetable farms |
| Rossi, Johnson, and Hendrickson (2017) | MO/NE local producers |
| and Mansury (2016) | NY apple industry |
| Swenson (2010) | Upper Midwest local F&V farms |
| Tuck et al. (2010) | MN farm to school |
| Watson et al. (2017) | ID local farms |

However, there are important acknowledged limitations of I-O modeling, some of which are inherent to I-O and some specific to investigations of food systems efforts. Some of these may be addressed through integration of new data sets, such as the EPA and IMPLANs environmentally extended I-O model, which enable governments and others to consider potential trade-offs to investments. However, data limitations make it difficult to incorporate other types of capital to evaluate potential impacts or trade-offs of food systems efforts using this type of method.

Also, while Thilmany McFadden et al. (2016) includes a module on community outreach, it is possible to undertake I-O modeling without engaging and empowering stakeholders. This implies that while I-O model outputs may be resonant with policymakers (Rahe, Van Dis, and Gwin, 2019), they could be less helpful to practitioners if not approached in a systematic way. Community-based food assessment processes in which stakeholders are brought in as project partners on the front-end, instead of having a more passive role (like supplying data to a researcher), implies that decisions on model type and hypotheses to be tested will be determined in a collaborative fashion between researchers and stakeholders. This is one of the reasons we highlighted the Incentive Programs and Nourish Colorado case study: There was a unique mix of program-driven goals, program-specific data, and important community context to inform assumptions informing the analysis and metrics most relevant to the stakeholders. Although this cannot address all limitations of I-O and IMPLAN, for example shedding light on racial justice issues, it improves our confidence in the estimates and helps to develop a richer story than just numbers to inform communities, programs, and policy leaders.

## Acknowledgements

The authors would like to acknowledge the Project Leads for The Economic Contributions of Healthy Food Incentives project, Holly Parker with Fair Food Network, Ronit Ridberg with UC Davis and Eli Zigas with SPUR. Special thanks also to Project Assistant Jennifer Sanchez with UC Davis and the project Advisory Committee of community-based partners for their insights, data and guidance in framing the case study presented in this chapter: Ecology Center (CA), Field & Fork Network (NY), Iowa Healthiest State Initiative (IA), Nourish Colorado (CO), Reinvestment Partners (NC), Sustainable Food Center (TX), The Food Basket (HI), University of California San Diego (CA), Vouchers 4 Veggies – EatSF (CA), and Washington State Department of Health (WA). In addition, we want to acknowledge support from the Colorado Agricultural Experiment Station, Cornell's Agriculture Experiment Station, the USDA Agricultural Marketing Service and NE1749, the multistate regional research committee on Enhancing Rural Economic Opportunities, Community Resilience, and Entrepreneurship for their support of the team's research in this topic. Finally, we thank the student peer reviewers from Tufts University, and Ashley McCarthy for support with the Fig. design. While Dr. O'Hara currently works at USDA's Office of the Chief Economist, his contribution to this manuscript occurred when he was employed at USDA's Agricultural Marketing Service.

## Disclaimer

## References

Bauman, A., Thilmany, D., 2017. Exploring localized economic dynamics: methods driven case studies of transformation and growth in agricultural and food markets. Economic Development Quarterly 31 (3), 244–254. https://doi.org/10.1177/0891242417709530.

Canning, P., Stacy, B., 2019. *The Supplemental Nutrition Assistance Program (SNAP) and the economy: New estimates of the SNAP multiplier, ERR-265*. U.S. Department of Agriculture, Economic Research Service Available at: https://www.ers.usda.gov/webdocs/publications/93529/err-265.pdf?v=1289.2 .

Cantrell, P., Conner, D., Erickcek, G. & Hamm, M.W. (2006). *Eat fresh and grow jobs, Michigan*. Available online at: https://www.canr.msu.edu/foodsystems/uploads/files/eatfresh.pdf (accessed May 5, 2020).

Christensen, L., Jablonski, B.B.R., Stephens, L., Joshi, A., 2019. Evaluating the economic impacts of farm-to-school procurement: an approach for primary and secondary financial data collection of producers selling to schools. Journal of Agriculture, Food Systems, and Community Development 8 (Suppl. 3), 73–94.

Conner, D.S., Knudson, W.A., Hamm, M.W., Peterson, H.C., 2008. The food system as an economic driver: strategies and applications for Michigan. Journal of Hunger & Environmental Nutrition 3 (4), 371–383.

Duval, D., Bickel, A.K., Frisvold, G.B., 2019. Farm-to-school programs' local foods activity in Southern Arizona: local foods toolkit applications and lessons. Journal of Agriculture, Food Systems, and Community Development 8 (Suppl. 3), 53–72.

European Commission, 2019. Natural capital accounting: Overview and progress in the European Union, *6th edition* Publications Office of the European Union, Luxembourg https://ec.europa.eu/environment/nature/capital_accounting/pdf/MAES_INCA_2019_report_FINAL-fpub.pdf.

Goetz, S.J., Partridge, M.D., Stephens, H.M., 2018. The economic status of rural American in the president trump era and beyond. Agricultural Economics, Sociology and Education 40 (1), 97–118.

Henneberry, S.R., Whitacre, B.E., Agustini, H.N., 2009. An evaluation of the economic impact of Oklahoma farmers markets. Journal of Food Distribution Research 40 (3), 64–78.

Hughes, D.W., Isengildina-Massa, O., 2015. The economic impact of farmers markets and a state level locally grown campaign. Food Policy 54, 78–84.

Hughes, D., Brown, C., Miller, S., McConnell, T., 2008. Evaluating the economic impact of farmers markets using an opportunity cost framework. Journal of Agricultural and Applied Economics 40 (1), 253–265.

IMPLAN. (2020). *Environmental data*. https://implanhelp.zendesk.com/hc/en-us/articles/1500001270601-Environmental-Data

Ingwersen, W., 2018. An Introduction to USEEIO. The Strategic Analysis team for the Advanced Manufacturing Office, U.S. Environmental Protection Agency https://cfpub.epa.gov/si/si_public_record_report.cfm?Lab=NRMRL&TIMSType=&count=10000&dirEntryId=344412&searchAll=&showCriteria=2&simpleSearch=0.

Jablonski, B.B.R., Bauman, A.G., Thilmany McFadden, D., 2020. Local food market orientation and labor intensity. Agricultural Economics, Sociology and Education. https://doi.org/10.1002/aepp.13059.

Jablonski, B.B.R., O'Hara, J.K., Thilmany McFadden, D., Tropp, D., 2016. Resource for evaluating the economic impact of local food system initiatives. Journal of Extension (6) 54.

Jablonski, B.B.R., Schmit, T.M, 2015. Differential expenditure patterns of local food system participants. Renewable Agriculture and Food Systems 31 (2), 139–147. https://doi.org/10.1017/S1742170515000083.

Jablonski, B.B.R., Schmit, T.M., Kay, D., 2016. Assessing the economic impacts of food hubs on regional economies: a framework that includes opportunity cost. Agricultural and Resource Economics Review 45 (1), 143–172.

Johnson, T.G., Raines, N., Pender, J.L., 2014. Comprehensive wealth accounting. In: Pender, J.L., Weber, B.A., Johnson, T.G., Fannin, J.M. (Eds.), Rural wealth creation. Routledge, New York.

Low, S.A., Vogel, S..

Miller, S.R., Mann, J., Barry, J., Kalchik, T., Pirog, R., Hamm, M.W., 2015. A replicable model for valuing local food systems. Journal of Agricultural and Applied Economics 47 (4), 441–461.

Mon, P.N., Holland, D.W., 2006. Organic apple production in Washington State: an input-output analysis. Renewable Agriculture and Food Systems 21 (2), 134–141.

O'Hara, J.K., Parsons, R.L., 2013. The economic value of organic dairy farms in Vermont and Minnesota. Journal of Dairy Science 96 (9), 6117–6126.

O'Hara, J.K., Pirog, R., 2013. Economic impacts of local food systems: future research priorities. Journal of Agriculture, Food Systems, and Community Development 3 (4), 35–42.

Partridge., M., 2013. America's job crisis and the role of regional economic development policy. The Review of regional studies 43, 97–110.

Pesch, R., Tuck, B., 2019. Developing a production function for a small-scale farm operator. Journal of Agriculture, Food Systems, and Community Development 8 (Suppl. 3), 27–36.

Rahe, M.L., Van Dis, K., Gwin, L., 2019. Communicating economic impact assessments: how research results influence decision-market attitudes toward the local food sector. Journal of Agriculture, Food Systems, and Community Development 8 (Suppl. 3), 95–105.

Rossi, J.D., Johnson, T.G., Hendrickson, M., 2017. The economic impact of local and conventional food sales. Journal of Agricultural and Applied Economics 49 (4), 555–570.

Schmit, T.M., Boisvert, R.N., 2014. Agriculture-based economic development in New York State: Assessing the inter-industry linkages in the agriculture and food system. Cornell University http://publications.dyson.cornell.edu/outreach/extensionpdf/2014/Cornell-Dyson-eb1403.pdf.

Schmit, T.M. & Jablonski, B.B.R. (2017). *A practitioner's guide to conducting an economic impact assessment of regional food hubs using IMPLAN: a systematic approach.* https://www.ams.usda.gov/sites/default/files/media/EB2017APractitionersGuide.pdf

Schmit, T.M., Jablonski, B.B.R., Mansury, Y., 2016. Assessing the economic impacts of 'local' food system producers by scale: a case study from New York. Economic Development Quarterly 40 (4), 316–328. https://doi.org/10.1177/0891242416657156.

Swenson, D., 2010. Selected measures of the economic values of increased fruit and vegetable production and consumption in the Upper Midwest. Leopold Center for Sustainable Agriculture.

The Economics of Ecosystems and Biodiversity (TEEB), 2018. TEEB for agriculture & food: Scientific and economic foundations. UN Environment, Geneva http://teebweb.org/agrifood/scientific-and-economic-foundations-report/.

Thilmany, D., Bauman, A., Love, E. & Jablonski, B.B.R. (2021). *The economic contributions of healthy food incentives.* https://www.spur.org/sites/default/files/2021-02/economic_contributions_incentives_2_2_21.pdf

Thilmany McFadden, D., Conner, D., Deller, S., Hughes, D., Meter, K., Morales, A., et al., 2016. *The economics of local food systems: A toolkit to guide community discussions, assessments and choices.* U.S. Department of Agriculture, Agricultural Marketing Service Report https://www.ams.usda.gov/sites/default/files/media/Toolkit%20Designed%20FINAL%203-22-16.pdf.

Tuck, B., Haynes, M., King, R., Pesch, R., 2010. The economic impact of farm-to-school lunch programs: A Central Minnesota example. University of Minnesota Extension.

U.S. Department of Agriculture National Institute of Food and Agriculture (USDA NIFA). (n.d.). *Gus Schumacher Nutrition Incentive Program* (formerly FINI). https://nifa.usda.gov/program/gus-schumacher-nutrition-incentive-grant-program

Watson, P., Cooke, S., Kay, D., Alward, G., Morales, A., 2017. A method for evaluating the economic contribution of a local food system. Journal of Agricultural and Resource Economics 42 (2), 180–194.

Watson, P., Wilson, J., Thilmany, D., 2007. Determining economic contributions and impacts: what is the difference and why do we care? Journal of Regional Analysis Policy 37 (2), 1–7.

Winfree, J., Watson, P., 2017. The welfare economics of "buy local.". American Journal of Agricultural Economics 99 (4), 971–987.

Winfree, J., Watson, P., 2021. Buy local and social interaction. American Journal of Agricultural Economics. https://doi.org/10.1111/ajae.12186.

# Environmental Input-Output (EIO) Models for Food Systems Research: Application and Extensions*

*Patrick Canning[a], Sarah Rehkamp[b] and Jing Yi[c]*

[a]USDA, Economic Research Service, Food Economics Division, Washington, DC, United States [b]University of Hawai'i at Mānoa, Honolulu, Hawaii, United States [c]Dyson School of Applied Economics and Management, Cornell University, Ithaca, NY, United States

## 9.1 Introduction

Chapter 3 introduces life cycle assessment (LCA), which focuses on the environmental impacts of single products. In this chapter, we present the background, basic framework, and an application of the environmental input-output (EIO) model, which assess environmental impacts across whole economic sectors. In this chapter we emphasize the environmental impact of food systems along precisely defined stages of the supply chain (e.g., food production-processing-distribution-marketing, or farm-to-fork), and by the component of diets (e.g., meats, grains, dairy, and so on). The EIO approach can also be used to conduct what-if scenario analysis. For example, Rehkamp and Canning (2018) estimated types and quantities of foods in different dietary scenarios to meet the Dietary Guidelines for Americans and measured corresponding blue water consumption to meet the demand at each food supply chain stage: farm inputs, crop production, livestock production, processing, food retailing, etc.

This chapter also links to chapter 8 as EIO is based on the economic input-output (IO) framework. Chapter 8 discusses how IO models can be used to estimate the economic impacts of food system initiatives, and in this chapter, we discuss how EIO, as extended IO models,

---

* Disclaimer: The findings and conclusions in this publication are those of the authors and should not be attributed to the USDA or the Economics Research Service.

can be applied to estimate environmental impacts resulting from food systems. By extending the conventional input-output model in dollar units to physical units, EIO allows for the measurement of both direct and indirect environmental impacts by economic sector. Direct impacts account for production activities that provide direct outputs to meet a specified demand, or first-tier transactions. For example, the environmental impacts of a change in final demand for the consumption of fresh vegetables would include the direct environmental impacts due to the change in the level of farm production and retail services. There are also indirect environmental impacts associated with the change in food demand, which are inter-activity transactions or second-tier activities to support first-tier activities. For example, to produce meat, livestock farms purchase feed from grain farms, and so indirect environmental impacts would incorporate the blue water used on grain farms to produce the feed (inter-activity transactions), or such as the change of water use in the energy industry (second-tier activities) for generating electricity required by livestock farms. The EIO approach considers not only the directly related activities, such as farm and food production, transportation to points of purchase, and food wholesale and retail services, but it also incorporates indirect transactions supporting these activities to provide a food system wide environmental analysis. This is considered a top-down approach since the modeling is done at an economy-wide scale allowing for consistent and clear boundaries of analysis but providing less detail than a process-based LCA, for example.

The following sections provide background information and technical details regarding the development of EIO models. In closing, a U.S. food economy EIO model application is presented to illustrate the concept.

## 9.2 Background

Empirical studies linking environmental services to models of national or regional economies came into prominence in the late 1960's with the realization that environmental pollution and its control is a material balance problem inextricably linked to the production and consumption processes of these economies (Ayers and Kneese, 1969). For example, in the case of fossil fuels, net flows of carbon emissions into the atmosphere have grown and economic processes, including those associated with the food economy, contribute to this growth.

Environmental services refer to qualitative functions of the natural non-produced assets of land, water, air, and minerals (and related ecosystems), and the animal and plant life within. Qualitative functions include (i) disposal services of residuals (from production and consumption in the present context) by the natural environment, (ii) productive services of natural resource inputs including space for production and consumption, and (iii) consumption services of providing for physiological and recreational needs of human beings (OECD, 2008; Canning et al., 2020). Measurable changes to the stock or flow of services from any of these qualitative functions that is brought on by any EIO model transaction we call an *environmental metric* (*em*), and that which it measures we call an *environmental factor* (*ef*). For example, freshwater withdrawal from surface or groundwater sources is an environmental factor often referred to as 'blue water use'. The volume of blue water use linked to an EIO model transaction, such as annual gallons of blue water withdrawal for crop irrigation per unit of grain production, is an environmental metric. Table 9.1 presents a list of common

**TABLE 9.1** Examples of food-related eio hybrid research by environmental metric environmental metric.

| | Description | Related publications | Example of role in the food system |
|---|---|---|---|
| Energy | The use of fossil fuels (coal, natural gas, or petroleum) or non-fossil fuels as a source of heat or power or as a raw material input to a production process. | Canning et al. (2010); Canning et al. (2017); Cazcarro et al. (2012); Egilmez et al. (2014); Hirst (1974); Reynolds et al. (2015); Sherwood et al. (2017); Xiao et al. (2019) | Fuel for tractors on farm or trucks transporting product to final markets. |
| Eutrophication | Excess nutrients in a body of water due to runoff. This causes excessive plant and algae growth due to lack of oxygen. | Behrens et al. (2017); Scherer & Pfister (2016); Tukker et al. (2011); Wolf et al. (2011) | Impacts from the runoff of agricultural chemical application. |
| Greenhouse gases | Gases that are released into the atmosphere including carbon dioxide ($CO_2$), methane ($CH_4$), nitrous oxide ($N_2O$), and fluorinated gases. These emissions trap heat in the atmosphere and contribute to climate change. | Behrens et al. (2017); Boehm et al. (2018); Camanzi et al. (2017); Egilmez et al. (2014); Hendrie, Ridoutt, Wiedmann, & Noakes, 2014; Hitaj et al. (2019); Ivanova et al. (2016); Virtanen et al. (2011); Webber & Matthews (2008) | Emissions from the combustion of fossil fuels or through enteric fermentation of ruminant livestock. |
| Land | Land can be subdivided in many ways. Some papers study cropland exclusively while others consider cropland plus one or more of the following: pasture land, forest land, marine land, and built-up land. | Behrens et al. (2017); Bruckner et al. (2019); Costello, Griffin, Matthews, & Weber, 2011; Egilmez et al. (2014); Ivanova et al. (2016); Mulik & O'Hara (2015); Weinzettel, Hertwich, Peters, Steen-Olsen, & Galli (2013) | The land base used for growing crops or grazing livestock. |
| Water | Freshwater withdrawn from surface or groundwater sources. | Avelino & Dall'erba (2020); Egilmez et al. (2014); Ivanova et al. (2016); Lenzen & Foran (2001); Rehkamp & Canning (2018); Rehkamp et al. (2021); Reynolds et al. (2015); Sherwood et al. (2017); Xiao et al. (2019) | Irrigation water for crop production or processing raw commodities. |

environmental metrics studied in food-related EIO or EIO-LCA hybrid research, and some accompanying publications.

An important concept underlying the measurement of environmental metrics throughout an economy is the "conservation of embodied energy" (Bullard and Herendeen, 1975). Although originally stated with explicit reference to energy, it is conceptually the same to reference the more broadly defined "environmental factors." Such an adaptation of the Bullard & Herendeen (1975) concept states that *environmental factors used or dissipated by a process are passed on, embodied in the products of that process.* In EIO multiplier analysis, total final demand, or gross domestic product (GDP) plus imports,[1] is the input multiplier that induces all economic system processes.

We refer to the measure of annual economywide uses of specific environmental factors as the annual factor budget. For example, the annual primary energy budget (fossil fuels, nuclear power, renewable energy) for the United States was 100.2 quadrillion British thermal units (Btu) in 2019 (EIA, 2021). By developing *ef* output multipliers that allocate all measured annual factor budgets among all economic processes, *conservation of embodied environmental factors* implies that all such uses of an *ef* are entirely embodied in an economy's final demand. The appeal of this approach is the impossibility of double counting the attribution of any environmental metric among different components of GDP plus imports. Additionally, the model's boundaries are comprehensive and consistent when measuring embodied *ef*. It is a basic modeling best practice to first verify that the vector of all environmental factor budgets ($\sigma$) in the EIO model are fully recovered. The sum of the environmental factors for all economic processes equals the total budget for that environmental factor, which is measured by the product of the final demand vector ($\mathbf{x}$), the total requirement matrix ($\mathbf{M}$), and the *ef* multiplier matrix ($\mathbf{E}$): $\sigma = E \times \mathbf{M} \times \mathbf{x}$.

One such component of final demand is the annual food acquisition by or for all domestic food consumers. By multiplying this annual food acquisition through the EIO multiplier model, we identify that subset of all annual economywide transactions linked to this food acquisition and by extension the portions of all annual environmental factor budgets linked to this food acquisition.

This type of analysis was one the earliest applications of the EIO model (Hirst, 1974). Hirst examined total energy flows (Btu) and found that the domestic food system, defined as the set of all transactions induced to accommodate total domestic personal consumption expenditures on food, accounted for 12 percent of the 1963 national energy budget of the United States. More recently, Weber and Matthews (2008) investigate the "food-miles" concept, or the environmental impact of transporting food from where it is produced to where it is consumed. The authors find that the production stage contributes more greenhouse gas (GHG) emissions than transportation in the U.S. food supply chain. The authors also find variations in emission levels by commodities and that red meat is the most GHG-intensive.

Other recent studies include those that focus on linkages between diet outcomes and environmental flows. Examples of papers studying U.S. dietary patterns include Behrens et al. (2017), Boehm et al. (2019), Canning et al. (2020), Hitaj et al. (2019), and Rehkamp and Canning (2018) using an EIO approach, or by extension, a computable general equilibrium model (Mulik and O'Hara, 2015). Examples of papers studying international diets include Cazcarro et al. (2012), Hendrie, Ridoutt, Wiedmann, & Noakes (2014), Reynolds et al. (2015), Tukker et al. (2011), Virtanen et al. (2011), and Wolf et al. (2011).

It is not certain whether healthier diets lead to improved environmental outcomes. The *Dietary Guidelines for Americans* (DGA) are the National dietary recommendations published

---

[1] All recorded transactions in country (region) EIO models are bilateral involving at least one domestic (regional) actor, and all environmental metrics assigned to specific transactions in the model represent domestic (regional) use. Use attributed to imports only measure those induced by the absorption of imports into the domestic (regional) economy. All *ef*'s already embodied in imports prior to absorption into the economy must be determined outside the basic model.

every five years by U.S. federal government agencies. Among findings from the US-based studies who use the DGA as a yardstick for heathy dietary patterns, it is not certain whether the healthier diets aligning with these recommendations would lead to improved *ef* outcomes. The studies indicate that following the Healthy US-Style Dietary Pattern, which is omnivorous, may have mixed environmental impacts. Energy use may be reduced, but greenhouse gas emissions may show little change and water use could increase when compared to current American diets. Additionally, healthy vegetarian diets are associated with a lower environmental impact, except for water. There is evidence in the international literature that plant-based diets are less impactful on the environment (e.g., Tukker et al., 2011). In the extensions section, we provide an example showing how the EIO modeling framework can be used for this type of research. The example will focus on linkages between diet outcomes and the use of fossil fuels and other primary energy sources to accommodate these diets.

## 9.3 Adapting the EIO framework for modeling food systems

Our approach is an adaptation of a Type I *EIO* multiplier model (see Glossary) in which activity-to-commodity (A × C) transactions (defined in the box included with the supplemental information material for this chapter titled "*An Explanation of Activities, Commodities, and Mathematical Notation*") and vice versa are determined by the model. The multipliers developed to make these determinations provide a measurement of how one monetary unit (e.g., $1 million) injected into the economy as an expenditure on a specific good or service (commodity) changes total production or output of each activity and commodity across the entire economy. There is a multiplier for every possible activity-to-commodity transaction and vice versa. Multiplier calculations are initiated by acquisitions of either activity or commodity outputs which are specified by the modeler and may result in three types of impacts: direct, indirect, and induced. Type I multipliers only concern the direct and indirect impacts, where direct impacts consider the activity and commodity outputs that directly meet the modeler specified demand, and indirect impacts account for the activity and commodity outputs resulting from the direct modeler specified acquisitions.

Any payments to primary factors of production resulting from these direct and indirect output affects, such as labor or proprietor incomes, will induce a new round of spending. These are known as the induced impacts (Hughes, 2018). However, a Type I EIO multiplier will not measure these induced affects, since the modeler using this type of EIO model only wants to consider the purchases and/or acquisitions which they are specifying for analysis. For example, if a modeler is studying an economy for a recent time period and wants to study the environmental flows linked to observed total annual food expenditures of all domestic consumers, a Type I multiplier model is appropriate since allowing induced effects of observed food spending would lead to additional food spending beyond the observed amounts under study. An example can be found in Rehkamp and Canning (2018). The demand for food types and quantities is estimated outside of the EIO model. Next, the EIO model is applied to evaluate water use of individual food items, including direct and indirect water requirements for production by each activity, so that the study can assess the U.S. food system blue water use across different dietary patterns.

---

**An Explanation of Activities, Commodities, and Mathematical Notation**

In this chapter we present the EIO framework but use the social accounting matrix (SAM) notation convention. This is because our use of industry-by-commodity accounting is far more synonymous with SAM multiplier models. However, we employ the type I multiplier in which personal consumption expenditures and distributions of primary factor payments are external to the modeling process, so the model is best described as an IO model. Primary factors are inputs into production processes including land, labor, and capital.

Activities and Commodities are described as follows:

- An **activity** is a grouping of establishments that produce one or more types of products using a similar production process, instead of the IO convention of referencing 'industries' to describe the grouping of establishments producing only one product type. For example, tomato canning plants may also grow their own tomatoes. Fresh tomatoes and canned tomatoes are two products produced by the activity.
- A **commodity** results from the assembly of one type of product, acquired from one or more activities that produce this product, instead of the IO convention where this assembly occurs prior to incorporation into the model.

Mathematical notation in this chapter is as follows:

- A matrix is a rectangular array or table of numbers, arranged in rows and columns, and denoted with bold capitalized letters.
- A vector is a single column or row of numbers, denoted with bold lower-case letters.
- Sets are a pre-defined collection of elements inside of a matrix or vector, and are denoted with capitalized and italicized letters.
- Set elements are specific individual elements within a set and are denoted with lower case italicized letters.
- A scalar is a single number and is denoted with nonbold lower case letters.
- Letters are from either the English or Greek alphabet.
- A matrix or vector transpose is denoted with a prime (′).
- A diagonal vector (square diagonal matrix) has zeros anywhere not on the main diagonal and is denoted with a double prime (″).
- A matrix inversion is indicated by its placement inside brackets as $\{matrix\}^{-1}$

---

Once the model determines the levels of activity and commodity production to accommodate the modelers specified level of food demand, the EIO modeler employs an environmental factors (*ef*) multiplier matrix to translate all model derived production levels into environmental factor outcomes linked to these production levels.

The four components of the basic EIO multiplier model for food systems research are:

1. the total (activity and commodity) requirement matrix **M**,
2. the monetized value-added multiplier vector **v**,

**3.** the quantized environmental factors ($Q$) multiplier matrix $\mathbf{E}$ ($\{\mathbf{e}^q\} \; \forall \; q \in Q$), and
**4.** the final demand food system scenarios ($F$) matrix $\mathbf{X}^F$ ($\{\mathbf{x}^f\} \; \forall f \in F$).

There are clear advantages to using the $A \times C$ EIO model approach in the present context. First, it internalizes the multi-product nature of production activities along commodity supply chains throughout the food economy, providing more opportunities to pinpoint environmental multipliers. Second, although the dimensions and complexity of this modeling approach are greater, the mathematics (matrix algebra) is virtually identical and one can derive both activity and commodity multipliers, thus achieving the strengths of both conventional alternative IO model approaches (Industry-by-Industry, or Commodity-by-Commodity). Third, we view the environmental factors multiplier matrix ($\mathbf{E}$) approach as preferable to the alternative convention of using a hybrid EIO model in which all internally determined model transactions involving the use of an environmental factor are recorded in physical units (such as the purchase of electricity measured in British thermal units) and all other transactions are recorded in monetary units. Whereas both approaches are interchangeable for many applications, the $\mathbf{E}$ multiplier matrix approach is more versatile for use in cases where specific transactions have implications across multiple metrics. In addition, the $\mathbf{E}$ multiplier matrix approach allows the model to produce both market value and environmental flow implications across the sequence of activities that are induced to accommodate the modelers specified level of food demand.

A key feature of our EIO modeling approach is the incorporation of supply chain analysis (Leontief, 1967; Rehkamp and Canning, 2018). This provides a clean and precise method to organize the decomposition of both value added, or workers' wages plus returns to capital, and environmental impacts into supply chain stages, which captures the organized sequencing of multiple activities that take place to assemble food commodity value chains that collectively comprise the entire domestic food economy. Supply chain analysis facilitates the application of life cycle assessment scenario analysis.

Four important assumptions about the economic setting for the period of analysis (both the pre- and post-scenario simulation economy) in this type of model inform our interpretation of the multiplier approach. They are that:

**1.** the supply of primary factors (labor, capital, natural resources) exceed the demand for these production inputs,
**2.** any additional use of these primary factors is equally productive as what is already in use,
**3.** the new scenario being studied does not change existing relative prices in factor and commodity markets or existing production technologies such as factor productivities and material discharge rates, and
**4.** all proceeds accruing to primary factor owners from the scenario induced production outcomes do not induce further spending by factor owners in the period of analysis.

In many contexts, these assumptions pose important limitations on the appropriate use of this model. For example, when the modeler is interested in how their final demand scenario, $\mathbf{x}^f$, induces further changes to final demand, or induces changes in the use of primary factors per unit of output, this model is not appropriate since these changes are ruled out by the four assumptions just discussed. However, there are many strengths and advantages to this model. Most notably, it is an ideal tool for developing decomposition measures of environmental factors using historical data, both in terms of supply chain stages (e.g., farm to fork) and across different food categories (e.g., meats, grains, dairy, fresh produce, and so on). The type I EIO model can also be effectively combined with other models when considering scenarios where the economic setting discussed above is not violated. These scenarios can be wide ranging. For example, a separate study to forecast real (inflation-adjusted) food demand scenarios can be traced through an EIO model. Provided real wages and average interest rates remain stable across these food demand scenarios (consistent with the slack factor market assumptions of the standard type I EIO model), this type of analysis can be justified. In other situations, modelers can make caveats to their analysis, such as 'upper-bound', 'lower-bound', and 'realistic distribution of impacts'.

Perhaps as important as any other strength of the type I EIO approach is the quality and timeliness of model input datasets used to compile the EIO model. Most countries worldwide follow the guidelines of the United Nations Statistical Commission (UN et al., 2009) by maintaining an official government data system of national accounts (SNA). It is common that these SNA programs are maintained by large government agencies and staffed with highly skilled economists and statisticians, with many participating in a global community of practice that strives to employ statistical best practices and harmonize cross-country methodologies. Foundational to SNA data systems are the National *Supply* and *Use* Tables (UN, 2018), and these Tables produce two of the four main components of the EIO model.

### 9.3.1 Supply and use tables

The supply Table (**S**) shows availability of goods and services by type of commodity (J) and by type of activity (I) distinguishing between supply by domestic activity and imports of goods and services (United Nations, 2018). The first Table depicted in Fig. 9.1 presents an example of a *Supply* Table for a fictional economy that classifies all activities into three groups ($I = 3$) and categorizes all products into 3 commodity varieties ($J = 3$). Totals in the last column represent total output by activity (also called outlays) and totals in the last row represent total supply by commodity plus total imports.

As depicted (top Table in Fig. 9.1), commodities are listed above the **S**′ matrix (a transposed Supply Table is also called a 'Make' Table). The element $s_{ij}$ in **S**′ shows the value of the output of commodity $j$ that is produced by activity $i$. Reading down the 'crop and animal products' column shows that these products are assembled from output of the domestic 'agriculture' activity totaling 23 monetary units (e.g., $billion), 4 units of output from the domestic 'manufacturing' activity, and 2 units of crop and animal products are imported from other countries or regions with no indication of what activities produced these imported products. Total availability of crop and animal products is 29 units. The other two commodities, 'goods' and 'services', are also assembled from two of the three domestic activities plus imports. In the case of goods, it is mostly assembled from output of the manufacturing activity with a small amount from agriculture, plus 20 units from imports. For services, three-quarters of total units assembled come from the 'other' activity, about a quarter from imports and a small amount from manufacturing.

A *Use* Table (**U**) shows both intermediate uses of commodities by type of activity, and final market uses by category, typically broken out into personal (household) consumption, private gross capital formation (investment), government consumption and investment expenditures,

**Supply of commodities by activity**

| | | Commodity | | | |
|---|---|---|---|---|---|
| | | Crop and animal products | Goods | Services | Total |
| **Activity** | Agriculture | 23 | 2 | | 25 |
| | Manufacturing | 4 | 128 | 3 | 135 |
| | Other | | | 86 | 86 |
| Imports | | 2 | 20 | 31 | 53 |
| Total | | 29 | 150 | 120 | |

**Use of commodities by activity and GDP components, and use of primary factors by activity**

| Commodity | Activity* | | | Final Uses* | | | | Total |
|---|---|---|---|---|---|---|---|---|
| | Agriculture | Manufacturing | Other | PCE | PDI | Govt | Exp. | |
| Crop and animal products | 5 | 10 | 1 | 5 | 1 | 2 | 5 | 29 |
| Goods | 5 | 40 | 25 | 50 | 15 | 5 | 10 | 150 |
| Services | 5 | 30 | 25 | 50 | 4 | 1 | 5 | 120 |
| Salary and wages | 2 | 25 | 23 | | | | | |
| Operating surplus | 8 | 30 | 12 | | | | | |
| Total | 25 | 135 | 86 | | | | | |

\* PCE = personal consumption expenditures; PDI = private direct investment; Gov = government consumption and investment expenditures; Exp = exports

**Activity by commodity input-output table**

| | Commodity | Activity | | | Commodity | | | Final uses | Total |
|---|---|---|---|---|---|---|---|---|---|
| | | Agriculture | Manufacturing | Other | Crop and animal products | Goods | Services | PCE+PDI+Govt+Exp | |
| **Activity** | Agriculture | | | | 23 | 2 | | | 25 |
| | Manufacturing | | | | 4 | 128 | 3 | | 135 |
| | Other | | | | | | 86 | | 86 |
| **Commodity** | Crop and animal products | 5 | 10 | 1 | | | | 13 | 29 |
| | Goods | 5 | 40 | 25 | | | | 80 | 150 |
| | Services | 5 | 30 | 25 | | | | 60 | 120 |
| **Factors+imports** | Salary/wages+Operating surplus+Imports | 10 | 55 | 35 | 2 | 20 | 31 | | 120 |
| | Total | 25 | 135 | 86 | 29 | 150 | 120 | 120 | |

FIGURE 9.1 Supply, use, and activity by commodity IO table.

and export sales. The Table also shows the components of gross value added by activity (here, the salary/wages and operating surplus). Totals by row represent the total intermediate and final uses by commodity. Totals by column represent total intermediate outlays and value added, or gross output by economic activity, and total final market outlays by category (United Nations, 2018).

The second Table from the top in Fig. 9.1 depicts a *Use* Table for the same fictional economy depicted in the *Supply* Table. $u_{ij}$ indicates the purchase value of commodity $i$ by activity $j$. The three activities are listed across the top above columns of the **U** matrix. Reading down the 'agriculture' column shows that this activity acquires all three commodities with each intermediate input valued at 5 monetary units (e.g., $billion), and pays 2 monetary units in salary and wages to hired labor. The column can be thought of as this activity's recipe for production. Total proceeds from output sales of the agriculture activity are reported in the far-right column of the top data row in Fig. 1's supply Table, and this value of 25 is also entered in the last row of the agriculture column of the *Use* Table. By deducting the intermediate input and labor costs, totaling 17, this leaves an operating surplus of 8, as reported in the Use Table. Although not depicted in this example, values for operating surplus can be negative when total sales proceeds do not fully cover intermediate plus labor costs. The other two activity columns, representing 'manufacturing', and 'other' respectively, show outlays for all three commodities plus payments to hired labor, and both also have a positive operating surplus after intermediate input and labor costs are deducted from their sales proceeds, also reported in the *Use* Table (Fig. 9.1). The last four columns to the left of the 'Total' column in the *Use* Table report total outlays among the four final market categories for the three commodities.

The third Table at the bottom of Fig. 9.1 combines the **S** and **U** Tables. In this consolidated 'S&U' Table, reading down any column reports outlays by category, with the sum of column outlays (bottom row) representing total supply. Each column in this Table has a corresponding row and reading across each row reports all uses of total supplies with the total uses reported in the far-right column. This Table shows that total supply by activity and commodity equals their corresponding total uses. It also shows that total final uses of commodities are exactly equal to total value added (payments to labor + operating surplus) by activity plus total value of commodity imports.

Fig. 9.1 also shows, with rectangles and arrows, the method of consolidation. The arrows descending from the **S** and **U** Tables respectively, identify their exact placement in the consolidated Table. Collapsing the final uses submatrix to a single column (PCE + PDI + Gov + Exp) and collapsing the value-added submatrix (salary/wages + Operating surplus) to a single row—both from the *Use* Table—ensures the consolidated Table is both square and balanced. There is no reason, *a priori*, to expect the 'Final uses' and 'GDI + imports' submatrices to have the same number of columns and rows respectively, much less for each column and corresponding row to have the same column and row sums respectively. However, by collapsing both to one column and row respectively, we know from Walras law (Nicholson and Snyder, 2012) that these column and row sums are equal. Specifically, Walras law states that the market value of all excess supply (supply − use) across products in a market economy equals 0. In the consolidated **S&U** Table at the bottom of Fig. 9.1, values for all activity and commodity column and row totals are equal by construction in their source **S** and **U** Tables. Although remaining columns and rows in this Table are not equal in number or in values for corresponding row and column totals, Walras law ensures that by collapsing all remaining columns and rows in the consolidated Table, their corresponding column and row totals must be equal. As a square and balanced matrix, this consolidated Table can be transformed into a multiplier model.

The discussions of the total requirement matrix and value-added multipliers are presented in the Supplementary Information (SI) material of this chapter. This SI material takes you through the mathematical derivations of the total requirement matrix and value-added multipliers with a simple numerical example using a fictional country's input output data. The data and computer code for this numeric example are also available in open source format with the supplemental materials for this chapter.

## 9.3.2 Environmental multipliers and food system scenarios

A parallel accounting structure to the *Supply* and *Use* Table accounting presented above exists for measurement of physical flows of environmental factors. Described in the SEEA (system of environmental-economic accounting) Central Framework (UN et al., 2014) and called the physical supply and use Tables (PSUT), it records flows from the environment to the economy, within the economy, and from the economy. In theory, the incorporation of data from PSUTs into the compilation of an environmental factors' matrix would follow a precise process analogous to that depicted above in Fig. 9.1 for compilation of the monetized A × C IO Table.

In practice, there is a scarcity of available PSUTs within official SEEA among countries worldwide. Because this chapter is focused on applied food systems research, our approach focuses on methods of working with available environmental statistics and adjusting those statistics when necessary to align with the concepts of the SEEA Central Framework, and with the specific structure of the modelers' EIO model. Examples of adjustments are too numerous to explain in detail here. A comprehensive discussion of this topic can be found in the handbook publication of the UN (2018, chapters 13 and 19).

For a single country (or region), the compilation of an environmental factor's matrix should proceed by carrying out the following 3 steps for each environmental factor to be included in the EIO model:

- Step 1: Define the economywide boundaries of use to be included.
- Step 2: Identify the appropriate data source(s) and compile a measure of the annual use budget that has comprehensive coverage within the boundaries defined in step 1—preferably with a detailed breakout for types of users across groupings of activities, commodities, and institutions.
- Step 3: Compile the appropriate allocation metrics to completely assign the annual use budget among all activities, commodities, and institutions.

For step 1, the geographic boundary of use is the physical borders of the country or region. Time period must also be included in the boundary definition, for example annual, quarterly, or monthly. But the definition of boundaries can include more than space and time. It could also be defined as a subset of a broader source or use category. Source subsets can be, for example, primary energy use from fossil fuel sources (excluding nuclear power and renewable sources), or freshwater use withdrawn from ground and surface waterbodies (excluding rainwater and other non-freshwater sources). Use subsets can be, for example, land use for all agricultural activities (excluding non-agricultural land uses), or greenhouse gas emissions from fossil fuel use (excluding biogenic and reactive chemical emissions).

Once boundaries are defined, the appropriate data source(s) must capture total uses within its boundaries (step 2). If, for example, data only exists for 11 months of the annual boundary defined in step 1, then either the modeler must have a validated method and information to impute use for the 12th month or must redefine the time boundary. If data measuring use comes from multiple sources, or from different parts of a single source, such that more than one metric is reported for subsets of uses, then a precise method for converting all measures to a single metric must be available. For example, when some data is reported in metric units and other data in non-metric units, a precise conversion of non-metric units to metric, or vice versa must be possible such that all sources and uses of an environmental factor are measured by the same metric and a summation of all uses recovers the known total factor budget. Another example is where some data is reported as annual use and others as average daily use. It must be verified that daily use can be converted to annual use by multiplying the former by the number of use days in the study period.

The allocation metric in step 3 is necessary when the source data in step 2 does not provide specific use data for all candidate processes occurring in any EIO model simulation. For example, every intermediate or final market transaction in an EIO simulation involves an activity outlay on a commodity, a commodity outlay on an activity, or a final market outlay on a commodity. It is very unlikely that the source data developed in step 2 explicitly allocates uses to each specific bi-lateral transaction in the EIO model, so the allocation metric is used for this purpose. Once the use category subtotals in the step 2 dataset are linked to subgroups comprising all potential users in the EIO model, the step 3 allocation metric dataset is used to allocate the use subtotals to specific model transactions. As was noted above, adjustments to those allocations may become necessary to align use with the concepts of the SEEA Central Framework, and with the specific structure of the modelers EIO model. The following example demonstrates the most common approaches.

Returning to the hypothetical economy and IO Table depicted in Fig. 9.1 and Table S1, suppose we are seeking to compile an environmental factor use Table and accompanying multiplier matrix (**E**) for two specific factors; annual U.S. blue water withdrawals, and annual U.S. fossil fuel use (Step 1). Each factor has a single data source, as summarized in Table 9.2 (Step 2).

The data in Table 9.2 depicts two of many possible scenarios with environmental factors data. In the scenario the data is complete but more aggregated, in terms of user details, than required for assignment to the EIO model. Knowledge of statistical coverage for each source dataset should inform the EIO modelers approach. With a growing number of applied EIO research publications, such as those listed in Table 9.1, practitioners can glean methods for environmental factors data development from these sources.

For the scenario presented in Table 9.2, agriculture does not have an aggregation problem since the user (crops and livestock) aligns exactly with the same commodity in the EIO model. Note, however, that this user is a commodity enterprise whereas the other user (nonfarm business) is an activity enterprise. This scenario is not uncommon. For example, every 5-years in the United States an Economic Census is enumerated (USDOC-CB, 2021b). This census compiles business statistics of most non-farm businesses, broken out by major industry groups such as a census of manufacturing and a census of retail trade. Also every 5-years

**TABLE 9.2** Fictional economy fossil fuel and blue water use by region.

| Source ↓ | Use → | Activity | Commodity | Institution |
|---|---|---|---|---|
| | | Nonfarm Business | Crops and livestock | Consumption |
| | | | **North Region** | |
| | | | *trillion Btu (tBtu)* | |
| **Fossil fuels** | Coal | 20 | 2 | 11 |
| | Petroleum products | 30 | 14 | 21 |
| | Natural gas | 20 | 14 | 18 |
| | | | *million gallons per day (mgal/day)* | |
| **Blue water** | Public supply | 20 | 10 | 25 |
| | Self supply | 180 | 140 | 25 |
| | | | **South Region** | |
| | | | *trillion Btu (tBtu)* | |
| **Fossil fuels** | Coal | 20 | 2 | 8 |
| | Petroleum products | 5 | 6 | 8 |
| | Natural gas | 10 | 7 | 9 |
| | | | *million gallons per day (mgal/day)* | |
| **Blue water** | Public supply | 10 | 5 | 12 |
| | Self supply | 90 | 70 | 13 |

in the United States, an Agricultural Census is enumerated ((USDA-NASS) U.S. Department of Agriculture, National Agricultural Statistics Service, 2021) in which most statistics in are reported on a commodity basis, such as planted soybean acreage and dairy cattle inventories. For the fictional economy in Table 9.2, we assume the fossil fuel and blue water use data for crops and livestock do not include any water use by nonfarm business that produce crops and livestock commodities. This knowledge will be accounted for later in calculations of energy and water use by category.

Environmental data in Table 9.2 for the non-farm activity and institutional uses does have an aggregation problem. Nonfarm businesses include both manufacturing and other (non-farm) activities, and the institutional consumption includes both personal and government consumption—investment (or accumulation) is not considered use and exports are outside the defined study boundary. With manufacturing and other production activities potentially exhibiting a range of environmental factor use intensities, their aggregation in these use measures reported in Table 9.2 can be problematic. This problem can be overcome through careful choice of data used for the distribution metrics (discussed below). It can also often be addressed with the use of regional data, as is reported in Table 9.2 with breakouts of both North region and South region statistics. Although the target model for this fictional economy is a National model, breaking out both the environmental and economic data by subnational regions can dramatically improve the distribution of environmental factor use measures of more aggregated data sources. To demonstrate, we first focus on compiling the distribution metrics data, or step 3 in the 3-step process.

Table 9.3 presents the compiled distribution metrics data for our fictional economy. By itself, the use of employment data as a distribution metric for fossil fuel use between manufacturing and other production has limitations. Employment is a good indicator of scale and, other things equal, scale is a strong predictor of use levels. However, employment is not necessarily a good indicator of factor use intensity—for example Btu of primary energy use per unit of output—since different activity classifications are likely to employ different technologies (e.g., a 'goods' producing activity verses a 'services' producing activity). Fortunately for our fictional economy, the regional breakout statistics reported in Tables 9.2 and 9.3 provide a layer of information to refine our distribution metric. As it turns out for our fictional economy, over 95 percent of nonfarm business employment in the North region (486k ÷ 511k) is working in the 'other' production activity, so applying the same use-per-worker ratio for all nonfarm business in this region will overwhelmingly reflect the use intensity of the 'other' activity worker. Since this activity will claim over 95 percent of that regions use, it is likely to be a reliable distribution. The manufacturing allocation of resource use in the North region may not be reliable. However, it will be getting most of its resource use allocation from the South region where manufacturing is the dominant employer (89 percent or 425k ÷ 479k) and so the source data is likely to be a reliable representation of manufacturing activity resource use. For institutional consumption of environmental factors such as households and government agencies, population is likely to be a reliable distribution metric. Given how statistics are enumerated in this fictional economy (see footnote to Table 9.3), the institutional consumption use data in Table 9.2 can be allocated between households and governments based on the population (employment for government agencies) of each category.

For environmental factors that are purchased, such as energy commodities, this allocation metric can be refined. For example, using national level data on employment by activity, the national *Use* Table data on energy commodity purchases by each activity can be converted to a per-worker measure and this measure can be multiplied into each regional employment metric. This would refine the allocation metric to reflect factor use intensities of different activities. Although this approach is not used in Table 9.3, our application in the next section does employ this technique.

To translate the information in Tables 9.2 and 9.3 into an **E** matrix, we denote the 5-by-3 North and South region data matrices in Table 9.2 as $\mathbf{EF}^n$ (environmental factors, North) and $\mathbf{EF}^s$. We convert the 2-by-6 South and North region data matrices from the first 2 data rows and 6 data columns in Table 9.3 as $\mathbf{DM}^s$ (distribution metrics, South) and $\mathbf{DM}^n$. We translate each regional distribution matrix into row share matrices (each element converted to its value share of corresponding row total):

$$\mathbf{D}^{s(n)} = \mathbf{DM}^{s(n)} \times ((\mathbf{DM}^{s(n)} \times i)'')^{-1} \tag{9.1}$$

We can obtain the environmental multiplier matrix as follows:

$$\mathbf{E} = (\mathbf{EF}^n \times \mathbf{D}^n + \mathbf{EF}^s \times \mathbf{D}^s) \times \{\mathbf{y}''\}^{-1} \tag{9.2}$$

**TABLE 9.3** Fictional economy factor distribution metric by region.

| User → | Activity | | | Commodity | | | Institution | |
| --- | --- | --- | --- | --- | --- | --- | --- | --- |
| | Agriculture | Manufacturing | Other | Crops and livestock | Goods | Services | PCE | Gov |
| User proxy ↓ | *employment (full time equivalent)* | | | | | | *population* (persons)* | |
| **North Region** | | | | | | | | |
| A_nonfarm business | – | 25,000 | 486,000 | – | – | – | – | – |
| C_crop and livestock | – | – | – | 6,700 | – | – | – | – |
| I_consumption | – | – | – | – | – | – | 1,750,000 | 67,000 |
| **South Region** | | | | | | | | |
| A_nonfarm business | – | 425,000 | 54,000 | – | – | – | – | – |
| C_crop and livestock | – | – | – | 3,300 | – | – | – | – |
| I_consumption | – | – | – | – | – | – | 750,000 | 33,000 |

* In this fictional economy, government employment is subsumed in the 'Other' activity and the population statisitcs refer to that portion of the total population of 2.5 million involved in the two institutional sectors.

**TABLE 9.4** Fictional Economy Environmental Multilier Matrix and Envrionmental Metrics Final Demand Vector.

| Environmental metric | Activity | | | Commodity | | | Institutions |
| --- | --- | --- | --- | --- | --- | --- | --- |
| | Agriculture | Manufacturing | Other | Crops and livestock | Goods | Services | $\sigma^x$ |
| | *Environmental multiplier matrix* (**E**) | | | | | | *Final demand EM vector* |
| Coal | – | 0.13869 | 0.24740 | 0.13793 | – | – | 19 |
| Petroleum products | – | 0.04373 | 0.33833 | 0.68966 | – | – | 29 |
| Natural gas | – | 0.07297 | 0.23429 | 0.72414 | – | – | 27 |
| Blue water: public supply | – | 0.07297 | 0.23429 | 0.51724 | – | – | 37 |
| Blue water: self supply | – | 0.65674 | 2.10860 | 7.24138 | – | – | 38 |

Table 9.4 presents the environmental multiplier matrix compiled in Eq. (9.2) by dividing through the activity and commodity column elements of the **EF** matrix (from the data in Tables 9.2 and 9.3) by their corresponding gross outputs (**y**)... Table 9.4 also reports the environmental metrics (EM) final demand vector, $\sigma^x$, compiled from the last two data columns of Table 9.2, summed over both regions.

From the data in Table 9.4, readers can verify that the national environmental factor budgets vector, $\sigma$, can be recovered from the Eq.:

$$\sigma = \mathbf{E} \times \mathbf{y} + \sigma^x \tag{9.3}$$

where the vector **y** represents the total activity and commodity output, and its numerical values are reported in Table S1 of the supplemental materials to this Chapter. The environmental factor budgets vector can also be compiled by adding the five North and South region row sums in Table 2—63 tBtu for coal (0.13869 * 135 + 0.24740 * 86 + 29 * 0.13793 + 19 = 63), 84 tBtu for petroleum products, 78 tBtu for natural gas, 82 mgal/day public supplied blue water, and 518 mgal/day for self-supplied blue water. Data from Tables 9.2 and 9.3, plus the code to compile the **E** matrix reported in Table 9.4 are provided with the supplemental materials to this Chapter.

### 9.3.3 Supply chain modeling

To further characterize environmental flows in the food system, it is important to conduct a supply chain analysis. Supply chains for specific food commodities are variants of the food supply chain depicted in Fig. 9.2.

To represent this structure in the EIO model, a matrix reorganization is developed as an alternative to aggregation (Leontief, 1967). This involves reorganizing the data into supply chain activities, $SA \subset J$, non-supply chain activities, $NA \subset J$ (where $SA \cup NA = J$), supply chain

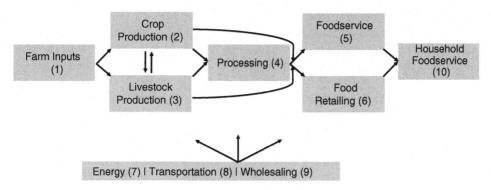

**FIGURE 9.2** Example of an agri-food supply chain.

commodities, $SC \subset I$, and non-supply chain commodities, $NC \subset I$ (where $SC \cup NC = I$). A section titled *"Mathematical Derivations of Supply Chain Modeling"* in the supplemental materials for this chapter presents the mathematical derivations of this method.

For the example in Fig. 9.2, all activities and commodities directly linked to each of the 10 supply chain stages are associated with sets $SA$ and $SC$ respectively, and all other activities and commodities are associated with sets $NA$ and $NC$ respectively. It is constructive to think of all non-supply chain enterprises, such as legal services and the chemical industry, as subcontractors such that each supply chain enterprise procures all the supply chain outputs and environmental flows required by their subcontractors.

### 9.3.4 Special topics for food system modeling

To study the use of environmental factors along food commodity supply chains, it is important that attributes of the food system that are known to drive the use of such factors as fossil fuels, freshwater withdrawals, and greenhouse gas emissions are not obscured by the representations of activities and commodities in the EIO model. For example, for many years the retail trade services depicted in the national **S** and **U** Tables for most countries worldwide have been represented by an aggregate 'retail trade' activity that provides trade services for all retail transactions. Purchases ranging from shoes, automobile tires, and fresh produce are all facilitated by a generic retail trade activity. This is problematic when, for example, one is studying energy use in the food system. The main reason being that the electricity used per dollar of product sales to maintain the quality of a grocery stores inventory of highly perishable products such as ice cream and fresh produce is far greater than that used to store shoes or automobile tires. If the EIO model doesn't capture this distinction, the analysis is going to understate the use of electricity throughout the food cold chain of a national economy. A similar scenario applies to wholesale trade services, and workarounds have been developed (e.g., see (Canning, 2011)) whereby primary data sources measuring business expenses (including electric and natural gas utility costs) are incorporated into specific commodity group measures of utility costs per dollar of retail services. More recent versions of country **S** and **U** Tables have included a breakout of wholesale and retail trade activities by broad commodity groupings, including a 'grocery and related products' grouping.

The need to address the obscuring of food supply chain activities in **S** and **U** Tables varies widely across countries and regions. However, two special topics that are universally reflected in **S** and **U** Tables merit special consideration. They are (i) the need to decouple the purchase of food and food services when representing food away from home expenditures, and (ii) the need to ensure that the treatment of imports in multiplier analysis are exclusively utilized domestically as opposed to being transshipped to other countries or regions.

By convention, retail expenditures on food are treated as several transactions in the standard EIO model: expenditures on food commodities (*food*) and simultaneous expenditures on retail and freight 'margin' services (*margin*). This treatment makes the calculation of environmental flows by supply chain stage routine. However, also by convention in the standard EIO model, when food is acquired through a foodservice establishment, the only transaction recorded is the purchase of a non-traded good (*foodservice*). In this case, the acquisition of the *food* is recorded as an intermediate transaction of the foodservice industry. The implication of this accounting practice is that multiplier analysis will fail to net out any imported *food* purchases by the *foodservice* industry. The work-around for this is to decouple the foodservice activity into two activities: 'foodservices' (*fs*) and 'food away from home' (*fafh*). This change aligns the treatment of food away from home purchases with those for food at home purchases. As demonstrated in Yi et al. (2021), where this decoupling is carried out with industry by industry input output matrices (see Eqs. 7 to 16 in Yi et al., 2021), no additional data is required and the source data is not modified, but is simply reorganized.

Another convention in **S** and **U** Table accounting is to exclude international transshipments (where shipments into a country represent an intermediate destination) from both the import and export accounts. This implies exports are entirely sourced from domestic production and imports are entirely absorbed by domestic uses. To ensure this is represented in the EIO multiplier model, an export assembly industry is added to consolidate all production for the export market. Again, no additional data is required, and the source data is not modified, but is simply reorganized (see Eqs. 17 to 20 in Yi et al., 2021).

## 9.4 Application: a U.S. food economy EIO model

To demonstrate application of the EIO model presented in the previous section, we compile an activity by commodity input output Table according to the layout reported in Fig. S1 for the 2012 annual economy of the United States. The activity and commodity aggregations we adopt are designed to facilitate the maximum food system detail while also grouping many of the non-food activities and commodities in a manner that facilitates an approach to producing annual updates to the **S** and **U** Tables using primary data sources that are annually updated from numerous data products of various U.S. Federal Government agencies. Efficient information-processing approaches for updating **S** and **U** Tables using numerous updated primary data sources is a powerful tool for EIO modeling practitioners. This topic is not covered here. However, a selected survey of relevant methods is presented in Miller and Blair (2009, see Chapters 7 and 8).

The 2012 U.S. EIO model we compile is comprised of 229 activities (A001 to A229), 231 commodities (C001 to C231), four leakage matrix elements (L01 to L04), four non-food expenditure related institutional matrix elements (X01 to X04), and 32 food commodity and home kitchen operation expenditure matrix elements (XF01 to XF32). Table S2 in the supplemental material to this chapter provides short descriptions of all 500 row and column addresses.

An accompanying environmental multiplier matrix ($\mathbf{E}$) is also compiled with non-zero data entries in most activity columns. We build this $\mathbf{E}$ matrix up from a U.S. State level environmental flows ($\mathbf{EF}$) matrix (see Table 9.2) reporting activity and institutional uses of 11 distinct primary energy commodities. We also compile a State level distribution metrics ($\mathbf{DM}$) matrix (see Table 9.3).

*Data Sources*

***Supply and Use Tables***: Three data sources are used to construct $\mathbf{S}$ and $\mathbf{U}$ Tables for the 2012 U.S. national economy: the Benchmark Tables published every five years by the United States Department of Commerce, Bureau of Economic Analysis (USDOC-BEA, 2021), annual summary $\mathbf{S}$ and $\mathbf{U}$ Tables published every year by USDOC-BEA, and the annual $\mathbf{S}$ and $\mathbf{U}$ Tables published by the U.S. Department of Labor, Bureau of Labor Statistics (USDL-BLS, 2020). More details can be found in the "Data Sources of Supply and Use Tables" section in the SI.

***Food-related personal consumption expenditures***: From Eq. S6 we see that food-related expenditures must be partitioned from all other final demand spending to carry out a food economy analysis. Our approach is to use the USDOC-BEA Table providing the National Income and Product Accounts (NIPAs) and is available in the 'underlying-estimates' section of the *Input-Output Accounts* data product (USDOC-BEA, 2021). Using various secondary data sources, we can identify expenditures and uses of goods and services for U.S. households to run their home kitchens, including travel to food retail establishments to obtain groceries. We are unable to isolate travel to foodservice establishments. A description of the data and methods to develop these kitchen operation expenditures is found in Canning et. al. (2017). A sample of the vectors for six of the 26 Tables (non-zero elements only) is depicted in Table 9.5 which is a composition of personal consumption expenditures from the NIPAs.

***Environmental Factors Data (Primary Energy Consumption)***. To demonstrate an actual EIO model application, we focus on one environmental factor category—primary energy consumption by type of energy commodity. The fictional data depicted in Table 9.2 is replicated here for the real U.S. 2012 economy. In the United States, a principal data source for primary energy consumption is the State Energy Data System (SEDS), published annually by the U.S. Department of Energy, Energy Information Administration (USDOE-EIA, 2021). SEDS reports a historical time series of energy production, consumption, prices, and expenditures by state that are defined as consistently as possible over time and across sectors for analysis and forecasting purposes. Table 9.6 reports an environmental factors Table with SEDS data covering 11 primary energy sources and 4 end user categories. The Table reports three States among the 51 regions used in this analysis, plus the U.S. totals across all 50 States plus Washington, DC.

For our purposes, we seek to assign the entire annual primary energy budget of the United States used as a source of heat or power, or as a raw material input to a manufacturing process. Use is measured in million British thermal units (mBtu). This use can be assigned to specific activities, such as a manufacturing or commercial activity, or can be assigned to a final market purpose such as the operation of home kitchens or for climate control of a private residence. As indicated by the far-right column in the bottom row of Table 9.6, the U.S. national energy budget in 2012 was roughly 94.366 quadrillion Btu's. To replicate the computations described above around Eqs. (9.1) and (9.2) with matrices compiled from the fictional data in Tables 9.2 and 9.3, we must compile an accompanying matrix of distribution metrics (the $\mathbf{DM}$ matrix) to assign state level use data reported in Table 9.5 to the detailed activity and institutional users represented in the EIO model.

**TABLE 9.5**  Select EIO model food expenditure vectors; annual 2012 personal consumption in the united states.

| | Food and beverages away from home | | | | Food at home | | | | | |
| ROW | ROW_DESC | Food | Beverages | Bakery products | Beef, pork and other meats | Fish and seafood | Fresh milk | Processed dairy products | Eggs | Fresh Fruits | Fresh vegetables |
| | | | | | | $million U.S. Dollars | | | | | |
| C002 | Grain farming | 29.0 | | | | | | | | | |
| C003 | Vegetable and melon farming | 313.0 | | | | | | | | | 16,974.0 |
| C004 | Fruit and tree nut farming | 73.0 | | | | | | | | 16,809.0 | |
| C005 | Greenhouse nursery and floriculture production | | | | | | | | | | 1,199.0 |
| C006 | Other crop farming | | | | | | | | | | 2,773.0 |
| C009 | Animal production except cattle and poultry and eggs | | | | | 132.0 | | | | | |
| C010 | Poultry and egg production | 195.0 | | | | | | | 4,757.0 | | |
| C012 | Fishing hunting and trapping | 136.0 | | | 435.0 | 2,455.0 | | | | | |
| C025 | Flour milling and malt manufacturing | 359.0 | | | | | | | | | |
| C026 | Wet corn milling | 190.0 | | | | | | | | | |
| C027 | Soybean and other oilseed processing | | | | | | | | | | |
| C028 | Fats and oils refining and blending | 440.0 | | | | | | | | | |

*(continued on next page)*

**TABLE 9.5** Select EIO model food expenditure vectors; annual 2012 personal consumption in the united states—cont'd

| | Food and beverages away from home | | | Food at home | | | | | | | |
| | Food | Beverages | Bakery products | Beef, pork and other meats | Fish and seafood | Fresh milk | Processed dairy products | Eggs | Fresh Fruits | Fresh vegetables |
|---|---|---|---|---|---|---|---|---|---|---|
| | | | | | $million U.S. Dollars | | | | | | |
| **ROW** | **ROW_DESC** | | | | | | | | | | |
| C029 | Breakfast cereal manufacturing | 223.0 | | | | | | | | | |
| C030 | Sugar and confectionery product manufacturing | 492.0 | | | | | | | | | |
| C031 | Frozen food manufacturing | 696.0 | | | | | | | | | |
| C032 | Fruit and vegetable canning pickling and drying | 731.0 | | | | | | | | | |
| C033 | Cheese manufacturing | 904.0 | | | | | | 13,361.0 | | | |
| C034 | Dry condensed and evaporated dairy product manufacturing | 252.0 | | | | | | 3,618.0 | | | |
| C035 | Ice cream and frozen dessert manufacturing | 175.0 | | | | | | 1,845.0 | | | |
| C036 | Fluid milk and butter manufacturing | 437.0 | | | | | 13,738.0 | 8,281.0 | | | |
| C037 | Animal (except poultry) slaughtering rendering and processing | 2,911.0 | | | 64,015.0 | | | | | | |
| C038 | Poultry processing | 1,422.0 | | | | | | | | | |

*(continued on next page)*

| Code | Description | | | | | | | | | |
|------|-------------|---|---|---|---|---|---|---|---|---|
| C039 | Seafood product preparation and packaging | 261.0 | | | 5,336.0 | | | | | |
| C040 | Bread and bakery product manufacturing | 1,064.0 | 41,058.0 | | | | | | | |
| C041 | Cookie cracker pasta and tortilla manufacturing | 612.0 | 8,837.0 | | | | | | | |
| C042 | Snack food manufacturing | 761.0 | | | | | | | | |
| C044 | Flavoring syrup and concentrate manufacturing | 29.0 | | | | | | | | |
| C045 | Seasoning and dressing manufacturing | 403.0 | | | | | | | | |
| C046 | All other food manufacturing | 418.0 | | | | | | 2,365.0 | | |
| C047 | Soft drink and ice manufacturing | 994.0 | | | | | | | | |
| C126 | Grocery and related product wholesalers | 1,760.9 | 7,668.7 | 5,580.7 | 398.5 | 1,560.7 | 3,197.2 | 510.0 | 3,491.6 | 3,998.8 |
| C128 | Food and beverage stores | | 16,975.9 | 21,838.1 | 1,782.7 | 4,899.1 | 9,517.2 | 2,386.6 | 3,884.7 | 5,590.2 |
| C129 | General merchandise stores | | 3,407.9 | 4,384.0 | 357.9 | 983.5 | 1,910.6 | 479.1 | 779.9 | 1,122.2 |
| C131 | Air transportation | 1.7 | 4.3 | 6.1 | 0.8 | 1.2 | 2.5 | 2.1 | 4.6 | 7.1 |
| C132 | Rail transportation | 65.8 | 169.1 | 243.8 | 30.6 | 48.8 | 98.1 | 82.1 | 180.8 | 283.7 |
| C133 | Water transportation | 6.6 | 16.9 | 24.3 | 3.1 | 4.9 | 9.8 | 8.2 | 18.0 | 28.3 |

*(continued on next page)*

**TABLE 9.5**  Select EIO model food expenditure vectors; annual 2012 personal consumption in the united states—cont'd

| | Food and beverages away from home | | Food at home | | | | | | | | |
|---|---|---|---|---|---|---|---|---|---|---|---|
| | Food | Beverages | Bakery products | Beef, pork and other meats | Fish and seafood | Fresh milk | Processed dairy products | Eggs | Fresh Fruits | Fresh vegetables |
| | | | | | $million U.S. Dollars | | | | | | |
| **ROW** | **ROW_DESC** | | | | | | | | | | |
| C134 | Truck transportation | 271.4 | | 697.3 | 1,005.4 | 126.3 | 201.3 | 404.6 | 338.5 | 745.9 | 1,170.2 |
| C203 | Service at full service restaurants | 154,532.6 | 42,112.0 | | | | | | | | |
| C204 | Service at limited service restaurants | 243,744.6 | 6,130.4 | | | | | | | | |
| C205 | Service at all other food and drinking places | 60,001.6 | 26,095.8 | | | | | | | | |
| C226 | Food at full service restaurants | 13,309.4 | 3,627.0 | | | | | | | | |
| C227 | Food at limited service restaurants | 21,812.4 | 548.6 | | | | | | | | |
| C228 | Food at all other food and drinking places | 1,697.4 | 738.2 | | | | | | | | |

Source: USDOC-BEA, 2021.

**TABLE 9.6** Primary energy use by fuel source and type of end user in U.S. states, 2012.

| | Type of End Use | | | | |
|---|---|---|---|---|---|
| | Industrial | Comercial | Transportation | Residential | Total |
| | Alabama | | | | |
| Primary fuel source | billion British thurmal units | | | | |
| All petroleum products | 77,598 | 8,752 | 445,613 | 4,326 | 536,289 |
| Biofuel | 2 | – | – | – | 2 |
| Coal | 72,938 | – | – | – | 72,938 |
| Electricity | 330,523 | 213,478 | – | 299,980 | 843,981 |
| Geothermal | 42 | – | – | 99 | 141 |
| Hydroelectric | – | – | – | – | – |
| Natural gas | 193,790 | 21,896 | 25,951 | 28,022 | 269,659 |
| Solar | – | 9 | – | 69 | 78 |
| Wind | – | – | – | – | – |
| Wood | – | – | – | 5,794 | 5,794 |
| Wood and biomass waste | 160,622 | 784 | – | – | 161,406 |
| Total | 835,515 | 244,919 | 471,564 | 338,290 | 1,890,288 |
| | California | | | | |
| All petroleum products | 411,479 | 31,630 | 2,839,268 | 23,366 | 3,305,743 |
| Biofuel | 9,526 | – | – | – | 9,526 |
| Coal | 30,679 | – | – | – | 30,679 |
| Electricity | 451,767 | 1,171,872 | 6,590 | 867,034 | 2,497,263 |
| Geothermal | 1,217 | 634 | – | 287 | 2,138 |
| Hydroelectric | – | 26 | – | – | 26 |
| Natural gas | 805,491 | 258,277 | 28,136 | 487,614 | 1,579,518 |
| Solar | 5,247 | 9,714 | – | 26,313 | 41,274 |
| Wind | 1 | – | – | – | 1 |
| Wood | – | – | – | 32,410 | 32,410 |
| Wood and biomass waste | 31,652 | 16,807 | – | – | 48,459 |
| Total | 1,747,059 | 1,488,960 | 2,873,994 | 1,437,024 | 7,547,037 |

(continued on next page)

**TABLE 9.6** Primary energy use by fuel source and type of end user in U.S. states, 2012—cont'd

| | Type of End Use | | | | |
|---|---|---|---|---|---|
| | Industrial | Comercial | Transportation | Residential | Total |
| | | | Alabama | | |
| Primary fuel source | | | | | |
| | | | billion British thurmal units | | |
| | | | Wyoming | | |
| All petroleum products | 65,966 | 6,004 | 100,675 | 2,783 | 175,428 |
| Biofuel | 600 | – | – | – | 600 |
| Coal | 31,064 | 458 | – | – | 31,522 |
| Electricity | 108,314 | 45,941 | – | 29,396 | 183,651 |
| Geothermal | 65 | 528 | – | 70 | 663 |
| Hydroelectric | – | – | – | – | – |
| Natural gas | 118,077 | 10,840 | 17,250 | 11,895 | 158,062 |
| Solar | 1 | 2 | – | 12 | 15 |
| Wind | – | – | – | – | – |
| Wood | – | – | – | 980 | 980 |
| Wood and biomass waste | 112 | 133 | – | – | 245 |
| Total | 324,199 | 63,906 | 117,925 | 45,136 | 551,166 |
| | | | United States | | |
| All petroleum products | 8,054,349 | 550,001 | 25,272,426 | 869,788 | 34,746,564 |
| Biofuel | 710,871 | – | – | – | 710,871 |
| Coal | 1,516,013 | 43,650 | – | – | 1,559,663 |
| Electricity | 10,287,524 | 13,640,682 | 73,842 | 14,149,504 | 38,151,552 |
| Geothermal | 4,200 | 19,702 | – | 39,600 | 63,502 |
| Hydroelectric | 22,393 | 261 | – | – | 22,654 |
| Natural gas | 8,822,590 | 2,968,401 | 781,762 | 4,252,794 | 16,825,547 |
| Solar | 7,196 | 33,300 | – | 78,844 | 119,340 |
| Wind | 182 | 513 | – | – | 695 |
| Wood | – | – | – | 438,094 | 438,094 |
| Wood and biomass waste | 1,621,149 | 105,929 | – | – | 1,727,078 |
| Total | 31,046,467 | 17,362,439 | 26,128,030 | 19,828,624 | **94,365,560** |

Source: USDOE-EIA, 2021.

*Multiplier Analysis of the Baseline U.S. Diet*

In the United States, virtually all food acquisition for domestic consumption is achieved through market transactions, ranging from grocery and convenience store purchases, fast and full-service restaurants, food trucks, various unconventional foodservice and food retail points of purchase, and through farmers markets. Each of these points of food acquisition are represented in the list of food and beverage expenditures categories of the EIO model we compiled, as depicted in Table 9.7, including food produced and consumed on farms and food purchased as a business expense. With the notable exception of food consumption from residential home gardens, total food acquisition to accommodate all domestic annual diets in the United States can undergo an analysis of energy use along the different stages of the commodity supply chains depicted in Fig. 9.2 with the EIO model described in this section and the computations outlined in Eq. (S11). This analysis can occur across the 23 food and beverage categories listed in Table 9.7. Further, for total food at home consumption the energy flows linked to the operations of home kitchens, including travel to points of food purchases, can also be analyzed across the five categories, $x^{f24}$ to $\sigma^{f28}$, also listed in Table 9.7.

Here, we restate the computations of Eq. (S11) from the supplemental material in the context of our application to the 2012 U.S. economy using the EIO model described in this section:

$$\sum{}^{f} = \left[ \mathbf{E}^* \times (\mathbf{y}^{f1}|\mathbf{y}^{f2}|...|\mathbf{y}^{f23})|\mathbf{E} \times (\mathbf{y}^{f24}|\mathbf{y}^{f25}|\mathbf{y}^{f26})|\sigma^{f27}|\sigma^{f28} \right] \tag{9.4}$$

where $\mathbf{E}^*$ is the special supply chain complied environmental multiplier (Eq. (S10) from the supplemental material), $\mathbf{E}$ indicates the conventional environmental multiplier in Eq. (9.2), $\mathbf{y}$ is the gross output, and $\sigma$ represents the environmental matrices final demand vector measured in physical units. In Eq. (9.4), final demand vectors $x^{f1}$ to $x^{f23}$ comprise all food expenditures that are subject to the special supply chain compiled environmental matrix ($\mathbf{E}^*$). Final demand vectors $x^{f24}$ to $x^{f26}$ comprise the acquisition of non-energy commodities for running home kitchens that are subject to the conventional environmental matrix ($\mathbf{E}$). Final demand vectors $\sigma^{f27}$ and $\sigma^{f28}$ comprise direct household purchases of energy commodities for home kitchen operations.

The results of this computation are summarized across a selection of the food, beverage, and food-related categories in Table 9.8 . The embodied energy use of the total diet, from farm to fork, is depicted in Fig. 9.3 for the overall food-related energy budget.

Our goal in this chapter is to explain and demonstrate the method of EIO food system analysis. Our findings in this section merit an extensive discussion, but this is beyond the scope of this chapter. However, several questions stand out in viewing these sample results which point to the value of extending the basic analysis. In the final section of this chapter, we highlight one of these extensions: how might energy use differ among subpopulations whose diets differ from the national average?

## 9.5 Extensions: linear programming and comparative diets analysis

A question often posed by researchers in the field of sustainable nutrition is whether healthier diet outcomes and resource conservation outcomes are synergistic or competing goals. In the context of EIO analysis, we rely on mathematical optimization models to formulate healthy diet scenarios. For example, suppose the complete baseline diet behind the outcome depicted

**TABLE 9.7** Food and food-related final demand expenditure categories.

| Final Demand Vector | Description |
| --- | --- |
| $x^{f1}$ | Food away from home |
| $x^{f2}$ | Beverages away from home |
| $x^{f3}$ | Food at home: Cereals |
| $x^{f4}$ | Food at home: Bakery products |
| $x^{f5}$ | Food at home: Beef, pork and other meats |
| $x^{f6}$ | Food at home: Poultry |
| $x^{f7}$ | Food at home: Fish and seafood |
| $x^{8}$ | Food at home: Fresh milk |
| $x^{f9}$ | Food at home: Processed dairy products |
| $x^{f10}$ | Food at home: Eggs |
| $x^{f11}$ | Food at home: Fats and oils |
| $x^{f12}$ | Food at home: Fresh Fruits |
| $x^{f13}$ | Food at home: Fresh vegetables |
| $x^{f14}$ | Food at home: Processed fruits and vegetables |
| $x^{f15}$ | Food at home: Sugar and sweets |
| $x^{f16}$ | Food at home: Other foods |
| $x^{f17}$ | Beverages at home: Nonalcoholic beverages |
| $x^{f18}$ | Food at home: salt and chemical additives |
| $x^{f19}$ | Food at home: Consumed on farms |
| $x^{f20}$ | Beverages at home: Beer |
| $x^{f21}$ | Beverages at home: Wine |
| $x^{f22}$ | Beverages at home: Spirits |
| $x^{f23}$ | Food at work: Vouchers |
| $x^{f24}$ | Home kitchen operations: fleet |
| $x^{f25}$ | Home kitchen operations: appliances |
| $x^{f26}$ | Home kitchen operations: equipment and supplies |
| $\sigma^{f27}$ | Home kitchen operations: Utilities |
| $\sigma^{f28}$ | Household transportation (grocery): petroleum use |

**TABLE 9.8** Annual U.S. food system energy use by supply chain stage, energy source, and selected food commodity groups or activity, 2012.

| | Renewable Energy | Coal | Electricity | Natural Gas | Petroleum | Total |
|---|---|---|---|---|---|---|
| | *million Btu* | | | | | |
| **Supply Chain Stage** | *Food away from home* | | | | | |
| Total | 50,119,982 | 35,575,537 | 1,598,549,814 | 286,835,054 | 242,571,786 | 2,213,652,172 |
| Agribusiness | 11,185,440 | 4,254,190 | 25,333,546 | 44,221,729 | 25,885,030 | 110,879,935 |
| Crops | 4,897,426 | – | 3,323,173 | 10,367,367 | 7,871,663 | 26,459,629 |
| Livestock | 4,327,447 | 3,462,969 | 9,153,499 | 13,937,083 | 11,076,972 | 41,957,970 |
| Other ag forestry fisheries | 1,119,400 | 847 | 608,420 | 1,109,765 | 2,290,133 | 5,128,566 |
| Food processing | 11,415,187 | 11,097,922 | 89,573,133 | 72,348,413 | 16,726,776 | 201,161,430 |
| Transportation and storage | 708,170 | 683,379 | 8,467,439 | 6,759,543 | 114,520,605 | 131,139,136 |
| Food wholesalers | 256,355 | 163,777 | 11,606,022 | 1,504,722 | 1,974,351 | 15,505,227 |
| Food retailers | 598,321 | 414,242 | 61,968,834 | 3,457,042 | 2,099,281 | 68,537,719 |
| Foodservice | 15,612,234 | 15,498,211 | 1,388,515,749 | 133,129,390 | 60,126,976 | 1,612,882,561 |
| | *Food at home: Bakery products* | | | | | |
| Total | 30,277,882 | 14,297,969 | 239,826,884 | 121,970,878 | 91,957,563 | 498,331,177 |
| Agribusiness | 6,956,247 | 997,703 | 11,529,837 | 30,228,083 | 12,623,631 | 62,335,500 |
| Crops | 5,469,204 | – | 2,342,333 | 11,700,900 | 8,276,157 | 27,788,594 |
| Livestock | 433,416 | 654,911 | 1,159,382 | 1,745,777 | 1,233,994 | 5,227,482 |
| Other ag forestry fisheries | 486,614 | 208 | 162,231 | 154,157 | 166,489 | 969,699 |
| Food processing | 15,361,526 | 11,464,271 | 107,857,455 | 67,950,751 | 19,888,063 | 222,522,066 |
| Transportation and storage | 306,509 | 304,190 | 3,768,645 | 2,723,974 | 43,883,953 | 50,987,271 |
| Food wholesalers | 330,502 | 211,146 | 14,962,855 | 1,939,935 | 2,545,397 | 19,989,835 |
| Food retailers | 915,091 | 647,902 | 96,543,328 | 5,375,624 | 3,271,842 | 106,753,787 |
| Foodservice | 18,773 | 17,638 | 1,500,817 | 151,677 | 68,036 | 1,756,942 |
| | *Food at home: Beef, pork and other meats* | | | | | |
| Total | 65,112,875 | 13,035,827 | 238,866,130 | 212,742,653 | 226,256,911 | 756,014,396 |
| Agribusiness | 28,703,525 | 2,038,418 | 41,641,251 | 88,689,885 | 63,057,178 | 224,130,258 |
| Crops | 4,842,864 | – | 1,968,440 | 10,293,383 | 7,336,751 | 24,441,438 |
| Livestock | 21,591,134 | 1,019,248 | 17,553,529 | 52,929,198 | 47,310,685 | 140,403,794 |
| Other ag forestry fisheries | 939,219 | 136 | 385,343 | 415,911 | 625,308 | 2,365,917 |
| Food processing | 7,414,763 | 8,707,777 | 70,712,124 | 48,785,076 | 5,920,326 | 141,540,065 |

(continued on next page)

**TABLE 9.8** Annual U.S. food system energy use by supply chain stage, energy source, and selected food commodity groups or activity, 2012—cont'd

| | Renewable Energy | Coal | Electricity | Natural Gas | Petroleum | Total |
|---|---|---|---|---|---|---|
| | | | *million Btu* | | | |
| **Supply Chain Stage** | | | *Food away from home* | | | |
| Transportation and storage | 521,342 | 502,408 | 6,325,895 | 5,110,727 | 96,992,311 | 109,452,681 |
| Food wholesalers | 259,150 | 165,562 | 11,732,521 | 1,521,122 | 1,995,871 | 15,674,225 |
| Food retailers | 822,055 | 584,639 | 87,048,958 | 4,845,145 | 2,950,311 | 96,251,108 |
| Foodservice | 18,825 | 17,640 | 1,498,069 | 152,206 | 68,170 | 1,754,910 |
| | | | *Home kitchen operations* | | | |
| Fleet | 2,448,154 | 3,266,374 | 30,520,569 | 11,212,102 | 12,732,150 | 60,179,348 |
| Appliances | 3,884,861 | 2,708,431 | 38,267,203 | 9,697,918 | 12,893,167 | 67,451,581 |
| Kitchenware | 7,471,963 | 7,461,853 | 113,402,201 | 43,422,079 | 36,168,084 | 207,926,180 |
| Utilities for kitchen | 110,951,078 | – | 2,820,836,533 | 202,435,236 | (725,956) | 3,133,496,891 |
| Petroleum for grocery trips | – | – | – | – | 524,455,924 | 524,455,924 |

below in Fig. 9.3 (ignoring the kitchen operation stages) is denoted $x^{f0}$. Further, assuming that the per capita nutritional content of this baseline diet is not aligned with the set of USDA dietary guidelines, we seek to find an alternative diet outcome that does align with those guidelines, $x^{f1}$, and that minimizes the change in current dietary patterns. In other words, we are creating a hypothetical healthy diet scenario that aligns with prevailing dietary guidelines and is as close as possible to how Americans are currently eating (Canning et al., 2017).

A straightforward linear programming representation of this problem is stated as follows:

$$\text{Min Z1} = \left(x^{f1} - x^{f0}\right) \times \left(x^{f1} - x^{f0}\right) \tag{9.5}$$

Subject to

a) $N \times x^{f1} \geq n^+$ (dietary goals constraints)
b) $N \times x^{f1} \leq n^-$ (dietary limit constraints)
c) $x^{f0}, x^{f1} \geq 0$ (non-negative consumption constraint)

In this example, the matrix '$N$' is developed in such a way to translate elements of the food consumption vector into an array of nutrition outcomes, and the $n^+$ and $n^-$ vectors represent the acceptable nutrition outcomes according to the dietary guidelines begin considered. Should baseline diets ($x^{f0}$) already adhere to all dietary guidelines, then clearly the optimal solution to the minimum change objective function in Eq. (9.5) is to keep the baseline diet. The purpose of this minimum change objective is to recognize that baseline diets reflect current tastes and preferences, so if consumers seek to better align their food choices, minimizing

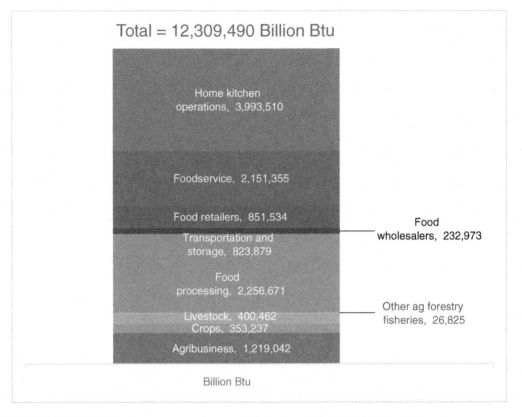

**FIGURE 9.3**   Total food system energy use by supply chain stage in the United States, 2012.

the change increases the likelihood that the solution healthy diet will be palatable to food consumers.

Examples of EIO analysis of this type include recent studies by Hitaj, et. al. (2019), and Canning et al. (2020). In Canning et al. (2020), variations of the objective function in Eq. (9.5) included budget constraints that ensure costs of the healthy diet alternatives do not exceed food budgets of observed diets. Additionally, in Hitaj et al. (2019), alterative healthy diets are considered, such as healthy vegetarian diets, and minimum resource use diets that disregard the minimum change objective and replace this with a minimum resource use (and perhaps less palatable) healthy diet outcome.

Fig. 9.4 reproduces findings from Canning et al. (2020). It depicts an EIO application of a comparative diet analysis like the one described in Eq. (9.5), and measures how a minimum change from average baseline diets of Americans in 2007 to meet the 2010 USDA Dietary Guidelines or Americans (USDA-HHS, 2010) would have both synergistic and competing outcomes among human nutrition and resource conservation/sustainability goals. Specifically, a shift to healthier American diets can lead to reduced fossil fuel consumption, land use, and forest product use, while having little or no change on greenhouse gas emissions and leading to substantial increases in freshwater withdrawals.

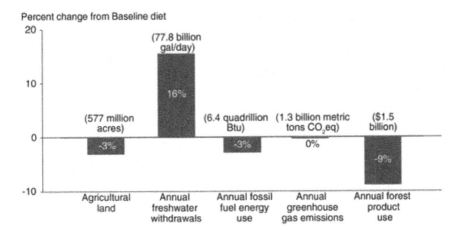

Notes: Btu = British thermal units; $CO_2$eq = carbon dioxide equivalent. Baseline diet values are in parentheses above the bars. Freshwater withdrawals refer to freshwater removed from the ground or diverted from a surface-water source.
Source: USDA, Economic Research Service.

**FIGURE 9.4**  Freshwater use could be affected the most by a shift from the baseline diet to the healthy american diet.

## 9.6 Conclusion

In this chapter, we discuss the background and methodology of EIO models for food systems research. A simplified fictional economy is used throughout this chapter to assist modelers in following up the technical details and confirming their understanding. Additionally, in this chapter we present an application of the U.S. food economy EIO model. We conclude this chapter by providing a research example of a comparative diets study that measures the relative use of five different environmental factors between the baseline diet and a representative healthy American diet.

The EIO models discussed in this chapter can be applied to measure both direct and indirect environmental impacts by economic sector, supply chain stage, and industrial group. It provides important opportunities for models to measure the demand for natural resources required by food systems. It can also be linked to other models to provide a more comprehensive analysis as we discussed in the extension section.

The EIO models have several limitations that should be considered by researchers. Four important assumptions are discussed in the section titled *"Adapting the EIO Framework for Modeling Food Systems"*. Secondly, the Type I EIO model does not incorporate the induced environmental impact which is also explained in the above-mentioned section. Researchers should examine the four assumptions and the limitations before developing an EIO model.

Lastly, identifying the appropriate data sources of supply and use Tables, food-related personal consumption expenditures, and environmental factors and developing the necessary matrices could be challenging. Moreover, to comply with the confidentiality law, there are often suppressions in publicly available data, and it increases the difficulty of compiling the measures. Fortunately, new data imputation techniques have opened more opportunities for addressing these limitations.

## Supplementary materials

Supplementary material related to this chapter can be found on the accompanying CD or online at doi:10.1016/B978-0-12-822112-9.00014-X.

## References

Avelino, A.F.T., Dall'erba, S., 2020. What Factors Drive the Changes in Water Withdrawals in the U.S. Agriculture and Food Manufacturing Industries between 1995 and 2010? Environ. Sci. Technol. 54 (17), 10421–10434.

Behrens, P., Jong, JeCK-d, Bosker, T., Rodrigues, J.F.D., Koning, A.D., Tukker, A., 2017. Evaluating the environmental impacts of dietary recommendations. Proc. Natl Acad. Sci. 114 (51), 13412–13417.

Boehm, R., Wilde, P.E., Ver Ploeg, M., Costello, C., Cash, S.B., 2018. A Comprehensive Life Cycle Assessment of Greenhouse Gas Emissions from U.S. Household Food Choices. Food Policy 79, 67–76.

Bruckner, M., Wood, R., Moran, D., Kuschnig, N., Wieland, H., Maus, V., et al., 2019. FABIO—The construction of the food and agriculture biomass input–output model. Environ. Sci. Technol. 53 (19), 11302–11312.

Camanzi, L., Alikadic, A., Compagnoni, L., Merloni, E., 2017. The impact of greenhouse gas emissions in the EU food chain: a quantitative and economic assessment using an environmentally extended input-output approach. J Clean Prod 157, 168–176.

Canning, P., 2011. A Revised and Expanded Food Dollar Series: A Better Understanding of Our Food Costs. U.S. Department of Agriculture, Economic Research Service, Economic Research Report, ERR-114.

Canning, P., Charles, A., Huang, S., Polenske, K.R., 2010. Energy Use in the Food System. U.S. Department of Agriculture, Economic Research Service, Economic Research Report, ERR-94.

Canning, P., Rehkamp, S., Waters, A., Etemadnia, H., 2017. *The Role of Fossil Fuels in the U.S. Food System and the American Diet*, ERR-224. U.S. Department of Agriculture, Economic Research Service.

Canning, P., Morrison, R.M., 2020. A Shift to Healthier Diets Likely To Affect Use of Natural Resources. Amber Waves: The Economics of Food, Farming, Natural Resources, and Rural America 2020(1490-2020-1032).

Cazcarro, I., Duarte, R., Sánchez-Chóliz, J., 2012. Water flows in the Spanish economy: agri-food sectors, trade and households diets in an input-output framework. Environ. Sci. Technol. 46 (12), 6530–6538.

Costello, C., Griffin, W.M., Matthews, H.S., Weber, C.L., 2011. Inventory development and input-output model of US land use: relating land in production to consumption. Environ. Sci. Technol. 45 (11), 4937–4943.

Egilmez, G., Kucukvar, M., Tatari, O., Bhutta, M.K.S., 2014. Supply Chain Sustainability Assessment of the U.S. Food Manufacturing Sectors: a Life Cycle-Based Frontier Approach. Resour. Conserv. Recycl. 82, 8–20.

Hendrie, G.A., Ridoutt, B.G., Wiedmann, T.O., Noakes, M., 2014. Greenhouse gas emissions and the Australian diet—Comparing dietary recommendations with average intakes. Nutrients 6 (1), 289–303.

Hirst, E., 1974. Food-related energy requirements. Science 184 (4133), 134–138.

Hitaj, C., Rehkamp, S., Canning, P., Peters, C.J., 2019. Greenhouse Gas Emissions in the U.S. Food System: current and Healthy Diet Scenarios. Environ. Sci. Technol..

Ivanova, D., Stadler, K., Steen-Olsen, K., Wood, R., Vita, G., Tukker, A., et al., 2016. Environmental impact assessment of household consumption. J. Ind. Ecol. 20 (3), 526–536.

Lenzen, M., Foran, B., 2001. An Input-Output Analysis of Australian Water Usage. Water Policy (3) 321–340.

Leontief, W., 1967. An alternative to aggregation in input-output analysis and national accounts. Rev. Eco. Stat. 49 (3), 412–419.

Miller, R.E., Blair, P.E., 2009. Input-Output Analysis: Foundations and Extensions, Second Edition. Cambridge University Press, New York.

Mulik, K., O'Hara, J.K., 2015. Cropland implications of healthier diets in the United States. J. Hunger. Environ. Nutr. 10 (1), 115–131.

Nicholson, W., Snyder, C., 2012. Microeconomic Theory: Basic Principles and Extensions, 11th Edition South-Western, Cengage Learning, USA.

(OECD) Organisation for Economic Co-operation and Development, 2008. OECD Glossary of Statistical Terms. OECD Publishing, Paris https://doi.org/10.1787/9789264055087-en.

Rehkamp, S., Canning, P., 2018. Measuring Embodied Blue Water in American Diets: an EIO Supply Chain Approach. Ecol. Econ. 147, 179–188.

Rehkamp, S., Canning P., and Birney C. (2021). Tracking the U.S. Domestic Food Supply Chain's Freshwater Use Over Time, Economic Research Report, ERR-288, U.S. Department of Agriculture, Economic Research Service.

Reynolds, C.J., Piantadosi, J., Buckley, J.D., Weinstein, P., Boland, J., 2015. Evaluation of the environmental impact of weekly food consumption in different socio-economic households in Australia using environmentally extended input–output analysis. Ecol. Econ. 111, 58–64.

Scherer, L., Pfister, S., 2016. Global biodiversity loss by freshwater consumption and eutrophication from Swiss food consumption. Environ. Sci. Technol. 50 (13), 7019–7028.

Sherwood, J., Clabeaux, R., Carbajales-Dale, M., 2017. An Extended Environmental Input-Output Lifecycle Assessment Model to Study the Urban Food-Energy-Water Nexus. Environ. Res. Lett. 12.

Tukker, A., Goldbohm, R.A., De Koning, A., Verheijden, M., Kleijn, R., Wolf, O., et al., 2011. Environmental impacts of changes to healthier diets in Europe. Ecol. Econ. 70 (10), 1776–1788.

(UN) United Nations, 2018. Handbook on Supply and Use Tables and Input-Output Tables with Extensions and Applications. New York Series F, No. 74, Rev. 1.

United Nations, 2014. European Commission, International Monetary Fund. Organisation for Economic Co-operation and Development, and World Bank *System of environmental-economic accounting 2012—Central framework*, Series F, No. 109 (ST/ESA/STAT/Ser.F/109).

(USDA-NASS) U.S. Department of Agriculture, National Agricultural Statistics Service, 2021. Census of Agriculture, 2017 Publications Web Data product last updated January 22, 2021.

(USDA-HHS) U.S. Department of Agriculture, U.S. Department of Health and Human Services, 2010. Dietary Guidelines for Americans, 7th edition U.S. Government Printing Office, Washington, DC.

(USDOC-BEA) U.S. Department of Commerce, Bureau of Economic Analysis, 2021. Input-Output Accounts Data Webpage last updated April 2, 2021.

(USDOC-CB) U.S. Department of Commerce, Census Bureau, 2021. Economic Census, Tables: 2017 Webpage last accessed July 1, 2021.

(USDOE-EIA) U.S. Department of Energy, Energy Information Administration. 2021. State Energy Data System (SEDS: 1960-2019 (complete). Webpage last updated June 25, 2021.

(USDOL-BLS) U.S. Department of Labor, Bureau of Labor Statistics. 2020. Inter-industry relationships (Input-Output matrix). Webpage last updated September 4, 2020.

Xiao, Z., Yao, M., Tang, X., Sun, L., 2019. Identifying critical supply chains: an input-output analysis for Food-Energy-Water Nexus in China. Ecol. Modell. 392, 31–37.

Virtanen, Y., Kurppa, S., Saarinen, M., Katajajuuri, J.-.M., Usva, K., Mäenpää, I., et al., 2011. Carbon footprint of food–approaches from national input–output statistics and a LCA of a food portion. J Clean Prod 19 (16), 1849–1856.

Weber, C., Matthews H.S., 2008. Food-miles and the relative climate impacts of food choices in the United States. 42 (10): 3508-3513.

Weinzettel, J., Hertwich, E.G., Peters, G.P., Steen-Olsen, K., Galli, A., 2013. Affluence drives the global displacement of land use. Global Environ. Change 23 (2), 433–438.

Wolf, O., Pérez-Domínguez, I., Rueda-Cantuche, J.M., Tukker, A., Kleijn, R., de Koning, A., et al., 2011. Do healthy diets in Europe matter to the environment? A quantitative analysis. Journal of Policy Modeling 33 (1), 8–28.

Yi, J., Meemken, E.M., Mazariegos-Anastassiou, V., et al., 2021. Post-farmgate food value chains make up most of consumer food expenditures globally. Nature Food **2**, 417–425. https://doi.org/10.1038/s43016-021-00279-9.

# Modeling biophysical and socioeconomic interactions in food systems with the International Model for Policy Analysis of Agricultural Commodities and Trade (IMPACT)

*Keith Wiebe[a], Timothy B. Sulser[a], Shahnila Dunston[a], Richard Robertson[a], Mark Rosegrant[a] and Dirk Willenbockel[b]*

[a] International Food Policy Research Institute, Washington, DC 20005, USA
[b] Institute of Development Studies, University of Sussex, Brighton, BN1 9RE, UK

## 10.1 Food system challenges

Food systems have generated remarkable achievements over the past half century. Yields and production of basic staple food crops have roughly tripled, driven by genetic innovation and improvements in farming practices (Fischer, Byerlee, and Edmeades, 2014). Continued growth in agricultural productivity in most of the world today is driven primarily by improvements in knowledge and efficiency rather than increases in input use (Fuglie, Gautam, Goyal, and Maloney, 2019). The share of the world's population that is hungry has declined to around 9 percent in 2019 (FAO, IFAD, UNICEF, WFP, WHO, 2020). Rising incomes and increasingly efficient value chains mean that a dazzling abundance and variety of foods are available in

*Food Systems Modelling.*
DOI: https://doi.org/10.1016/B978-0-12-822112-9.00008-4

supermarkets around the world today, including in developing countries (Reardon, Timmer, and Minten, 2012).

Nevertheless, food systems today contribute to and are confronted by an all-too-familiar and daunting array of challenges. Poverty prevents hundreds of millions of people from being able to afford nutritious diets. The number of people at risk of hunger remains unacceptably high and has risen by nearly 60 million over the past five years to 690 million in 2019 (FAO, IFAD, UNICEF, WFP, WHO, 2020). At least 2 billion people suffer from micronutrient deficiencies, and 2 billion more are overweight or obese, in developing and developed countries alike (Ingram, 2020). Pressure on land and water resources is increasing. Food safety concerns are rising, and COVID-19 highlights the risks of widespread health and socioeconomic consequences of zoonotic diseases originating in ever-more intensive and tightly interlinked food systems. Finally, food systems are major contributors to greenhouse gas emissions, and climate change is beginning to undermine the natural resource foundation on which agriculture and food systems fundamentally depend.

Each of these challenges is difficult and complex, and deservedly the subject of focused and detailed analysis in its own right. Other chapters of this book explore some of these challenges in greater detail. But links and feedbacks among the various challenges are also critical – across space, over time, and across a variety of biophysical and socioeconomic dimensions. These interactions may dampen the impact of shocks originating elsewhere in a system, or amplify them, often in unpredictable ways. Given the analytical tools available today, it is not possible to capture all this richness of breadth and depth simultaneously, so we necessarily face a tradeoff when choosing our level of focus and scope. Different approaches are complementary. In this chapter, we focus on a relatively macro scale using as an example the International Model for Policy Analysis of Agricultural Commodities and Trade (IMPACT). IMPACT is an integrated system of modeling tools that allows us to explore interactions between biophysical and socioeconomic drivers and impacts in food systems at national, regional and global levels, over a period of several decades. Better understanding of these interactions and impacts can help food producers, consumers, businesses, and policymakers make better-informed decisions today that will help steer food systems towards better social, economic, nutritional and environmental outcomes in the future.

## 10.2 How do we think about the future?

Thinking about the future comes naturally to us as humans, and we do it every day. Individual decisions about what to wear or how to get to work may depend on uncertain weather or traffic conditions, but are so routine as to become second nature. Decisions about food systems are of a completely different order of magnitude, and fundamentally more complex, not just because of their scale but precisely because of the types of interactions and feedback effects alluded to above (Barrett et al., 2021). Such decisions require more formal systematic foresight analysis, in which scenario development and modeling can play a key role (Wiebe and Prager, 2021).

Regardless of the level of complexity of a system or decision, we can think about the future in several basic ways. One option is to assume that the future will be much like the past. This may be a reasonable guide when underlying conditions are changing little or not

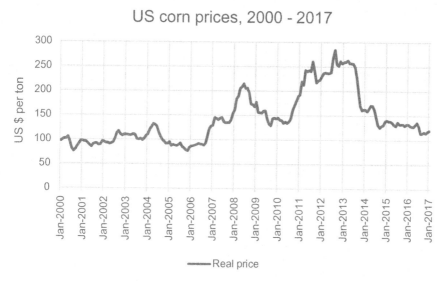

FIGURE 10.1   US corn prices, 2008–2017. Note: Prices are reported in real (inflation-adjusted) terms. Source: The authors, based on data from OECD-FAO (2017).

at all, or if we are looking only into the very near future. But what historical period is the appropriate reference? There is no single measure of food system performance, but let's say we are interested in food commodity prices as an indicator of food sector revenue or food availability and access, for example (see Fig. 10.1). We might look at recent trends and conclude that corn prices were declining (since 2013), or rising (since 2000), or fluctuating with no clear trend (over the past decade). In this case, our choice of the relevant reference period would give us quite different expectations of how prices might change in the next few years.

Looking at an even longer historical period clarifies some questions and raises others (see Fig. 10.2). Large fluctuations in the US price for corn since the early 19th century show the effects of world wars, economic depression, and other geo-political events. Even so, the underlying downward trend over the past century is clear, driven by fundamental changes in demographics, income and technology. The seemingly large fluctuations over the past decade that were apparent in the previous Fig. are now seen as relatively small variations around a clear longer-term trend. Given this context, if we are wondering how prices will change in the future, we need to distinguish whether we are asking about the next few years to a decade (in which case we would need to consider drivers of near-to-medium-term fluctuations), or whether we are asking about the next few decades (in which case we would need to consider the drivers of longer-term trends, such as changes in demographics, income, technology, and increasingly, climate). In the particular case of corn, for example, changes in those longer-term drivers will determine whether the trend in prices continues to decline, levels off, or changes direction and begins to rise in the coming decades.

Using the past as a guide is often the easiest and sometimes the best guide to the future, at least the near future. But in some cases, the past is not a reliable guide, particularly if we are looking more than a few years into the future in a complex and changing environment.

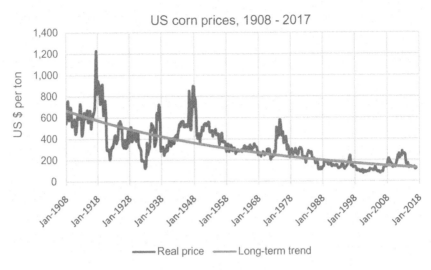

**FIGURE 10.2** US corn prices, 1908 - 2017. Note: Prices are reported in real (inflation-adjusted) terms. Source: The authors, based on data from OECD-FAO (2017).

Climate change presents such a case. The Earth's climate has changed drastically over its 4.5-billion-year history, based on changes in geology and orbital dynamics, but it is striking to note that global average temperatures have been relatively steady within a range of about 1 °Celsius over the past 10,000 years (Schellnhuber, Rahmstorf, & Winkelmann, 2016). Not coincidentally, this timeframe closely matches the entire period during which humans have domesticated plants and animals, settled on farms and then in villages and towns and cities, developed writing, expanded commerce and technology and industry, and in general developed into the cultures and societies we know today.

Over the past century and a half, however, average global land surface temperatures have risen about 1.5 °Celsius above pre-industrial levels, and the scientific consensus is that these increases will continue at least for the next several decades, even if ambitious mitigation efforts are successful, given inertia in the climate system based on greenhouse gases already emitted (IPCC, 2019). This gives rise to growing concerns both about long-term changes in temperature and precipitation as well as nearer-term increases in the frequency and severity of extreme events such as hurricanes, floods, and droughts.

Based on these changes and prospects, it is reasonable to question how useful the past can continue to be in guiding decisions about the future of food systems. The insurance sector provides an interesting and important example, since it is their core business to carefully assess the likelihood and magnitude of risks. Traditionally they have done so with rigorous analysis of historical data, but that same historical data may no longer adequately capture the risks that they will face in the future. Indeed, to explore that question, the insurance firm Lloyds recently commissioned an analysis of the potential impacts and costs associated with the simultaneous occurrence of multiple climate-related breadbasket failures and other disruptions to global trade, drawing on simulations using the IMPACT model. They concluded that

"There is little doubt that a systemic production shock to the world's most important food crops as described in this scenario would generate a cascade of economic, political and social impacts. What is striking about the scenario is that the probability of occurrence is estimated as significantly higher than the benchmark return period of 1:200 years applied for assessing insurers' ability to pay claims against extreme events." (Lloyds, 2015)

The combination of uncertainty, complexity, changing underlying conditions, and potentially significant impacts is what makes it important to systematically examine alternative future scenarios and explore ways to achieve (or avoid) them. This clearly applies to the case of food systems.

Numerous modeling tools have been developed to explore different aspects of food systems through simulation of alternative future scenarios. These include models that focus on climate, water, land, crops, livestock, fish, biodiversity, pests and diseases, nutrition, health, demographics, and economics. Within the subset of economic models, models vary in spatial scale (including household, community, national, regional, and global), in temporal scale (including historic, near-term future, medium-term future, and long-term future), and sectoral scope (including general equilibrium models of the whole economy and partial equilibrium models of agriculture and food systems); see for example Nicholson and Gomez (Chapter 7 in this volume), Jablonski et al. (Chapter 8 in this volume), and Canning, Rehkamp and Yi (Chapter 9 in this volume). We focus in this chapter on IMPACT, a partial equilibrium model of agriculture and food systems over the medium-to-longer term at national, regional and global scales.

## 10.3 The IMPACT model

This section briefly summarizes the development and evolution of the IMPACT model over the past 30 years and its structure today, drawing on full documentation of the model that has been published previously in Robinson et al. (2015); Rosegrant & IMPACT Development Team, 2012; and Willenbockel et al., 2018. Interested readers can find more technical detail in those sources, which are freely available online.

### 10.3.1 Development and evolution of the model

The International Model for Policy Analysis of Agricultural Commodities and Trade (IMPACT) was developed at the International Food Policy Research Institute (IFPRI) at the beginning of the 1990s to do medium- to long-term scenario analysis. The first results using IMPACT were published in a 1995 report on *Global Food Projections to 2020: Implications for Investment* (Rosegrant, Agcaoili-Sombilla, and Perez, 1995), which analyzed the effects of population, investment, and trade scenarios on food security and nutrition status, especially in developing countries.

Since 1995 IMPACT has gone through a process of constant expansion and improvement to add new components and modules that better capture the inter-relationships between people, natural resources, markets, and outcome indicators. First, water and aquaculture were added in the first half of the 2000s. The full integration of an explicit water module in the modeling framework was an important innovation and was a focus of several IMPACT studies investigating how water demand and availability affect future food production, demand, and trade. The water model consists of three separate modules: (1) a global hydrology model (focusing on rainfall and runoff), (2) water basin management models (which allocate water between agriculture and other uses), and (3) water stress models (which determine the impact of changes in water supply on crop yields). Later, links were added to food security modules to estimate changes in the number of undernourished children, and to crop models to allow for systematic analysis of climate change impacts on agricultural productivity and food security. New features currently in development include improved treatment of land use, livestock, fish, and links to nutrition and health.

Improving availability of data and greater computing capacity has allowed for increasing coverage of commodity markets, expanding from the original 17 commodities and 35 countries to the current 62 commodities and 158 countries. The model has increased not only the breadth of coverage but also the depth of the commodity markets. For example, the first version started with two aggregate processed commodities: food oils and oil meals. IMPACT 3 (the latest version) now simulates six oilseed complexes (groundnut, palm, rapeseed, soybean, sunflower, and other oilseeds). The model also includes the value chain for livestock, from feed grains to dressed meat and dairy.

## 10.3.2 The IMPACT model today

IMPACT consists of a core multimarket economic model linked to a variety of other models (see Fig. 10.3). The multimarket model simulates the operation of national and international commodity markets, solving for production, demand, and prices that equate supply and demand across the globe. The multimarket model is linked to several modules that include climate models, water models (hydrology, water basin management, and water stress models), crop simulation models (for example, Decision Support System for Agrotechnology Transfer [DSSAT]), value chain models (for example, sugar, oils, livestock), land use models (pixel-level land use, cropping patterns by regions), nutrition and health models, and welfare analysis. The IMPACT model system integrates information flows among the component modules in a consistent equilibrium framework that supports longer-term scenario analysis. Some of the model communication is one way, with no feedback links (for example, climate scenarios to hydrology models to crop simulation models), while other links require capturing feedback loops (for example, water demand from the core multimarket model and water supply from the water models must be reconciled to estimate water stress impacts on crop yields). IMPACT's modular structure supports integrated analysis of complex biophysical, and socioeconomic trends and phenomena of interest to policymakers.

IMPACT simulates the operation of commodity markets and the behavior of producers and consumers who determine supply and demand for agricultural commodities in those markets. In particular, IMPACT provides a detailed specification of production technology and factors affecting productivity (for example, water shortages and changes in temperature). It is a partial equilibrium (PE) model in that it deals only with agricultural commodities and so covers only a share of overall economic activity, while allowing for linkages with other sectors. Computable general equilibrium (CGE) models, another class of long-run simulation models, cover the entire economy and hence are "complete" in the sense that they specify all economic flows

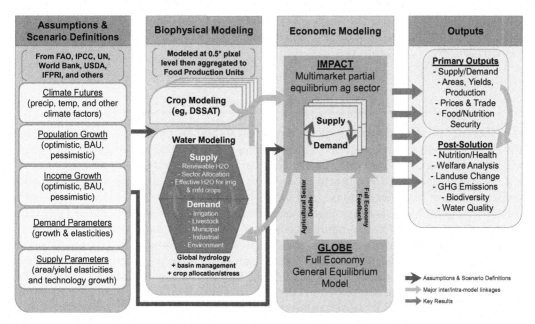

**FIGURE 10.3** The extended IMPACT modeling framework. Source: The authors, reproduced from IFPRI (2021).

and include all commodity markets and usually all factor markets (for example, labor and capital markets) – although such models typically include less detail on the agricultural sector than the IMPACT model does. PE and CGE models have different strengths and weaknesses for scenario analysis and have proven to be complementary in analysis of long-run trends under climate change and other drivers of change in food systems (see for example Nicholson and Gomez, Chapter 7 in this volume).

In IMPACT, projections for gross domestic product (GDP) growth by country enter the model as external factors that drive household food demand. Due to the partial equilibrium nature of IMPACT, feedback effects from agriculture to real GDP are necessarily neglected in stand-alone applications of IMPACT. Ignoring such feedbacks is not particularly problematic for high-income regions where the contribution of agricultural and food processing activities to aggregate GDP is small, but for low-income regions with a large share of agriculture in total GDP, the omission of these feedback effects may potentially lead to simulation results that miss an important part of the story. In some applications, IMPACT is thus linked with the dynamic CGE model GLOBE (Willenbockel et al., 2018). This linked modelling framework enables analysis of the wider implications of agricultural sector scenario projections generated by IMPACT by taking systematic account of linkages between agriculture and the rest of the economy.

This provides indications of the direction and order of magnitude of the effects on non-agricultural and macroeconomic variables, such as changes in factor prices and household incomes as well as changes in relative commodity prices throughout the global economy. These simulated changes in turn allow an internally consistent assessment of the associated general equilibrium welfare impacts. The aggregate real income effects generated by GLOBE are then passed back to IMPACT to analyze the detailed implications for agricultural variables, water and food security.

## 10.4 Scenario analysis using IMPACT

The IMPACT model is designed for scenario analysis and projections of alternative futures, rather than forecasting or prediction of the most likely future. Analysis of global agricultural markets several decades into the future requires a flexible, scenario-based approach that involves specification of the impacts of long-run drivers (such as changes in population, income, consumer behavior, climate, and technology development) whose nature is highly uncertain. Scenario analysis is a powerful analytical tool that allows policymakers to explore plausible futures in a systematic manner, considering future uncertainties (Wiebe et al., 2018). Scenario analysis is distinct from forecasting analysis in that the objective is not to predict the most likely outcome (usually extrapolating from historical experience). Instead, scenario analysis focuses on system dynamics, generating logically consistent future pathways that include trends and nonlinear interactions that may deviate significantly from past experience or from a "business-as-usual" (BAU) future scenario.

Looking more than a few years into the future requires considering how the broader context within food systems function will also change – including changes in population, income, technology, and climate, as indicated in Fig. 10.4. To do that, we (the IMPACT team) use assumptions about changes in these factors from the global modeling community, developed for the Intergovernmental Panel on Climate Change (IPCC) and its periodic Assessment Reports. Using a common set of assumptions for basic drivers makes it easier to compare results with those from other models, to better understand sources of differences in projections of future trends.

On the climate side, we use four scenarios called Representative Concentration Pathways (RCPs), expressed in terms of greenhouse gas concentrations and their associated levels of radiative forcing, ranging from 2.6 W per meter squared to 8.5 W per meter squared in 2100 (representing warming potential of roughly 1 to 4 °Celsius over the course of the 21st century, respectively) (Moss et al., 2010; Stocker et al., 2013; van Vuuren et al., 2011). These in turn are associated with changes in temperature and precipitation at daily and annual scales, as well as varying by location.

In the IMPACT modeling framework, these changes in temperature and precipitation are used as inputs in crop models (see Fig. 10.5), which determine how specific crops will grow and how much they will yield with different patterns of temperature and precipitation in different locations with different soils. Changes in temperature and precipitation also serve as inputs in water models, which determine how much water will be available for irrigation in each of around 150 water basins around the world, taking into account competing demands for residential and industrial uses.

On the socioeconomic side (see Fig. 10.4), we use scenarios called Shared Socioeconomic Pathways (SSPs), which differ in their assumptions about changes in population, income and other factors (O'Neill et al., 2014, O'Neill et al., 2017). These include the relatively optimistic SSP1 scenario (with slower population growth and faster growth in sustainable technologies,

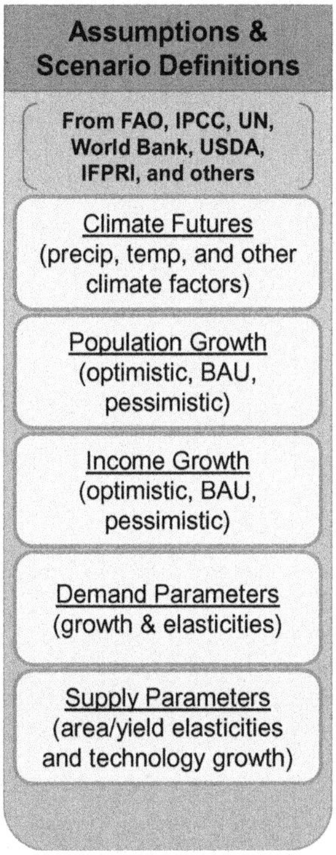

**FIGURE 10.4** Assumptions and scenario definitions in the IMPACT modeling framework. Source: The authors, reproduced from IFPRI (2021).

productivity and income), the "middle-of-the-road" SSP2 scenario, and the more pessimistic SSP3 scenarios (with faster population growth, slower growth in productivity, and more political fragmentation), among others.

SSPs make different qualitative assumptions about technology change through their respective narrative storylines, but they do not specify changes in technology or productivity explicitly. To account for future changes in technology in IMPACT, we need to make explicit assumptions about how technology will change productivity for each commodity in the model, as well as for each country and year. These assumed rates of technology-induced productivity growth are the starting point for the final yield changes that are estimated in the model, which we will come back to in a moment. Changes in climate, crop productivity (for any given technology) and water availability combine with changes in technology to drive changes in total production for each commodity in each country and year.

On the demand side, changes in income (from the SSPs) drive changes in per-capita demand for different agricultural commodities, following patterns that are empirically well-established. Per-capita demand for most commodities rises with incomes – typically more slowly for basic commodities like starchy staples, and more rapidly for nutrient-dense food groups like fruits, vegetables, and animal-sourced foods – but in all cases, at rates that eventually decrease as consumption levels reach satiation. Changes in per-capita income combine with changes in population to drive changes in total demand for each commodity, country and year.

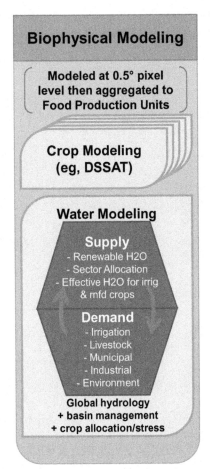

**FIGURE 10.5** Biophysical modeling in the IMPACT modeling framework. Source: The authors, reproduced from IFPRI (2021).

A key contribution of the IMPACT model (and other similar models) is to explore how changes in production and demand interact (see Fig. 10.6). This interaction is reflected in changes in prices as the balance between supply and demand changes between commodities, between countries, and over time. As changes in population and income combine to increase demand for fruits and vegetables and animal-source foods faster than demand for staple cereals, for example, prices of the former food groups rise faster than those of the latter. These price changes trigger further adjustments on both the demand and supply sides.

On the supply side, price increases for certain commodities create incentives for farmers to shift resources into the production of those commodities, either by increasing the area planted to those crops or by intensifying management of land already devoted to those crops. Changes in management practices may include increases in fertilizer application as crop prices rise, for example. (Application of fertilizer and other inputs is not modeled explicitly in IMPACT, but rather captured through supply response coefficients or "elasticities" for each commodity.)

On the demand side, changes in prices will induce some consumers to adjust their consumption patterns, reducing consumption of food items that have become more expensive

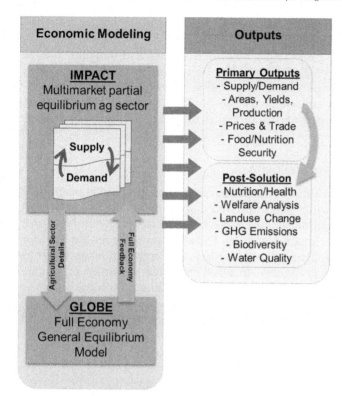

**FIGURE 10.6** Economic modeling and outputs in the IMPACT modeling framework. Source: The authors, reproduced from IFPRI (2021).

in favor of items that are more affordable. (These adjustments are typically smaller for high-income consumers, for whom food represents a relatively small share of their total income and expenditures, and larger for low-income consumers, for whom food can represent as much as 50 percent or more of their total expenditures.)

Adjustments on both the supply and demand sides continue until an equilibrium solution is reached for all commodities and countries, given the particular assumptions specified in the scenario of interest.

It is these adjustments between supply and demand, between commodities, and between countries, based on the interactions between biophysical and socioeconomic factors at multiple scales, that are the key contribution of models like IMPACT. Looking at an individual commodity or country or driving factor in isolation will yield interesting insights into the nature of that particular commodity or country or driving factor, implicitly (or explicitly) assuming that everything else remains unchanged. If it is reasonable to assume that everything else remains unchanged, it is unnecessary and even undesirable to use a more complicated model. But if those broader factors, linkages and interactions are central to the question at hand, it is essential to consider them. This then is the critical factor in deciding what methodological approach or scale of analysis is needed to address any particular question.

## 10.5 Examples of IMPACT-based applications

Since its first application in 1995 to projecting global food supply and demand in 2020 (Rosegrant, Agcaoili-Sombilla, Perez, 1995), IMPACT has been used in collaboration with many different partners to explore a wide variety of questions relating to production, consumption, markets, trade, prices, land and water use, technology adoption, diets, hunger, nutrition, health, climate change, environmental quality, agricultural policy, and investment.

### 10.5.1 On the impacts of specific technologies

Rosegrant et al. (2014) examined how different agricultural technologies and management practices, including improved crop varieties and improved management of soils, water, and pests, could help improve crop yields (for maize, rice and wheat), food security and resource quality under different economic and climate scenarios. Focusing on varieties for a wider range of crops, Islam et al. (2016) examined the potential of heat- and drought-tolerant varieties of maize, wheat, potato, sorghum and groundnut to offset the adverse impact of climate change on yields while accounting for changing socioeconomic conditions. Both studies show how use of a multi-commodity and multi-regional modeling framework allows consistent comparison of the projected impacts of improved technologies across diverse crops and regions as well as by changing climate and economic conditions.

### 10.5.2 On the importance of incorporating both biophysical and socioeconomic drivers and interactions

Nelson et al. (2014) and Wiebe et al. (2015) used IMPACT and other global economic models to generate projections of crop yields and other food and agricultural outcomes through 2050 under a range of socioeconomic and climate scenarios. They found that economic feedback effects (such as higher prices) trigger adjustments in behavior (such as cropping choices) that offset some – but not all – of the adverse impacts of climate change. This illustrates the importance of considering these socioeconomic effects along with the direct biophysical effects captured by crop models and other approaches.

### 10.5.3 On impacts of different investment strategies and tradeoffs across outcomes

Rosegrant et al. (2017) compared impacts of three categories of investment (agricultural R&D, improved soil and water management practices, and improved infrastructure) on a range of social, economic and environmental outcomes, finding that each strategy offered particular advantages but also tradeoffs in terms of outcomes and costs. For example, investment in improved access to markets increased productivity and reduced prices, but also increased pressure on land and water resources. By contrast, investment in a comprehensive package of agricultural research, infrastructure and resource management offered the highest benefits across a wide range of outcomes, but was the most costly option.

Other studies have explored similar questions with greater focus on particular geographies or commodities, but still within a wider food system context. Mason-d'Croz et al. (2019) applied the Rosegrant et al. (2017) results to a more focused analysis of the cost of different approaches to ending hunger in Africa, while others looked in greater depth at particular commodities and development issues, such as livestock (Enahoro et al., 2019), fish (Chan et al., 2019), roots, tubers and bananas (Petsakos et al., 2019), cereals (Kruseman et al., 2020), and employment (Frija, Chebil, Mottaleb, Mason-D'Croz, & Dhehibi, 2020). Sulser et al. (2021a) further extended this analysis to estimate the cost of agricultural adaptation to climate change globally and by region. In each application, use of a model such as IMPACT allowed comparison of impacts and tradeoffs across multiple outcomes that would not be evident in an analysis that did not capture interactions across commodities, resources and geographies.

## 10.5.4 On comparing impacts of multiple drivers on desired outcomes

Nelson et al. (2018) compared the impacts of climate change and socioeconomic factors on the availability of macro nutrients (calories and protein) and micronutrients (including iron, zinc and vitamin A) over the next three decades. They found that climate change will have an adverse impact on the availability of many nutrients, but socioeconomic factors (such as changes in population, income and technology) will play a larger role, at least through 2050. (Climate change is likely to play a progressively larger role later this century.)

Of course, climate change is a complex process with multi-dimensional impacts, even for a particular crop in a particular location. Beach et al. (2019) looked more closely at different ways in which changes in climate, technology and markets will interact to affect nutrient availability. By distinguishing effects of improved technology from the impacts of increased $CO_2$ on crop productivity and nutrient quality, they found that the positive effects of increased $CO_2$ on crop yields are projected to be outweighed by the negative effects of increased $CO_2$ on levels of protein, iron and zinc in the harvested crops, but productivity gains from improved technology will be able to offset these losses through 2050.

## 10.5.5 On examining impacts of future trends on multiple outcomes

Springmann et al. (2018) used IMPACT with health models and other tools to explore options for keeping the food system within planetary resource boundaries. They found that no single strategy will suffice; rather, a combination of changes in diet, improvements in technology and management practices, and reductions in food losses and waste will be needed to keep the projected future effects of the food system within safe environmental limits. Dietary changes (specifically a shift towards more plant-based diets) were found to be particularly effective at reducing greenhouse gas emissions, while improvements in technology and management were more effective at easing pressure on land and water resources.

Sometimes it is useful to see how multiple factors combine to affect a composite outcome measure, such as human well-being. Sulser, Beach, Wiebe, Dunston, and Fukagawa (2021b) used the metric of "disability-adjusted life years" (DALYs) to capture the multiple impacts of climate and socioeconomic change on diets and health, finding that DALYs associated with protein-energy undernutrition and micronutrient deficiencies are projected to increase in total through 2050 (particularly in Africa south of the Sahara) but decrease in per-capita terms.

These studies reflect an on-going process of model development, improvement and application in collaboration with other researchers, including the CGIAR foresight community of practice and the Agricultural Model Intercomparison and Improvement Project (AgMIP). While the topics covered are diverse, these analyses share the common goal of informing decision making by national governments, international organizations, funding agencies, and other development partners to reduce poverty, end hunger and malnutrition, and protect the natural resources on which people and food systems depend.

## 10.6 Use of modeling to inform decision making

Previous sections have described how models like IMPACT are used to explore complex interactions between changes in food demand, climate, technology, resource use, and other factors that characterize food systems today. If the results of such analyses are to be useful in informing decision making, equal care is needed in understanding the questions faced by decision makers and in communicating analytical results in an accessible way. This is particularly important when the issues being addressed involve trade-offs between the interests of different stakeholders.

For these reasons, in order to be successful, modeling exercises like these need to be seen as part of a larger process (Wiebe et al., 2018). This process generally includes

- identifying a set of questions or challenges that confront a particular group of people;
- dialog to clarify primary concerns and involve key stakeholders;
- consideration of alternative methodological approaches;
- development of a discrete set of scenarios that reflect a range of desired goals and strategies to reach them;
- analysis of the likely impacts of those alternatives; and
- decisions about the best course of action.

In some cases, a single sequence of steps like these (or even a subset of these steps) will suffice to provide the information needed for a particular set of choices. More generally, and particularly for complex challenges involving multiple dimensions and diverse stakeholders (as is the case with food systems), the process requires an iterative cycle, with multiple rounds of analysis and dialog before sufficient understanding is reached to support effective decision making. Even after decisions are made – say, about policies to improve resource management or investments to accelerate innovation – a complete process will involve monitoring and impact assessment to guide course corrections and consider new needs and opportunities as conditions continue to evolve. In practice, an ideal process such as this poses its own challenges to decision makers who face time pressures and competing demands, so a balance needs to be struck between completeness and expedience.

## 10.7 Lessons learned and conclusions

Addressing food system challenges requires careful analysis to understand complex interactions and impacts across multiple scales and dimensions. Models like IMPACT are tools that can help us structure our thinking about these challenges and interactions, and about our goals and options to achieve them. Models are simplifications of reality, which implies both limitations and strengths. Models are limited because they do not capture the full richness of the real world. When models are developed and used carefully, however, simplification is also their advantage, because it allows us to experiment in a simulated environment and explore impacts of alternative policy and investment strategies under a variety of future scenarios for which we don't have data but still need to understand the range of possible outcomes.

As with any toolkit, it is important to choose the proper tool for the job at hand. IMPACT is well-suited to explore biophysical and socioeconomic drivers and outcomes related to food production, consumption, and natural resource use at national, regional and global levels, over a medium- to longer-term time horizon of several decades. For other questions and scales – for example at the household or local community scale, or for shorter time horizons, or in relation to other parts of the food system or the wider economy, other modeling tools will be needed. Many other examples are described in this book. As part of a larger toolkit, these different models can be complementary, and help form a more complete understanding of the challenges we face.

Finally, it is essential to keep in mind that food system challenges and the changes needed to address them are deeply embedded in complex social and political systems, and often involve contentious tradeoffs. As noted throughout this volume, modeling thus needs to be undertaken as part of a systematic, iterative and on-going process of analysis and dialog with stakeholders and decision makers at multiple scales, in order to effectively inform changes in decisions and behavior today that will lead to improved food systems in the future.

## Acknowledgments

This chapter draws on work supported by the CGIAR Research Program on Policies, Institutions, and Markets (PIM) and the CGIAR Research Program on Climate Change, Agriculture and Food Security (CCAFS).

## References

Barrett, C.B., Beaudreault, A.R., Meinke, H.R., Ash, A., Ghezae, N., Kadiyala, S.,…, Torrance, L., 2021. Foresight and trade-off analyses: tools for science strategy development in agriculture and food systems research. Q Open 1 https://doi.org/10.1093/qopen/qoaa002.

Beach, R.H., Sulser, T.B., Crimmins, A., Cenacchi, N., Cole, J., Fukagawa, N.K.,…, Ziska, L.H., 2019. Combining the effects of increased atmospheric carbon dioxide on protein, iron, and zinc availability and projected climate change on global diets: a modelling study. Lancet Planetary Health 3, e307–e317.

Chan, C.Y., Tran, N., Pethiyagoda, S., Crissman, C.C., Sulser, T.B., Phillips, M.J., 2019. Prospects and challenges of fish for food security in Africa. Global Food Security 20, 17–25.

Enahoro, D., Mason-D'Croz, D., Mul, M., Rich, K.M., Robinson, T.P., Thornton, P., Staal, S.S., 2019. Supporting sustainable expansion of livestock production in South Asia and Sub-Saharan Africa: scenario analysis of investment options. Global Food Security 20, 114–121.

FAO, IFAD, UNICEF, WFP, WHO, 2020. The State of Food Security and Nutrition in the World 2020. Transforming food systems for affordable healthy diets. FAO, Rome.

Fischer, T., Byerlee, D., Edmeades, G., 2014. Crop yields and global food security: will yield increase continue to feed the world? Monograph No. 158. Australian Centre for International Agricultural Research, Canberra.

Frija, A., Chebil, A., Mottaleb, K.A., Mason-D'Croz, D.M., Dhehibi, B., 2020. Agricultural growth and sex-disaggregated employment in Africa: future perspectives under different investment scenarios. Global Food Security 20, 100353 https://doi.org/10.1016/j.gfs.2020.100353.

Fuglie, K., Gautam, M., Goyal, A., Maloney, W.F., 2019. Harvesting prosperity: Technology and productivity growth in agriculture. World Bank, Washington DC.

Ingram, J., 2020. Nutrition security is more than food security. Nature Food 1   https://doi.org/10.1038/s43016-019-0002-4.

International Food Policy Research Institute (IFPRI), (2021). Foresight modeling with IFPRI's IMPACT model. <https://www.ifpri.org/project/ifpri-impact-model> (accessed 21.06.2021).

IPCC, 2019. Climate Change and Land: An IPCC Special Report on Climate Change, Desertification, Land Degradation, Sustainable Land Management, Food Security, and Greenhouse Gas Fluxes in Terrestrial Ecosystems. Intergovernmental Panel on Climate Change, Geneva.

Islam, S., Cenacchi, N., Sulser, T.B., Gbegbelegbe, S., Hareau, G., Kleinwechter, U.,…, Wiebe, K., 2016. Structural approaches to modeling the impact of climate change and adaptation technologies on crop yields and food security. Global Food Security 10, 63–170.

Lloyds, 2015. Food System Shock. Emerging Risk Report – 2015.

Kruseman, G., Mottaleb, K.A., Tesfaye, K., Bairagi, S., Robertson, R., Mandiaye, D., Frija, A.,…, Prager, S., 2020. Rural transformation and the future of cereal-based agri-food systems. Global Food Security 26, 100441 https://doi.org/10.1016/j.gfs.2020.100441.

Mason-d'Croz, D., Sulser, T.B., Wiebe, K., Rosegrant, M.W., Lowder, S.K., Nin-Pratt, A.….., Robertson, R.D., 2019. Agricultural investments and hunger in Africa: modeling potential contributions to SDG2 – Zero Hunger. World Development 116, 38–53.

Moss, R.H., Edmonds, J.A., Hibbard, K.A., Manning, M.R., Rose, S.K., van Vuuren, D.P.,…, Wilbanks, T.J., 2010. The next generation of scenarios for climate change research and assessment. Nature 463, 747–756.

Nelson, G., Bogard, J., Lividini, K., Arsenault, J., Riley, M., Sulser, T.B.,…, Rosegrant, M., 2018. Income growth and climate change effects on global nutrition security to mid-century. Nature Sustainability 1, 773–781.

Nelson, G.C., Valin, H., Sands, R.D., Havlik, P., Ahammad, H., Deryng, D.,…, Willenbockel, D., 2014. Climate change effects on agriculture: economic responses to biophysical shocks. Proceedings of the National Academy of Sciences of the United States of America 111 (9), 3274–3279.

OECD/FAO, 2017. OECD-FAO Agricultural Outlook 2017-2026. OECD Publishing, Paris.

O'Neill, B.C., Kriegler, E., Riahi, K., Ebi, K.L., Hallegatte, S., Carter, T.R.,…, van Vuuren, D.P., 2014. A new scenario framework for climate change research: the concept of shared socioeconomic pathways. Climatic Change 122 (3), 387–400.

O'Neill, B.C., Kreigler, E., Ebi, K.L., Kemp-Benedict, E., Riahi, K., Rothman, D.S.…, Solecki, W., 2017. The roads ahead: narratives for shared socioeconomic pathways describing world futures in the 21st century. Global Environmental Change 42, 169–180.

Petsakos, A., Prager, S.D., Gonzalez, C.E., Gama, A.C., Sulser, T.B., Gbegbelegbe, S.,…, Hareau, G., 2019. Understanding the consequences of changes in the production frontiers for roots, tubers and bananas. Global Food Security 20, 180–188.

Reardon, T., Timmer, C.P., Minten, B., 2012. Supermarket revolution in Asia and emerging development strategies to include small farmers. Proceedings of the National Academy of Sciences of the United States of America 109 (31), 12332–12337.

Robinson, S., Mason-D'Croz, D., Islam, S., Sulser, T.B., Robertson, R.D., Zhu, T,…., Rosegrant, M.W., 2015. The International Model for Policy Analysis of Agricultural Commodities and Trade (IMPACT): Model Description for Version 3. International Food Policy Research Institute, Washington DC IFPRI Discussion Paper 01483.

Rosegrant, M.W., IMPACT Development Team, 2012. International Model for Policy Analysis of Agricultural Commodities and Trade (IMPACT): Model description. International Food Policy Research Institute, Washington DC.

Rosegrant, M.W., Sulser, T.B., Mason-D'Croz, D., Cenacchi, N., Nin-Pratt, A., Dunston, D.…., Willaarts, B., 2017. Quantitative foresight modeling to inform the CGIAR research portfolio. International Food Policy Research Institute, Washington DC Project Report for USAID.

Rosegrant, M., Koo, J., Cenacchi, N., Ringler, C., Robertson, R.D., Fisher, M., Cox, C.M.,…, Sabbagh, P., 2014. Food Security in a World of Natural Resource Scarcity: The Role of Agricultural Technologies. International Food Policy Research Institute, Washington DC.

Rosegrant, M., Agcaoili-Sombilla, M., Perez, Nicos, 1995. Global Food Projections to 2020: Implications for Investment. 2020 Discussion Paper. International Food Policy Research Institute, Washington DC.

Schellnhuber, H.J., Rahmstorf, S., Winkelmann, R., 2016. Why the right climate target was agreed in Paris. Nature Climate Change 6, 649–654.

Springmann, M., Clark, M., Mason-D'Croz, D., Wiebe, K., Bodirsky, B.L., Lassaletta, L.,…, Willett, W., 2018. Options for keeping the food system within environmental limits. Nature 562, 519–525.

Stocker, T.F., Qin, D., Plattner, G.-.K., Alexander, L.V., Allen, S.K., Bindoff, …, Xie, S.-P., 2013. Technical Summary. Climate Change 2013: The Physical Science Basis. Contribution of Working Group I to the Fifth Assessment Report of the Intergovernmental Panel on Climate Change In Stocker, T.F., Qin, D., Plattner, G.-K., Tignor, M., Allen, S.K., Boschung, J., et al. (Eds.). Cambridge University Press, Cambridge, UK.

Sulser, T.B., Wiebe, K., Dunston, S., Cenacchi, N., Nin-Pratt, A., Mason-D'Croz, D.,…, Rosegrant, M.W., 2021a. Climate Change and Hunger: estimating costs of adaptation in the agrifood system. Food Policy Report. International Food Policy Research Institute, Washington DC.

Sulser, T.B., Beach, R.H., Wiebe, K.D., Dunston, S., Fukagawa, N.K., 2021b. Disability-adjusted life years due to chronic and hidden hunger under food system evolution with climate change and adaptation to 2050. American Journal of Clinical Nutrition 114, 550–563.

van Vuuren, D.P., Edmonds, J., Kainuma, M., Riahi, K., Thomson, A., Hibbard, K.,…, Rose, S.K., 2011. The representative concentration pathways: an overview. Climatic Change 109, 5–31.

Wiebe, K., Prager, S., 2021. Commentary on foresight and trade-off analysis for agriculture and food systems. Q Open 1 https://doi.org/10.1093/qopen/qoaa004.

Wiebe, K., Zurek, M., Lord, S., Brzezina, N., Gabrielyan, G., Libertini, J.,…, Westhoek, H., 2018. Scenarios Development and Foresight Analysis: exploring Options to Inform Choices. Annual Review of Environment and Resources 43, 545–570.

Wiebe, K., Lotze-Campen, H., Sands, R., Tabeau, A., van der Mensbrugghe, D., Biewald, A., Bodirsky, B.,…, Willenbockel, D., 2015. Climate change impacts on agriculture in 2050 under a range of plausible socioeconomic and emissions scenarios. Environmental Research Letters 10, 085010.

Willenbockel, D., Robinson, R., Mason-D'Croz, D., Rosegrant, M.W., Sulser, T.B., Dunston, S., Cenacchi, N., 2018. Dynamic computable general equilibrium simulations in support of quantitative foresight modeling to inform the CGIAR research portfolio: Linking the IMPACT and GLOBE models. International Food Policy Research Institute, Washington DC IFPRI Discussion Paper 01738.

# Using social network analysis to understand and enhance local and regional food systems

*Sarah Rocker[a], Jess Kropczynski[b] and Clare Hinrichs[a]*

[a] The Pennsylvania State University  [b] University of Cincinnati

## 11.1 Introduction

Many food systems researchers and practitioners have become interested in how knowledge about social networks can contribute to evaluating impact and developing more sustainable and equitable food systems. The ability to visualize and measure the structure and strength of social and economic connections can increase our understanding of otherwise hard to see 'soft infrastructure.' Such soft infrastructure is comprised of relationships, norms, and knowledge undergirding social and economic systems, including those related to food and agriculture. Modeling social networks makes these otherwise invisible ties demonstrable and measurable over a period of time. This approach is often regarded as complementary to systems thinking by providing the ability to investigate connections between diverse and often discreetly studied sectors, populations, and industries.

Social network analysis (SNA) can be used to investigate how food system entities, such as individuals, organizations, or businesses, are or are not connected to one another. This approach is different than other forms of modeling, such as Input-Output modeling, which focus on economic impact at the more aggregated sector level. Examples of types of network connections examined in food systems research include the exchange of data, information or advice, commercial transactions, resource sharing, collaboration on projects or initiatives. When network studies include information about different attributes of individuals and organizations, they can help shed light on critical questions of power and inclusion, such as, "Who is connected to whom, and what attributes appear to matter?" Social network analysis may serve as a starting point for qualitative research to inform deeper investigation about the patterns of relationships present or lacking and the associated conditions and consequences. As researchers around the globe are now designing and implementing studies

*Food Systems Modelling.*
DOI: https://doi.org/10.1016/B978-0-12-822112-9.00015-1

to understand impacts of the pandemic and social crises of the past year, we contend that social network analysis can be a useful method to understand multi-level systems for concurrent and overlapping crises, and provide the ability to design and evaluate interventions and planning efforts to rebuild community and economic systems. Examining connections between individuals, organizations and businesses across different sectors and domains can reveal potential strengths and weaknesses of current connectivity - as well as possible exclusions - and illuminate new possible levers of change for how to rebuild or reorient relationships of resources, information sharing and power for more equitable and resilient futures, as explored elsewhere in this book's examination of community food systems modeling (see chapter 12 (Glickman et al.) and Chapter 13 (Hennessy et al.)).

This chapter considers the use of social network analysis (SNA) to understand social and economic connections, or the lack thereof, particularly within food systems research. SNA is both a set of methods, as well as a methodology, comprised of underlying social science theories. Social networks have long been a focus in fields such as public health (Valente and Pitts, 2017), education (Carolan, 2014), supply chain management (Borgatti and Li, 2009), planning (Dempwolf and Lyles, 2012), and natural resource management (Bodin, Crona, and Ernstson, 2006). Such work complements and supports current efforts to measure and model social networks among business, governmental and non-governmental entities that structure the interconnected relations across production, distribution and consumption in food systems. In contributing to this book's emphasis on diverse approaches to food systems modeling and practice, this chapter focuses primarily on the methodological foundations of SNA that have been applied to food systems, as well as considerations for future food systems research. We highlight relevant theoretical foundations as they shape methodological choices for data interpretation and provide suggestions for new approaches that elevate the value of data interpretation in future applied research scenarios.

The chapter is organized as follows. We begin with a brief history and broad overview of SNA. We then illustrate the core components of network study design from conceptualization and data collection to analysis and interpretation. We discuss possibilities for network study design in the context of recent food systems network research, highlighting relevant examples. We also provide an in-depth look at one study examining regional food value chains in the US, to demonstrate more underutilized methods in network analysis including: an example of a primary data collection tool for collecting data at live events, visualization techniques for connected ego-network level analysis, and mixed methods data interpretation. We then present future research scenarios in which SNA could be applied to answer theoretically and pragmatically motivated research questions in the field of food systems. We conclude by discussing ethical considerations for SNA research, as well as new areas of inquiry that may be explored in network research in the context of emerging discussions to rebuild systems in the wake of global health and social crises.

## 11.2 A brief history and overview of social network analysis

It is widely understood that 'relationships matter' - whether these be positive influences of gaining social or material support, or negative impacts where opportunities are constrained by 'not having the right connections.' These relationships may be easy to recognize in one-on-one situations or small groups, but difficult to conceptualize or identify in large or overlapping systems. Social network analysis offers a methodology with specific analytic tools to quantify and visualize complex relational information that is otherwise difficult to perceive. The advent of digital methods to collect information about interactions has stimulated a turn in recent decades to using SNA as a way to study and understand human relationships. These methods include extracting information from friend networks made public on social media, examining linked information in open-access databases (such as co-citation networks of authors based on information available on the Web of Science), or other digital footprints of social infrastructure. While digital sources can simplify the data collection process, many researchers still use direct observation of in-person interactions, interviews, and surveys to collect data for SNA in order to quantify aspects of social relationships in a way that allows us to more fully describe, predict, and visualize networked interactions.

Networked interactions are most often depicted as *sociograms*, which visualize *nodes* (the actors in the network) connected by links (lines connecting the actors). Modern graphic artists have made these visualizations synonymous with technology and the internet age. Given this trend, some might be surprised to learn that these network visualizations have been used for some time, dating back to the classic Western Electric (Hawthorne Plant) observational study of bank wiring room employees (Roethlisberger and Dickson, 1939). Visual representations of social networks are useful to leaders interested in seeing at a glance the overall connectivity of a group and who is most or least connected to the whole. This information can be used to make strategic changes to connect isolated components, reduce reliance on a node that is overwhelmed, or leverage the popularity of a particular node to share information.

One reason that SNA has become more widely used is because online interactions offer digital footprints of connections between individuals and thus a rich, potential source of secondary data amenable to network analysis. Other methods of data collection include administering social network questionnaires or directly observing interactions. Due to the special nature of linked data, some methods of analysis require gathering data that represents as much of the whole network as possible in order to understand patterns in social structure. **Whole network** studies observe relationships that exist within a population or group with an identifiable boundary, such as membership in an organization or attendance at an event. Gathering whole network data requires that the researcher has an idea of the bounded population they intend to study, and a high response rate to achieve an accurate picture of the whole network or a whole subcomponent of a network. It also means that it is difficult to infer meaning from an incomplete social network or by sampling from a large dataset. That is not to say that sampling of network actors is not possible, but some social network measures cannot be calculated when whole network (or complete network) data is not available. Whole network studies are applicable when the research question is related to dynamics that describe the whole group, rather than the individuals within it. Alternatively, data related to dyadic relationships (i.e., relationships between pairs of individuals) or individual-level descriptors may be answered through an *ego network* approach.

An *ego network study* examines the personal network of individuals; it centers on a focal node (or *ego*) and their connected peripheral nodes (or *alters*). Each alter is directly connected to the ego through social relationships which could be role-based (e.g., business ties or support ties), perceived relationship status (e.g., trusted individuals), or action-based (e.g., gives resources to, receives resources from). The analysis of ego networks entails gathering data from a selection of individuals in a network and mapping out social opportunities and constraints experienced by individual egos and alters as a direct result of the nature of their relationships. However, understanding interactions through the perspectives of actors sampled in a network who are disconnected from one another can provide a limited view. Therefore, another approach to data collection may be called for when sampling actors from a network; this can take the form of asking sampled individuals (or egos) to describe the perceived connections among others they are directly connected to (or alter-alter connections). Throughout this chapter, we will discuss how the suitability of data analysis techniques depends on the type of data collection method employed.

## 11.3 Emerging applications of SNA in food systems research and practice

With broad, growing interest in SNA approaches, the 2010s saw more frequent and diverse applications of SNA specifically within food systems research and practice. The move toward SNA dovetailed with emerging interest in evaluating the impact and effectiveness of development interventions, and thereby, to guide planning and strategy for future food systems development. Looking beyond the commonly used metrics of job and revenue creation, researchers have called for more attention to other important dimensions of impact such as social welfare, health, equity, and environmental stewardship (O'Hara and Pirog, 2013), many of which are illustrated in other chapters in this book, notably Chapter 3 (Thoma et al.), Chapter 4 (Schyns et al.), Chapter 9 (Canning et al.) and Chapter 14 (Skonberg et al.). Previously in the literature, specific attention has been raised regarding dimensions of equity in the food system, such as the distribution of impacts, including the use of energy resources, social well-being of consumers and farmworkers (Gomez et al., 2011). Efforts have been made to build upon and innovate beyond traditional modeling methods, most notably exemplified in the USDA Economics of Local Food Toolkit (Thilmany McFadden et al., 2016) and a corresponding special edition of the *Journal of Agriculture, Food Systems and Community Development* (Jablonski and McFadden, 2019). The method of SNA received mention in the Toolkit as a nascent approach for use in food systems research, with these authors acknowledging the method deserves deeper investigation and integration into any food system assessment process a community may choose to undertake.

Social network analysis approaches have also been applied in research addressing long-standing questions about organization, governance, performance and change in food systems. Food systems related studies have used SNA approaches in evaluating collaboration within a regional food project (Christensen and O'Sullivan, 2015), characterizing and mapping geographically-bounded alternative food networks (Brinkley, 2017; 2018) including changes over time (Brinkley, Manser, & Peschi, 2021), assessing the "value proposition" of network participation in the Vermont Farm-to-Plate initiative (Koliba, S.Wiltshire, Turner, Zia, and Campbell, 2017), examining opportunities for innovation in multi-stakeholder platforms for agricultural research (Hermans, Sartas, van Schagen, vanAsten, & Schut, 2017), predicting small and minority farm economic performance (Khanal et al., 2020) and examining how issue alignment reveals sub-networks within the broader Denver, Colorado food movement (Luxton and Sbicca, 2021).

One particular interest in food systems research has been the use of SNA to map, understand and potentially enhance patterns of coordination and cooperation. It has also been used to illustrate flows of information and resources between the diverse actors, businesses and organizations comprising food systems. Studies have examined relationship dynamics between food systems' *support actors,* such as service providers and food systems' practitioners, with regards to collaboration (Christensen and O'Sullivan, 2015) and advice sharing and information (Meter, Goldenberg, and Ross, 2018). Other studies have engaged *business actors* in local and regional food supply chains, exploring, for example, commercial relationships between farmers and delivery sites (Goldenberg and Meter, 2019), business exchange between farmers and food retailers (Trivette, 2019), and development of the local grain economic network in Arizona over a period of seven years (Forrest and Wiek, 2021).

Application of SNA approaches in food systems related studies still remains relatively novel. While there is some diversity in approaches to the design of a network study, there are some common considerations that must be addressed. In the next section, we discuss sequential elements for designing, analyzing and interpreting a social network analysis study related to food systems. Drawing on examples of food systems network research through its emergence in the past 10 years, we discuss the tenets and contributions underlying the common approaches thus far.

## 11.4  Designing a food systems network study

### 11.4.1  Identifying a research question, theoretical concept and study context

A dynamic interplay often exists between a research question, the study context and the theoretical and conceptual underpinnings guiding research design. Which comes first will depend on many factors, including the presence or emergence of a particular network, whether the research has applied or theoretical aims, and the relationship of the investigator to the context to be studied, in other words, is the investigator embedded in the network of interest or a neutral, distanced observer.

The network research context is the situation in which the research question is being explored. The context includes the types of actors, organizations, or events to be studied, as well as other contextual circumstances such as regional location or timeframe. Some research may be more applied, posing research questions that are pragmatically selected for an evaluation or to inform planning an intervention in the network in the future. Other research may be more theoretically driven, pursuing a more traditional academic inquiry. Whether academic or applied, examining conceptual and theoretical drivers at the outset of your research design will help inform other decisions to be made about all other characteristics of the study, including level of analysis, attributes, specific measures of analysis and interpretation.

Theoretical concepts can be integrated into a study design to support the research hypothesis, and to aid in identifying processes and structures within the food system being studied, which affect anticipated conditions or outcomes. Theoretical concepts may be operationalized through one or more network measures of analysis, and can be utilized to guide the interpretation of network results. Examples of theoretical concepts that may be examined in network studies include "food security", "resiliency", "embeddedness" (Brinkley, 2017), or "entrepreneurial social infrastructure" (Rocker, 2019). A network study tests a theoretical concept by crafting a research question such as, 'Is this network demonstrating a structure that is contributing to regional food security?" In order to answer this question, the researcher must operationalize the theoretical concept into network measures that serve as indicators to demonstrate presence, absence or strength of the concept within the network. As an example, SNA measures that may indicate "food security" could include density (percentage of connected nodes relative to the whole network), centrality (number of ties incoming or outgoing from any node), or redundancy (the extent to which a tie reaches nodes that are connected to each other), between critical actors in a regional supply chain. For additional examples of how theoretical concepts may be operationalized into network measures, see the Case Study Spotlight, later in this chapter.

Practitioners who are designing a more applied study or evaluation will also benefit from integrating basic theoretical network concepts from the outset of their efforts, as doing so will assist in guiding the analysis and planning actionable and strategic steps for shaping future growth of the network. Some examples of network concepts that may be helpful to applied practitioners include "community wealth" building (Plastrik & Taylor, 2006), "adaptable" networks (Holley, 2012), and "regenerative" networks (Plastrik, Taylor, & Cleveland, 2014). For those interested in network development, facilitation, and evaluation, these three aforementioned books serve as excellent references and step-by-step guides.

## 11.4.2 Determining network research level of analysis

The study context, research question and theoretical concepts will influence the unit and level of analysis for your network study, or in network terms, whether you will be conducting an analysis of a **whole network**, such as the structure of a whole entity like an organization, or an analysis of an **ego-network**, which examines the personal networks of individuals. There is also a third option, which can also yield insight, that involves examining a group of **connected ego-networks** in relationship to each other. This option may be appropriate when you have a sample of individuals within a group, but not a high enough response rate to yield the accuracy of whole network measures.

As noted earlier, the focal units within a network are referred to as **nodes**. They may be linked individuals, organizations, businesses, sectors, events, or objects. Both whole network and ego-network studies can examine any of these types of nodes. In the field of food systems, whole network studies may, for example, examine organizations working on a shared project, individuals attending a common event, or co-membership in a formal organization. Ego network studies may examine support networks for producers, buyer-supplier ties between food businesses, or an individual's collaboration ties on a project.

The research question and the level of analysis must be aligned in a network research design. For example, if one wants to understand to what extent a community has a supply chain structure that can promote "regional food security," that would likely only be achievable by designing a whole network study of, for example, aggregated buyer-supplier relationships in a particular region. In contrast, a study seeking to investigate "household food security" could design an ego-network study that looks at individual heads of household's food access points, analyzing each household's personal network discretely. Levels of analysis could also be mixed; for example, a study could begin focusing on an individual household network, then compare networks or attributes of households in relationship to each other to explore the notion of regional household food security.

### 11.4.3 Selecting data collection methods

SNA can be performed on both primary and secondary data sources. Primary data sources most commonly involve interviews, surveys or observation or some combination thereof. Secondary sources for food systems research may include online marketplace databases, websites, social networking platforms and observable data, such as attendance at public events. Data collection methods must align with the level of analysis the researcher intends, which follows their core research hypothesis. Utilizing secondary data, when available, is much more time-efficient, because of the high burden of collecting sensitive and detailed relational data from participants. Yet with access to a willing group of participants, where sufficient trust and rapport is established between researchers and participants, primary collection may yield data that is more meaningful and relevant to the specific research question, and possibly, more accurate. Trivette (2019) utilized secondary data from an online retail database to examine buyer-seller relationships by farm and food business type across a region of New England. Primary data collection typically involves surveys or interviews with network actors, such as Aculo, Mexico cheese producers (Crespo, Réquier-Desjardins, and Vicente, 2014), local farms and their markets and product users in Chester County, Pennsylvania (Brinkley, 2017, 2018), participants in the Growing Opportunities regional food systems collaborative in North Carolina (Cumming, Kelmenson, and Norwood, 2019) and members of Cass Clay Food Partners (a network linking eastern North Dakota and western Minnesota) as they responded to the COVID-19 pandemic (Harden et al., 2021). Such data can be collected in-person or via mail, but is increasingly gathered via on-line questionnaires.

### 11.4.4 Collecting attribute data

Attributes are the characteristics that describe the nodes (individuals or organizations) and relationships being studied in a network. **Node attributes** may consist of traditional demographic information for individuals including age, race, nationality, gender, socioeconomic status or education or organizational information, such as business, government agency, nonprofit or educational institution. **Tie attributes** describe a characteristic of the relationship *between* nodes, rather than the node itself. You may ask individuals, for example, about the depth of acquaintance or trust or frequency of communication, measured on a Likert scale to describe dimensions of the strength of the tie. Attributes can be selected specifically to investigate a hypothesis the researcher has going into the study about what types of dimensions may affect the presence, absence or amount of connectivity. For example, if you were studying an acquaintance and collaboration network of members on a Food Policy Council, the member's primary occupation or sectoral affiliation may be a relevant node attribute to collect, as that could likely be related to who knew whom before joining the council, and may also be a mediating factor for why some individuals would be more likely to connect with others in the future. In their study of collaborative governance in a food policy council, Carboni, Siddiki, Koski, and Sadiq (2017) considered members' organizational affiliations and domain foci for assessing how the inclusive design of the council aligned with how its work was conducted in actual practice. Other studies that involve organizations may have attributes that describe an organization or business's mission, values areas, or target populations served. An example of tie attributes will be presented in greater detail in the case study below, where Rocker (2019) coded connections made at a convening event for local and regional farmers as either 'horizontal' (between transactional businesses along the value chain) or 'vertical' (between transactional businesses and support organizations).

You can get as creative as you would like thinking about attributes to collect as they can be another qualitative data point in your network. If you collect open-ended responses, you may end up working with a wide range of discrete responses. In turn, you may need to create larger buckets or categories for these open-responses (in effect, to recode them) to identify broader patterns of connectivity. You could also set your attributes and a list of categories at the outset when collecting data, but this approach may limit some participants' ability to adequately describe where they see themselves. Having an "Other: Write In" category is a good idea in this instance. Those with the same attribute category will essentially become a subgroup that can be identified and examined in the context of the overall network patterns.

This approach to studying networks using attribute data can inform understanding of how interdisciplinary teams conduct their science across disciplines and sectors, as discussed in Chapter 15 (Cross et al.) in this book. Whether you code the responses from network participants into categories before data collection or after is up to you, but having the ability to look at specific responses and broader categories will enable you to see smaller or larger patterns in who is connected and why, when it comes time to analyze the network. As one example, Luxton and Sbicca's (2021) social network study of the Denver food movement demonstrates how initial attribute data categories, which focused on the broad focal areas of organizations and initiatives in the network, can be further delineated and even regrouped later during the analysis stage to engender new insights.

Geo-spatial data constitutes another attribute that can describe the node (e.g., the location of a business) or the tie (e.g., distance between two individuals, organizations, etc.). Food systems studies have utilized geo-spatial network data to investigate organizational collaboration for state-wide food systems development goals, embeddedness of social and geographic relationships, and how different types of businesses across supply chains develop buyer-seller relationships over time within a particular region (Brinkley, 2017, 2018; Brinkley, Manser, & Peschi, 2021; Christiansen and O'Sullivan, 2015; Forrest and Weik, 2021).

Certain node attributes can also inform the growing interest in understanding equity, inclusion, justice and power in food systems. While there are few examples to date of utilizing SNA to explore common dimensions of marginalization, such as race, ethnicity, gender or nationality, we believe SNA can offer insights not yet provided by other research methods. If social network studies focus primarily on patterns in the sharing of information and resources, then visualization of the current situation regarding who is connected and resourced, and who is not, can become a useful tool for reflection and planning for future corrective action. In network research, relationship data demonstrates 'what' the picture of connectivity looks like, and attribute data can offer insights about the 'why.' As we look at a network picture, attributes can help reveal the strong core hubs of activity, as well as more isolated individuals or clusters. Attribute data describing such individuals or clusters can help illuminate possible patterns and answer questions of a more collective nature about which groups are more central or more peripheral and isolated within a given network.

## 11.4.5 Selecting SNA measures for analysis

Measures are the types of quantitative analysis indicators that describe a network's features. Some measures describe the entire structure of a network (whole network measure), while others describe the significance of certain individuals in the network (ego-level measure). Whole network measures can only be used when a whole network data collection approach is used, while ego-level measures can be used in both ego-network design and whole-network design. In other words, if you only have a 45 percent response rate from a network survey of members of a Food Policy Council, you cannot meaningfully generate whole-network results, because your data represents less than half of the entire group's structure. Generally, an 80 percent response rate (Stork and Richards, 1992) is required to meaningfully and appropriately utilize whole network measures to describe the whole structure of a network. One example of whole network study design comes from Christiansen and O'Sullivan (2015), who used degree centrality (the number of connections a node has), closeness (average distance between nodes), and betweenness (shortest path between nodes) to investigate changes in collaboration ties in a regional food systems project group over a three-year period.

Designing a research method around ego-level analysis can be easier, especially when gathering primary data, because there is greater flexibility on the number of participants needed to achieve meaningful results. In an example of ego-level analysis in food systems research, Scott and Richardson (2021) examined a farm incubator training program, collecting and analyzing ego-networks of social support for 11 participating farmers. They measured total number of ties (size of the farmers' personal advice network), as well as density and strength of the ties in personal advice networks for farmers. In the field of public health, one study had 3,148 ego observations looking at food choices among spouses, friends and siblings (Pachucki, Jacques, and Christakis, 2011). Researchers used correlation analysis and logistic regression to understand how social influences affect eating choices, identifying 7 categories of eating patterns among alter-ego ties within the network.

The case study highlighted in this chapter demonstrates how overlapping ego network data in a smaller dataset (under 40 participants) may be visualized in the same sociogram. While certain whole network measures (e.g., centrality, structural holes, isolates) cannot be ascertained from connected ego-network data, patterns of individuals' connectivity, such as those relationships formed at a convening event, can still be ascertained, particularly through attribute data. (See further detail below in the Case Study Spotlight). The number of network research participants will likely depend on the research question, analysis tools available, and how much time and research support are available to the project. At the outset, it is important to consider how you will be prepared to meaningfully and practically organize and analyze the number of observations you collect. For a comprehensive guide and further steps for ego-network research design, we recommend the book *Egocentric Network Analysis: Foundations, Methods, and Models* (Perry, Pescosolido, & Borgatti, 2018).

## 11.4.6 Interpretation: putting it all together

Social network analysis will yield data results in the form of a sociogram, which is the visual representation of the network. (See Fig. 11.1 for an example of a sociogram from Trivette [2019]). The measures of network, such as network density or closure, network size and degree

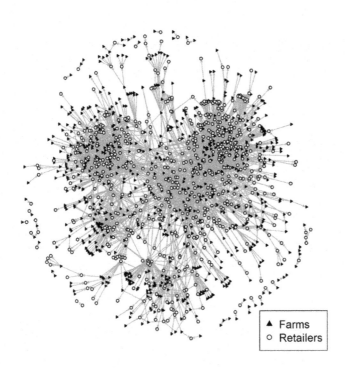

**FIGURE 11.1**  Example of a sociogram from Trivette (2019).

of clustering and isolates, convey the nuance of the data visualization on the sociogram. In studies with smaller networks, sometimes meaningful patterns can be observed with the naked eye. Indeed, visualizing only the sociogram is perhaps most useful for practitioners. For more on analog network analysis without the use of computer-aided technology, Holley (2012) is an excellent resource. However, traditional research studies, or any study with larger networks, will need to have the measures presented in tandem with the sociogram to fully understand the complexity of the data. (See Fig. 11.1 as an example). Whole network data will also provide outputs about how individual nodes function in the network, for example how many connections a node has going outward to others or is receiving from others (out and in-degree centrality), their ability to serve as a bridge or link between otherwise unconnected nodes (betweenness score), and how their attributes affect who they are connected with (homophily).

Network measure results don't stand alone. Rather, they must then be interpreted in the context of the original research question and any guiding concepts or theories ideas underlying the hypotheses. For example, a study about whether or not a network is growing towards greater or lesser 'resilience' will require researchers to examine how prior studies operationalized similar concepts, possibly from other methods and also network studies from other fields, that may have a longer history with SNA research. Such a study may also require some creativity on the part of the researcher to operationalize a concept using network measures in some way not yet explored. In the Case Study Spotlight, we show how Rocker (2019) operationalized the concept of 'entrepreneurial social infrastructure' for a network study on food value chain coordination. Other examples of operationalizing food systems concepts in network studies include 'embeddedness', 'social capital', and 'collective action' (Brinkley, 2017; Chiffoleau, 2009; Crespo et al., 2014).

## Case Study Spotlight:

### Collecting Primary Data from In-Person Food System Convening Events

Rocker (2019) conducted research on regional food value chain coordination, gathering primary data in a comparative case study method. SNA was implemented to evaluate connections formed among business and support actors associated with food value chains in two U.S. regions. This research was part of a national funding and training initiative called Food LINC (Leveraging Investment through Network Coordination) whose programmatic mission was to support the activities of regional food value chain coordination (United States Department of Agriculture, 2016). In each case, regional coordinators hosted buyer-supplier convening events for targeted food value chains, with these events serving as the opportunity for network data collection. The purpose of these convening events was to promote networking to foster business connections among food producers, processors and end-buyers, as well as new support connections for businesses with technical assistance providers such as lenders, non-profits, and Extension. Table 11.1 shows attributes of these events including the type of food value chains explored and types of attendees at each event.

**TABLE 11.1** Characteristics of regional food value chains and their convening events.

| Network feature | Mid-atlantic regional event | Southeastern regional event |
| --- | --- | --- |
| Lead Value Chain Coordination Organizers | Partnership between a mid-tier processor, consumer advocacy non-profit, and a university faculty | City government staff member (who is also a volunteer board member of a state-wide organic farming non-profit) |
| Commodity/Product Focus of Value Chain | Small grains for brewing, distilling, and baking | Mixed organics (produce, meat, grains, and dairy) |
| Number attending event | 80 Individuals 44 Businesses | 148 Individuals 82 Businesses |
| Number participating in network survey | 20 Individuals 16 Businesses | 31 Individuals 27 Businesses |
| Value chain categories in network | Input Supplier, Producer, Processor, Distributor, Retailer, Extension/Research, Other Support: Lender/ Policy, Coordinator | Input Supplier, Producer, Maltster, Miller, Brewer, Distiller, Baker, Brew Pub, Restaurant/Retail, End Consumer, Education/ Research Other Support: Lending/ Marketing/Consulting, Coordinator |

*Collecting Primary Network Data*

Collecting primary data can be one of the most challenging aspects of network analysis due to logistics of disseminating and receiving valid and sufficient responses. Table 11.1 shows total

*(Continued)*

## Case Study Spotlight: —cont'd

participants and response rate for the following primary data research example. Concerns about participant willingness to share information balanced with their understanding of benefits gained from participating in the research can be one challenge. Other challenges associated with data collection at a live event include capturing participant attention for a sufficient amount of time to complete the survey. Because network analysis is a new approach for evaluating food systems convenings and conferences, explaining the process to participants can take longer than a more traditional event evaluation survey. Rocker (2019) developed a data collection procedure to mitigate as many of these challenges as possible, including disseminating surveys in multiple formats (paper and electronic), using a free-recall method of only the "top priority" connections (as opposed to requiring an exhaustive list), and limiting the number of unique relationship types queried to only two (business exchange and business resource support) as shown in the network survey instrument (Fig. 11.3.)

*Post-Network Interview Guide Design and Data Collection Process*

Six months following the convening events, semi-structured qualitative phone interviews (see Interview Guide in Fig. 11.4) were conducted with survey participants to learn more about their perception of the event and their perspective on what had taken place subsequently with each of the connections they had indicated on their network survey. These interviews also served as a form of member checking to determine that the network surveys were reported accurately. These follow-up interviews also yielded important insights about the kinds of outcomes that resulted from connections made, such as resources gained, new or increased products to market, and innovations or efficiencies in practices.

*Utilizing Attributes in Analysis*

The instrument yielded both individual and business level data. However, to balance meeting the research aims with protecting participant anonymity, all businesses were coded into value chain "type" (see Table 11.2) and analysis was conducted focusing on the attribute. ZIP codes were collected to consider analysis of network data in relation to geographic location, however was not published publicly in order to protect respondent anonymity and potentially sensitive data regarding business partnerships. The analysis was conducted at the business level, using pre-selected and emergent value chain categories.

*Visualization*

A network visualization was created using UCINet and NetDraw (Analytic Technologies) (see Fig. 11.5). Instead of the standard network visualization in NetDraw, a more intentional visualization structure was created by the researcher, organizing the value chain nodes in relational space to each other horizontally and vertically, to demonstrate the transactional value chain flow and access to support actors outside of the economic chain. This type of visualization comes from the method of Qualitative Structural Analysis (QSA) in which 'sensitizing concepts' (Blumer, 1954; Herz, Peters, and Truschkat, 2015)- concepts used to guide the empirical analysis of data – are considered in how to shape the visualization structure of the sociogram. Additionally, two network images were placed side-by-side to show pre- and post- event network changes.

*Interpretation*: The type of purposive visualization demonstrated in Fig. 11.5 allows for key concepts to be represented within the sociogram. This sociogram revealed the surprising finding that

## Case Study Spotlight: —*cont'd*

VALUE CHAIN CONVENING
NETWORKING SURVEY

PENN STATE
College of
Agricultural
Sciences

Please check all that apply:

YOUR NAME: _____

BUSINESS: _____

BUSINESS ZIP:_____

☐ Farmer        ☐ Educator / Extension
☐ Processor     ☐ Lender
☐ Distributor   ☐ Consumer
☐ Wholesale Buyer ☐ Other_____

Your response to this survey is voluntary and appreciated. This survey will be used in a university research study on local food system development and the evaluation of this event. All data will be kept confidential and anonymous. Please feel free to contact us with any questions and thank you for your response!

**1a. Which NEW contacts did you make at the Conference that you plan to do business with in the future?**

Please list top 6:
Who might you sell to / buy from in the future
(ex. John Jones, Sunnydale Farm)

1 _____
2 _____
3 _____
4 _____
5 _____
6 _____

**1b. Of the attendees at this event, who did you do business with BEFORE today?**

Please list top 6:
Who do you sell to / buy from currently

1 _____
2 _____
3 _____
4 _____
5 _____
6 _____

**2a. Which NEW business resource contacts did you make at the Confernce?**

Please list top 3:
New connections with Lenders, Extension, etc.

1 _____
2 _____
3 _____

**2b. Which business resources have you worked with BEFORE today?**

Please list top 3:
Former connections with: Lenders, Extension, etc.

1 _____
2 _____
3 _____

We will be conducting a 3 month follow-up on connections made at this event. May we contact you?  Y / N
Phone:
Email:

If you have any questions about this survey or research please contact:

Sarah Rocker                (Name of Conference
Penn State University       Organizer Partner)
111 Armsby Building
University Park, PA 16802

**FIGURE 11.2**  Network survey instrument for regional value chain convening events.

(*Continued*)

Case Study Spotlight: —cont'd

---

### Value Chain Convening Event Participant Interview Guide – 6 Month Follow Up

**Connections made at event**

1. Several months ago you attended the _____ (VCC hosted convening event). What made you interested in attending this event? Prompt: Former relationship with VCC organization?

**Top Connections**

2. What were the most valuable kinds of connections made? Prompt: e.g., Sales connections? To resources? End customers?

**Benefits following the event**

3. What are the greatest benefits at this point (approx. 9-12 months after) attending the event with these connections? a. Did business contracts come to fruition? Did you make any new partnerships with resource agencies?

**Quality of Business Relationships**

4. Thinking about your top connections, what makes them so valuable?    a. How do you plan for the future seasons? Do you have a formal contract in place? b. What is the level of trust? Do you need a formal contract, or is a handshake agreement enough?

**Prices Received / Prices Paid**

5. Do you feel that you are receiving is fair? Is it meeting your operational needs? a. What are the best kinds of arrangements for you currently? Sales to distributors? Sales direct to processors? Sales direct to retail? Sales direct to consumer? For buyers: flip the question to prices paid, for each of these categories.

**For Farmers – Brand Identification**

6. When you sell to your buyers across different markets, do you know if your farm name / brand is being represented to the end consumer? a. Of these four categories, which tends to represent your farm brand identity the best? Sales to distributors?    Sales direct to processors?  Sales direct to retail?  Sales direct to consumer?

**Relationship with VCC Organization**

7. How long have you worked with (VCC organization)? a. What are the greatest benefits of working with this organization? What services do they provide you as a (farmer/buyer/distributor)? [Note: listening for matching of VCC roles: Convener, networker, matchmaker, facilitator, etc.]

---

FIGURE 11.3    Interview guide, value chain convening event participant follow up.

Case Study Spotlight: —*cont'd*

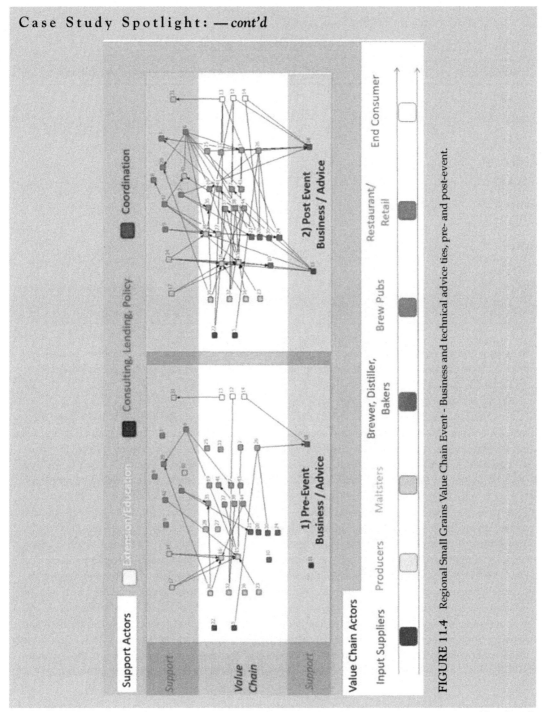

FIGURE 11.4  Regional Small Grains Value Chain Event - Business and technical advice ties, pre- and post-event.

(Continued)

## Case Study Spotlight: —*cont'd*

**TABLE 11.2**   Code scheme for network actors by value chain category attribute.

| Node color | Mid Atlantic small grain and malt event |
|---|---|
| Brown | Input Supplier |
| Green | Producer |
| Cyan | Maltster |
| Light Blue | Miller |
| Dark Blue | Brewer, Distiller, Baker |
| Orange | Brew Pub |
| Pink | Restaurant/Retail |
| White | End Consumer |
| Yellow | Education/Research |
| Purple | Other Support Lending, Marketing, Consulting |
| Red | Coordinator |

these convening events were not just fostering 'horizontal' business ties between buyers and sellers, but that 'vertical' support connections were likewise being fostered, including dyadic patterns (i.e., Producer-Lending, Processors-Extension).

*Operationalizing Theoretical Concepts in Network Studies*

To determine the direction of analysis of these value chain networks, this study drew from theoretical literature related to business and sociology and network indicators of social capital and social infrastructure to interpret network results. Theoretical concepts were drawn from relevant literature to inform research question about which network attributes have positive outcomes for economic and entrepreneurial development. They include social infrastructure impacts on economic action through 1) the presence of structural holes, 2) the presence of embedded ties, 3) density of the network, and 4) the presence of weak ties (Granovetter, 2005). In addition, the approach integrated entrepreneurial social infrastructure (ESI), designed for *community-level* indicators of endogenous economic activity which include 1) a diversity of ties with medium network density, 2) horizontal ties for lateral learning, and 3) vertical ties to outside institutions and resources (Flora and Flora, 1993;Sharp and Flora, 1997). Because the researcher hypothesized that social infrastructure development of regional food value chain coordination requires social infrastructure at multiple levels including individuals, businesses, and the community, a framework was created that integrates theories of social capital and economic and entrepreneurial capacity together to assess social capital creation at the individual and network/value chain level resulting from value chain convening events (see Table 11.3). This Table lists the network characteristics related to economic and entrepreneurial performance, linking them to corresponding network metrics and methods of observation for whole and ego-networks.

## Case Study Spotlight: —cont'd

**TABLE 11.3**  Social infrastructure network attributes and corresponding metrics and methods.

| Theoretical foundation | Network concepts | Whole network measures | Ego network measure |
|---|---|---|---|
| Social Infrastructure Impact on Economic Action (Granovetter, 2005) | Network Cohesion | Tie Count / Density | Tie Count |
| | Strength of Weak Ties | Density/ Clique Analysis | * |
| | Structural Holes | Structural Holes | * |
| | Embeddedness | Multiplex ties (economic and non-economic) | Qualitative Data |
| Entrepreneurial Social Infrastructure (Flora and Flora, 1993) | Horizontal Ties | Tie Count / Density | Tie Count/ Qualitative Data |
| | Vertical Ties | Tie Count / Density | Tie Count / Qualitative Data |
| | Diverse Ties | Homophily | ** |

\* Denotes attributes are unable to be examined in an ego-centric network \*\* Denotes attributes are unable to be examined due to limitations with the data Source: Rocker (2019) adapted from Granovetter (2005) and Flora and Flora (1993).

Tables 11.4 demonstrates a way to show patterns of ego-networks, as horizontal and vertical ties increased for attendees at mid-Atlantic small grains convening event (based on counts). Table 11.5 compares ego-network data patterns from two different networks, comparing average tie counts at two different events, and in this case demonstrating higher gains for horizontal ties than for vertical ties in both events.

*Interpretation:* This case is an example of mixed methods data collection and analysis between qualitative and network data. Consistent with sociological literature (Flora and Flora, 1993), this study found that 'horizontal', or 'lateral learning ties' foster innovations for new products, operations, and new ideas for research and development, regardless of whether they were formed between transactional businesses or support organizations. Building from previous qualitative research, this study was able to *visually* demonstrate the presence of horizontal and vertical ties in a network context, by analyzing qualitative data in tandem with network data and the broader network structure.

*Lessons from this Case*

In addition to demonstrating a mixed qualitative and network methods, this case has several unique features not commonly found in other food systems network studies, including primary, longitudinal data collection at an in-person event and a data visualization approach to modifying a standard sociogram by identifying a sensitizing concept to guide the network visualization. This case shows the creativity and adaptation that often needs to be deployed to make network approaches work for the unique context and constraints of food systems research, particularly in applied field-based settings. We conclude our cast study spotlight with a few tips to guide and enhance primary data collection success with SNA

*(Continued)*

## Case Study Spotlight: —cont'd

TABLE 11.4  Ego level horizontal ties in mid atlantic small grains network.

| ID | Value chain or ecosystem support | Pre-event ties | New ties made | Post-event ties | Increase |
|----|----------------------------------|----------------|---------------|-----------------|----------|
| Mid Atlantic, small grains, horizontal ties | | | | | |
| 1 | Value Chain | 1 | 1 | 2 | Yes |
| 2 | Value Chain | 0 | 3 | 3 | Yes |
| 3 | Value Chain | 4 | 2 | 6 | Yes |
| 4 | Value Chain | 2 | 2 | 4 | Yes |
| 5 | Value Chain | 0 | 7 | 7 | Yes |
| 6 | Value Chain | 0 | 3 | 3 | Yes |
| 7 | Support Ecosystem | 0 | 0 | 0 | No |
| 8 | Value Chain | 0 | 2 | 2 | Yes |
| 9 | Value Chain | 1 | 3 | 4 | Yes |
| 10 | Value Chain | 1 | 2 | 3 | Yes |
| 11 | Support Ecosystem | 0 | 0 | 0 | No |
| 12 | Support Ecosystem | 0 | 4 | 4 | Yes |
| 13 | Support Ecosystem | 0 | 2 | 2 | Yes |
| 14 | Support Ecosystem | 0 | 1 | 1 | Yes |
| 15 | Value Chain | 0 | 2 | 2 | Yes |
| 16 | Support Ecosystem | 0 | 0 | 0 | No |
| 17 | Value Chain | 0 | 1 | 1 | Yes |
| 18 | Value Chain | 2 | 0 | 2 | No |
| 19 | Value Chain | 1 | 0 | 1 | No |
| 20 | Value Chain | 3 | 1 | 4 | Yes |

- **Plan sufficiently for participant engagement and recruitment** - Be prepared to explain and answer questions about research purpose and SNA method *(option: create materials for engaging participants such as handouts, brochures, or link to project website)*
- **Be transparent and specific about the data being collected, analyzed and shared** - Since SNA is still a relatively new method used in this field, ensure the participant understands what the data will reveal *(option: include an example of sample sociogram results)*
- **Consider the participant experience** - do everything you can to make taking the survey an informed, stress-free process (e.g., *concise surveys, multiple and convenient formats*)

Case Study Spotlight: — *cont'd*

- **Make an adaptable plan for analysis** - Survey response will ultimately determine the measures you can use. Plan for several options prior to collecting data (e.g., *determine measures of interest from each category: whole network, connected ego-networks and individual ego network measures*)

TABLE 11.5 Average horizontal and vertical ties compared between convening events.

| Tie type | Case | Pre-event ties | New ties made | Post event ties |
| --- | --- | --- | --- | --- |
| Horizontal | Southeastern Organic Mixed Products Network | 2.13 | 2.77 | 4.90 |
| Horizontal | Mid Atlantic Small Grains Network | 0.75 | 1.80 | 2.55 |
| Vertical | Southeastern Organic Mixed Products Network | 0.77 | 1.45 | 2.23 |
| Vertical | Mid Atlantic Small Grains Network | 0.85 | 0.95 | 1.80 |

## 11.5 Future opportunities for SNA in food systems research

Having discussed considerations in designing a food systems network study and after presenting specifics of our spotlight case study, we present ideas for implementing social network analysis methods in future food systems research. Our purpose is to stimulate innovative research that leverages research goals, data collection methods, or promote analytical approaches that have not been central thus far in food systems research. Because context is critical to determining appropriate methodological procedures, we have prepared two example scenarios for whole network and ego network studies of food systems that will demonstrate possible data collection procedures, analysis measures and data outcomes.

### 11.5.1 Scenario 1 - Whole Network study: business exchange among local/regional food producers and buyers

- *Context*: Let's suppose there was an online regional food marketplace of buyers and sellers and one wanted to understand which businesses bought and sold to each other. This study could be undertaken by the organization that hosts the online buying platform or by a researcher, working in partnership with the online hub, who then has access to the database of sales.
- *Research Aim:* Understand patterns in business exchange interaction among buyers and sellers over time.
- *Data Collection:* Access to information on digital platforms typically takes one of two forms. The first may be the availability of a public Application Programming Interface

(API) that allows developers to query public information posted to the platform; this is commonly done with social media platforms such as Twitter. A second form would be coordination with the platform developers directly in order to access information stored in their database. Another advantage of data gathered from online interactions is that interactions are collected continuously over time, allowing for an examination of longitudinal trends. Therefore, a simple question that this type of data collection can answer is: What are the patterns of purchasing partnerships over time in this regional food marketplace? Network data could be recorded from sales transactions and could be gleaned to discover patterns of buying and selling and supply chain partnership development within a particular sector, region, etc.. Attribute information may correspond to characteristics of the business and could include size of business, gross sales, and geographic location or individual owner-operator demographics, such as gender or race/ethnicity.

- *Analysis:* To prepare data, time-based intervals that are meaningful to this type of network should be determined, for example, interactions corresponding to seasons, months, or pre- and post- a funding or policy initiative that is hypothesized to influence this network. For each time interval, all sales interactions should be recorded during that time. Changes in social cohesion among nodes that considers attribute data can inform patterns of which advanced hypothesis testing tools such as quadratic assignment procedure (QAP) (Dekker, Krackhardt, and Snijders, 2007) can offer a statistical correlation and significance between time periods or the relationship between particular attributes and sale interactions. Other statistical tools such as stochastic modeling (Snijders, 2001) can use longitudinal data to predict what future interactions may look like.
- *Possible Outcomes:* Results could help illuminate how transactional relationships between buyers and sellers persist or disappear over time, and which attributes may be related to whether connections are sustained. Node and tie level attributes could be explored to assess and inform network leaders on what makes long-lasting buyer-seller relationships sustainable over time.

### 11.5.2 Scenario 2 - Connected Ego-Network study: information networks of community food leaders with a focus on inclusion and access

- *Context:* Let's now suggest that one wants to conduct an assessment of a local community of food systems leaders in a particular region to better understand patterns of diversity, inclusion and access with regards to collaborative information sharing. Perhaps this effort contributes to a community's 'Asset Mapping' exercises for taking broader stock of the social capital in the community. This scenario is largely exploratory.
- *Research Aim:* To ascertain a baseline of the current information sharing network, including attributes of the most connected individuals and organizations. This research could illuminate equity in access to (and potentially control of) resources within the network, revealing both concentration and gaps in power and influence.
- *Data Collection:* In a network study on food systems actors in a particular region, there may be some awareness of key organizations and businesses that are currently working together on initiatives, but no known formal group or organization links them together in membership. In this case the study is intentionally exploring relationships in an *unbounded* context, meaning there are no pre-existing boundaries of who is in or out of the network. In this case, a free recall method of listing associations may give the most inclusive and

comprehensive idea of who is in the network. Here participants freely list all the other businesses and organizations they know or work with.

- ○ The alternative roster method would involve starting with a list of all possible organizations and businesses that would be pre-populated by the researcher. Attribute data to be collected could be at the individual (e.g. race, gender, age, education) or organizational level, (e.g. year organization was founded, number of paid employees and/or volunteers, annual revenue, types of clients served).

- *Analysis:* Ego-level analysis largely focuses on the placement of individuals with particular attributes within the overall network. One possibility would be to examine the network with attention towards core and periphery structure. This measure detects a core area of density within the network and creates a cut point within the network that creates a boundary between the core of the network and the periphery. A researcher could examine patterns such as individual attributes that relate to an actor's position as more core or peripheral to the network. In this analysis, we can examine whether members of the network periphery have connections to core members that can be leveraged to advance their placement within the network.

- *Possible Outcomes:* The value in conducting a network study on an unbounded network of food systems actors is perhaps as diverse as a community's intentions and aspirations. One possible outcome of collecting such data is the ability to collate a 'master list' of the relevant actors named as alters on the surveys. This list could be ranked by in-degree centrality, in other words, how many times someone else referred to them. Though this would be an unbounded network, patterns of cliques or subgroups may still emerge. These subgroups may be related to shared projects, proximity, values or some other reason meriting exploration. A researcher could also rank actors by out- and in-degree centrality to gain understanding of who is making the most outward ties and receiving the most connections. From an applied perspective, this information could be useful to identify leadership and future network weaving efforts.[1]

## 11.6 Implementing social network analysis going forward

It is our hope that this chapter inspires new ways of thinking about the significance and possibilities of networks and network analysis within food systems research and practice. We believe it is important to highlight a few final topics for those considering the use of SNA methods in future research or practice.

### 11.6.1 Ethical considerations

There are layers of interpretation that can be ascertained from network study results and sociogram visualizations, beyond what the researcher may recognize. In all human subjects' research, but with network research in particular because of its still novel use and the dynamic data it produces, it is critical to remember that there can be a gap between researcher intent and the possible impacts and implications of the study results. In some cases, removing names of actors alone will not be sufficient to anonymize results, given clear assumptions that can be

---

[1] Such as those described in the network practitioner guide, "Building Smart Communities through Network Weaving" by Krebs and Holley (2006).

made about networks. In other cases, a network may imply that some actors are unpopular, lacking power, or in possession of other negative social traits that they would prefer not to be revealed. In our experiences, IRB units have varying approaches to interpreting and approving network study protocols. Because of a lack of widely recognized and accepted standards, the onus falls on the researcher to take particular care and attention to account for all possible situations in which collecting and using SNA data could either benefit or potentially harm each of the different types of participants in the study.

In terms of data collection, and especially with primary data collection, researchers should consider how they can achieve accuracy and verifiability with regards to attribute data of participants. For example, if one were collecting racial demographic data, it would be advisable to have participants self-identify their race versus relying on a third party or researcher observation, to avoid making inaccurate assumptions. Using open responses, or other categories in addition to pre-selected categories will also help to provide greater accuracy with regard to assigning attributes. Because researchers often use attributes in identifying and sorting patterns of activity within network measures, verifying that the attributes are accurate should be a high priority.

Another ethical consideration involves researcher sensitivity and discernment around collecting and publishing select network results. For example, for network data that includes geographic identifiers in sectors or communities that are particularly tight-knit, small or otherwise have very well-known actors, like prominent businesses or organizations, publishing data that could identify their business details could bring unwanted exposure of internal details or trade secrets. We urge researchers planning primary data collection to consider how network data, in the wrong hands, could be marshalled competitively against those who have disclosed their information. Researchers should always weigh seriously the implications of publishing novel findings versus ensuring maximum risk protection for research participants.

This is especially true in the case of overlaying network data with geospatial data which can be highly identifying, given the public's ability to use Google maps and other social media platforms to identify virtually any public organization or business. In the case of network data, just because you can collect data, doesn't always mean that you should. Network data can be a very powerful decision-making tool, particularly when collected and shared with consideration about the welfare of participants and the intentions of audiences who will access the results. All contexts are nuanced, so this decision lies with the researcher who best knows the specifics of the scenario, but nonetheless here is a general rule of thumb to consider: Share the data with those who have the whole network's (all participants) best interests in mind, and be cautious about sharing data with any party who could use the data for individual gain. Finally, consider to what extent the research results could be useful to the participants themselves. Especially if the project has participatory or emancipatory research aims, sharing network data back to participants could itself be an intervention that influences the future direction and actions in the network.

## 11.6.2 Analysis and visualization tools

There are a number of network analysis and visualization tools ranging from paid software licenses to open-sourced and web-based options. UCINet and its companion visualizer, Net

Draw tend to be the academic standard and have in-depth training manuals and courses available.[2] NodeXL is commonly used for drawing public data from social networking sites such as Facebook and Twitter. Gephi is a visualization platform that works with Excel formatted network data, and produces high-quality sociogram images with a range of visualization editing options. Kumu is another web-based platform that has been embraced more by practitioners. Some tools, such as UCINet and the R statistical package can have a higher-learning curve that requires some dedication to time and training to become competent. An alternative option is to partner with researchers who specialize in the software programs for the analysis.

### 11.6.3 Integration of SNA with other methods

We have observed a high degree of mixed methods in food systems research, combining, for example, survey and interview data, or network data with geo-spatial data in food systems network research. Other studies utilize mixed methods sequentially, such as beginning with a network analysis and conducting qualitative interviews to investigate underlying influences that impact network connectivity (Luxton and Sbicca, 2021; Rocker, 2019). Some researchers have expressed the suggestion that SNA and Input-Output analysis could be seen as complementary methods, articulating multiple levels and dimensions of food systems phenomena from the economic sector level to the individuals, organizations and businesses (Goldenberg and Meter, 2019; Trivette, 2019). Future opportunities for multi-level systems research could be approached by interdisciplinary research teams to more deeply explore connections between sector-level economic data with individual and organizational level relational dynamics, as well as subjective accounts of particular practices and activities.

## 11.7 Conclusion

In this chapter, we have explained foundational concepts and methodological approaches in social network analysis, with particular attention to contexts in which SNA has been applied in existing food systems research. Throughout we have sought to offer ideas and open pathways for possible future uses of SNA in food systems research. We have identified some common characteristics of food systems research that has integrated SNA to date and we have highlighted the common patterns found in these early studies, as well as gaps and opportunities for ways that SNA could be integrated in future research design. We have also identified several possible research design ideas, as well as instruments and tools for data collection, analysis and visualization. Food systems studies at their core address the complex interactions of both coordination and contestation among people and resources through production, trade, and consumption as mediated by diverse influences including policy, funding, cultural and social practices, institutional structures, and much more. It is our hope that social network analysis methods, tools and concepts can help to illuminate and marshal different types of data that will contribute to a more dynamic, nuanced and constructive understanding of food systems today and in the future.

---

[2] Borgatti, S. P., Everett, M. G., & Johnson, J. C. (2018). Analyzing social networks. Sage. and annual LINKs training at University of Kentucky

## 11.8 Acknowledgement

Hinrichs's contribution to this work was supported by the USDA National Institute of Food and Agriculture and Hatch/Multi-State Appropriations under Project #PEN04630 and Accession # 1,014,133.

## References

Blumer, H., 1954. What is wrong with social theory? American Sociological Review *19* (1), 3–10.

Bodin, Ö., Crona, B., Ernstson, H., 2006. Social networks in natural resource management: what is there to learn from a structural perspective? Ecology and Society *19* (1), 3–7.

Borgatti, S.P., Li, X., 2009. On social network analysis in a supply chain context. Journal of Supply Chain Management *45* (2), 5–22.

Brinkley, C., 2017. Visualizing the social and geographical embeddedness of local food systems. J Rural Stud *54*, 314–325.

Brinkley, C., 2018. The small world of the alternative food network. Sustainability 10 (2921). doi:10.3390/su10082921.

Brinkley, C., Manser, G.M., Peschi, S., 2021. Growing pains in local food systems: a longitudinal social network analysis on local food marketing in Baltimore County, Maryland and Chester County, Pennsylvania. Agric Human Values 38, 911–927.

Carboni, J.L., Siddiki, S., Koski, C., Sadiq, A.-A., 2017. Using network analysis to identify key actors in collaborative governance processes. Nonprofit Policy Forum *8* (2), 133–145.

Carolan, B.J., 2014. Social Network Analysis and Education: Theory, Methods and Applications. SAGE, Thousand Oaks, CA.

Chiffoleau, Y., 2009. From politics to co-operation: the dynamics of embeddedness in alternative food supply chains. Sociol Ruralis *49* (3), 218–235.

Christensen, L.O., O'Sullivan, R., 2015. Using social networking analysis to measure changes in regional food systems collaboration: a methodological framework. J Agric Food Syst Community Dev *5* (3), 113–129.

Crespo, J., Réquier-Desjardins, D., Vicente, J., 2014. Why can collective action fail in Local Agri-food Systems? A social network analysis of cheese producers in Aculco, Mexico. Food Policy *46*, 165–177.

Cumming, G., Kelmenson, S., Norwood, C., 2019. Local motivations, regional implications: scaling from local to regional food systems in northeastern North Carolina. J Agric Food Syst Community Dev *9* (Suppl. 1), 197–213.

Dekker, D., Krackhardt, D., Snijders, T.A., 2007. Sensitivity of MRQAP tests to collinearity and autocorrelation conditions. Psychometrika *72* (4), 563–581.

Dempwolf, C.S., Lyles, L.W., 2012. The uses of social network analysis in planning: a review of the literature. J Plan Lit *27* (1), 3–21.

Flora, C.B., Flora, J.L., 1993. Entrepreneurial social infrastructure: a necessary ingredient. Annals of the American Academy of Political and Social Science *529* (1), 48–58.

Forrest, N., Wiek, A., 2021. Growing a sustainable local grain economy in Arizona: a multidimensional analytical case study of an alternative food network. J Agric Food Syst Community Dev *10* (2), 507–528.

Goldenberg, M.P., Meter, K., 2019. Building economic multipliers, rather than measuring them: community-minded ways to develop economic impacts. J Agric Food Syst Community Dev *8* (Suppl. 3), 153–164. https://doi.org/10.5304/jafscd.2019.08C.010.

Gomez, M.I., Barrett, C.B., Buck, L.E., DeGroote, H., Gao, H.O., McCullough, E., Yang, R.Y., et al., 2011. Research Principles for Developing Country Food Value Chains. Science 332 (June 3), 1154–1155.

Granovetter, M., 2005. The impact of social structure on economic outcomes. Journal of Economic Perspectives *19* (1), 33–50.

Harden, N., Bertsch, B., Carlson, K., Myrdal, M., Bobicic, I., Gold, A., et al., 2021. Cass Clay Food Partners: a networked response to COVID-19. J Agric Food Syst Community Dev *10* (2), 181–196.

Hermans, F., Sartas, M., van Schagen, B., vanAsten, P., Schut, M., 2017. Social network analysis of multi-stakeholder platforms in agricultural research for development: opportunities and constraints for innovation and scaling. Plos One 12 (2), 1–21. doi:10.1371/journal.pone.0169634.

Herz, A., Peters, L., Truschkat, I., 2015. How to do Qualitative Structural Analysis: the Qualitative Interpretation of Network Maps and Narrative Interviews. Forum: Qualitative Social Research (Vol. 16, No. 1). Freie Universitaet Berlin, Public Library of Science.

Holley, J., 2012. Network Weaver Handbook. Network Weaver Publishing, Athens, OH.

Jablonski, B.B., McFadden, D.T., 2019. IN THIS ISSUE: what Is a 'Multiplier' Anyway? Assessing the Economics of Local Food Systems Toolkit. J Agric Food Syst Community Dev 8 (C), 1–8.

Khanal, A.R., Tegegne, F., Goetz, S.J., Li, L., Han, Y., Tubene, S., et al., 2020. Small and minority farmers' knowledge and resource sharing networks, and farm sales: findings from communities in Tennessee, Maryland, and Delaware. J Agric Food Syst Community Dev 9 (3), 149–162.

Koliba, C., S.Wiltshire, S.S., Turner, D., Zia, A., Campbell, E., 2017. The critical role of information sharing to the value proposition of a food systems network. Public Management Review 19 (3), 284–304. doi:10.1080/14719037.2016.1209235.

Krebs, V., Holley, J., 2006. Building smart communities through network weaving. Appalachian Center for Economic Networks.

Luxton, I., Sbicca, J., 2021. Mapping movements: a call for qualitative social network analysis. Qualitative Research 21 (2), 161–180.

Meter, K., Goldenberg, M.P., Ross, P., 2018. Building community networks through community foods. Report to Maricopa County Food System Coalition Food Assessment Coordination Team. Crossroads Resource Center, Minneapolis Retrieved from http://www.crcworks.org/azmaricopa18.pdf .

O'Hara, J.K., Pirog, R., 2013. Economic impacts of local food systems: future research priorities. J Agric Food Syst Community Dev 3 (4), 35–42.

Pachucki, M.A., Jacques, P.F., Christakis, N.A., 2011. Social network concordance in food choice among spouses, friends, and siblings. American Journal of Public Health 101 (11), 2170–2177.

Perry, B.L., Pescosolido, B.A., Borgatti, S.P., 2018. Egocentric network analysis: Foundations, methods, and models (Vol. 44). Cambridge university press.

Plastrik, P., Taylor, M., 2006. Net gains: a handbook for network builders seeking social change. Innovation Network for Communities. version. 1. http://staging.community-wealth.org/sites/clone.community-wealth.org/files/downloads/tool-networks-social-change.pdf.

Plastrik, P., Taylor, M., Cleveland, J., 2014. Connecting to change the world. Location?. Island Press, Washingon DC.

Rocker, S., 2019. Value chain coordination: A new strategy for developing soft infrastructure in regional agri-food systems in the United States. (Doctoral Dissertation). OCoLC 1107171233. Pennsylvania State University, University Park, Pennsylvania.

Roethlisberger, F., Dickson, W., 1939. Management and the Worker. Cambridge University Press, Cambridge.

Scott, C., Richardson, R., 2021. Farmer social connectedness and market access. J Agric Food Syst Community Dev 10 (2), 1–23.

Sharp, J.S., Flora, J.L., 1997. Entrepreneurial social infrastructure and growth machine characteristics associated with industrial-recruitment and self-development strategies in nonmetropolitan communities. Community Development 30 (2), 131–153.

Snijders, T.A., 2001. The statistical evaluation of social network dynamics. Sociol Methodol 31 (1), 361–395.

Stork, D., Richards, W.D., 1992. Nonrespondents in communication network studies: problems and possibilities. Group & Organization Management 17 (2), 193–209.

Thilmany McFadden, D., Conner, D., Deller, S., Hughes, D., Meter, K., Morales, A. et al. (2016). The Economics of Local Food Systems: a Toolkit to Guide Community Discussions, Assessments and Choices U.S. Department of Agriculture, Agricultural Marketing Service Report. Posted at: https://www.ams.usda.gov/sites/default/files/media/Toolkit%20Designed%20FINAL%203-22-16.pdf

Trivette, S.A., 2019. The importance of food retailers: applying network analysis techniques to the study of food systems. Agric Human Values 36, 77–90.

United States Department of Agriculture, 2016, March 31. Federal, Philanthropic Partners Join to Strengthen Local Food Supply Chains, "Food LINC" to Boost Farm Sales, Grow Local Foods Sector in Ten Selected Regions [Press Release]. Retrieved from https://www.ams.usda.gov/press-release/federal-philanthropic-partners-join-strengthen-local-food-supply-chains-food-linc.

Valente, T.W., Pitts, S.R., 2017. An appraisal of social network theory and analysis as applied to public health. Annual Review of Public Health 38, 103–118.

# Participatory modeling of the food system: The case of community-based systems dynamics

*Alannah R. Glickman[a], Jill K. Clark[a] and Darcy A. Freedman[b]*

[a] The Ohio State University, United States of America [b] Case Western Reserve University, United States of America

## 12.1 Introduction

This chapter explores community-based systems dynamics (CBSD) as a mechanism for integrating members of the public into food systems modeling processes that ultimately aim to shape food policy and effect systemic change. Food systems present policymakers with many examples of wicked problems–food insecurity, food waste, diet-related chronic disease, access to healthcare for food sector employees, and more. Some scholars have argued that these types of problems should lead policymakers to reconsider the food system as a public commons (i.e. a shared and collectively governed resource) rather than a collection of privately held and managed commodities (Vivero-Pol, Ferrando, De Schutter, & Mattei 2018). Viewing food systems as a public commons implies a need for collective community-based decisions around food systems management and policy. As discussed in previous chapters, systems dynamics modeling offers a potential tool for policymakers to identify innovative tipping points and move their policy initiatives from incremental shifts to more comprehensive solutions. CBSD specifically seeks to incorporate and center the lived experiences and wisdom of stakeholders and community members into the modeling process. Whereas other types of modeling processes invite feedback from the public along the way, CBSD participants build capacity in systems dynamics and collectively create systems models themselves. In addition, CBSD

*Food Systems Modelling.*
DOI: https://doi.org/10.1016/B978-0-12-822112-9.00003-5

emphasizes centering communities that have been historically marginalized within research and policymaking processes. Integrating lay members of the public into a systems modeling process presents both practical challenges and the promise of using systems modeling to develop a shared vision that articulates actionable and community-based policy changes. This chapter provides a practical overview of the CBSD modeling process, and provides a case study of foodNest 2.0, a multiyear CBSD food system project based in Cleveland, OH that extends the CBSD model to incorporate a deliberative and situated system approach.

## 12.2 Community-based systems dynamics

Group model building (GMB) emerged in the 2000s in response to the failure of systems dynamics models to shape public policy (Hovmand, 2014). GMB entailed a collaborative process of model building, but often remained restricted to private sector and/or government employees (Rouwette, Vennix and van Mullekom, 2006). Community-based systems dynamics (CBSD) extended GMB to focus explicitly on the inclusion of community members throughout the modeling process. Rather than researchers and policymakers collaboratively using systems dynamics modeling to craft a top down policy approach, "CBSD is about building the public constituency to support the policy reversals that can address the root causes of dynamic problems from a feedback perspective. It is about engaging communities, helping communities cocreate the models that lead to system insights and recommendations, empowerment, and mobilizing communities to advocate for and implement changes based on these insights" (Hovmand, 2014, p. 6). In other words, CBSD focuses on the perspectives of community members embedded within the systems that the modeling process hopes to understand. Furthermore, community empowerment is an essential outcome of the process, in addition to knowledge creation and policy recommendations. The following subsections provide a brief overview of the main components of CBSD. This chapter builds on Hovmand (2014) book, *Community-Based Systems Dynamics*, an essential resource for anyone considering undertaking a CBSD project.

### 12.2.1 Project planning: problem identification and the importance of reflexivity

The first step of any CBSD project is confirming whether or not the problem of interest is indeed a systems problem. As stated in Chapter 2, systems issues involve feedback loops that lead to changes over time (Richardson, 2011).[1] Once the planning team confirms that systems dynamics is an appropriate approach, they must clarify what insights they anticipate emerging from the project and how these insights relate to the inclusion of community members in the modeling process. In other words, why use CBSD rather than other systems dynamics modeling?

There are several reasons that systems dynamics modeling might benefit from the inclusion of community members. First, systems dynamics aims to make mental models explicit. The absence of community members' perspectives may miss important information based on lived

[1] See chapters 2 and 10 for more detail on systems dynamics.

experiences and lead to biased models. This remains particularly true when modeling systems problems that affect communities that have been historically marginalized, racialized, and underrepresented in government leadership and research. A local initiative that uses systems modeling to understand the dynamics of food insecurity, yet excludes community members with experiences of food insecurity, may misrepresent or dismiss important dynamics. An incomplete and/or biased model may reveal potential "solutions" that are ineffective, at best, or harmful, at worst.

In addition, CBSD aims to build collective political power around shared visions of community-based solutions. Systems dynamics models created solely by experts may point to policy changes that, due to the exclusion of community members, are unlikely to garner political support. Context is important; the research question, study site(s), affected communities, and funding should all be considered when deciding to pursue a CBSD project.

As part of the initial planning process, the planning team must also consider how they define the relevant community (or communities) for the project. Hovmand (2014) writes, "What is critical to recognize in CBSD is that the drawing and redrawing of boundaries is a fluid process and something that is done by those within the community... In CBSD, we therefore begin to understand what is meant [by community] by asking those who already have ties to the community how they define community and continually trying to understand their use of the term 'community' within the context of their discussions" (p. 7). Thus, building connections with community members remains an important first step that then shapes the boundaries of the community from which the CBSD planners will recruit participants. In the context of food systems problems, a CBSD process focusing on school nutrition, for example, will involve different communities than one focusing on farm labor in the organic sector.

Before committing to the implementation of a CBSD project, project leaders should undertake an internal examination of values and goals. Clark et al. (2017) find that reflexivity among designers of public participation processes relates to perceptions of social equity and corresponding design choices, such as recruitment and facilitation. Reflexivity refers to the act of "examining critically the assumptions underlying our actions, the impact of those actions, and from a broader perspective, what passes as good" (Cunliffe, 2016). In the case of food systems research, this means individuals conducting food systems research must examine their social identities and lived experiences, the assumptions underpinning their worldviews, how those assumptions shape their work, and the underlying values and norms of food systems research at large.

In the U.S. context, structurally embedded systems of discrimination, such as racism, sexism, ableism, and classism, and a legacy of exclusionary policies has resulted in chronic underrepresentation of marginalized groups in academia and at higher levels within government (Griffin, 2019). CBSD planners, especially those who do not have lived experience within communities affected by the systems problem at hand, must thus reflect on their positionalities and implicit mental models to avoid implementing ineffective, and potentially harmful, group modeling processes. Adopting a stance of active reflexivity will also help CBSD planners build the long-term relationships of mutual respect required for the process.

Furthermore, those designing CBSD processes that investigate problems in historically marginalized communities must understand and acknowledge the role of historical power relations, and especially structural racism,[2] as a key building block of the project (Holley, 2016). Since CBSD elucidates shared mental models for the end goal of emancipatory and community-based policy change, CBSD planners must be able to embed such an historical analysis. Many resources exist to facilitate these two aspects of pre-CBSD preparation, and the specific needs of a given team will reflect the particular policy problem and communities they plan to engage through the CBSD process. As a starting point, some helpful resources include: the Kirwan Institute for the Study of Race and Ethnicity, Academics for Black Survival and Wellness, Duke Sanford World Food Policy Center's Report "Identifying and Countering White Supremacy Culture in Food Systems," Michigan State University Center for Regional Food Systems' annotated bibliography on structural racism in the food system, and the Rural Advancement Foundational International Food Justice Toolkit.

### 12.2.2 Core modeling team recruitment

The heart of CBSD is the Core Modeling Team (CMT). The CMT is a relatively small group of individuals composed of members of key stakeholders with lived experiences as well as roles and responsibilities related to the topic of interest. The CMT is tasked with designing, facilitating, and participating in the group model building activities that comprise the CBSD process. Though Hovmand (2014) recommends keeping the size of the CMT to five-to-seven people and allocating CMT members sole responsibility over facilitation, others conceptualize the CMT as a larger group (usually no more than 15–20) of community members participating in the CBSD process when needed to capture critical perspectives. In the latter case, the original planning team often plans and facilitates the first model building activities, and as CMT members grow capacity in systems thinking and modeling, they contribute to planning and facilitation. Since the case study discussed in this chapter utilizes the second approach, the following sections will define the CMT as the group of community participants and the planning team as the team that initiates the project, participates in the planning phases, receives funding for the project, and designs and implements the first several CMT meetings.

Who should the planning team seek to recruit for the CMT? Hovmand writes,

> It is important to have individuals who are in a position to anticipate how various community members from different stakeholder groups might respond to exercises, know the local language and customs, have some sense of what would be meaningful deliverables at the end of a workshop, and be able to identify potential issues related to power dynamics and cultural appropriateness of the activities. It is also essential that each person on the CMT is someone who will raise questions and challenge assumptions about the design and process, as opposed to deferring to the system dynamics experts in the room. It is often ideal to have representatives from various stakeholder groups from the community, but if this is not feasible, people who can serve a proxy such as direct service providers and community organizers are often a good choice. (2014, p. 55)

---

[2] Describing the connection between structural racism and racialized health inequities, (Gee and Ford, 2011) write, "Structural racism is defined as the macrolevel systems, social forces, institutions, ideologies, and processes that interact with one another to generate and reinforce inequities among racial and ethnic groups (Powell, 2008). The term *structural racism* emphasizes the most influential socioecologic levels at which racism may affect racial and ethnic health inequities. Structural mechanisms do not require the actions or intent of individuals (Bonilla-Silva, 1997). As fundamental causes, they are constantly reconstituting the conditions necessary to ensure their perpetuation (Link and Phelan, 1995). Even if interpersonal discrimination were completely eliminated, racial inequities would likely remain unchanged due to the persistence of structural racism (Jones, 2000)."

This perspective underscores a core value of CBSD: that the process requires shifting power over design and implementation from those on the planning team (often academic researchers or other organizational leaders) to CMT members to ensure that the approach remains relevant and valued by the community. The quote also highlights the importance of seeking CMT members with different perspectives on the systems problem at hand.

Since the planning phase includes defining the relevant community (or communities) through relationships with key stakeholders, building on these key stakeholders' knowledge and social capital through network-based targeted recruitment efforts is an efficient and effective strategy to reach potential CMT members. Recruiting CMT members requires an honest proposal that indicates the time commitment, potential compensation, and ultimate outcomes of the project. CBSD often requires attending multi-hour meetings monthly or biweekly for over a year, and potentially for several years. In addition, depending on funding, the project may not have the resources to financially support the potential actions identified as tipping point solutions for systems change. Participants hoping to implement solutions may be disappointed if this is not possible (or at least not guaranteed by project funding). Thus, it is important to offer potential CMT members a clear and frank picture of the benefits and outputs of the process. Finally, ensuring that CMT members receive fair compensation for the time they dedicate to the project aligns with CBSDs value of community empowerment.

### 12.2.3 Group model building: activities and outputs

Once the planning team has recruited all CMT members, group model building can commence. Group model building includes a series of formalized activities, noted in **bold** throughout the rest of this section. Scripts for these activities can be found via a wikibooks website dedicated to CBSD called Scriptapedia. Early group meetings often include **Hopes and Fears**, an activity that invites participates to name their hopes and fears in order to elicit and establish CMT members' expectations for the project (Scriptapedia). These expectations help shape the following meetings and group goals. The next activity, **Graphs Over Time**, involves CMT members hand-drawing graphs where the X-axis is time and the Y-axis is a variable related to the problem at hand. The resulting graphs can be used in framing the problem, generating variable ideas, and deciding on the reference mode, or starting point, for the modeling process (Scriptapedia). These two activities complement each other as CMT members may graph a variable they desire to change in the direction they hope or fear. Fig. 12.1 depicts an example graph over time that demonstrates a hope that fresh fruits and vegetables will become more affordable over time.

After **Graphs Over Time**, CMT members take multiple variables from the graphs over time and begin to draw causal relationships between variables to create **Causal Loop Diagrams** (CLDs). To indicate a causal relationship, CMT members draw arrows between variables. The arrows will be marked with a "+" or "-" depending on whether a change in the focal variable (the one where the arrow is drawn from) causes the other variable (the variable the arrow points to) to increase or to decrease. Two variables may both cause change in the other (resulting in two arrows between the variables, one in each direction). Over time, the CMT incorporates new variables in the diagram, drawing causal arrows to and from each new variable as they relate to existing variables. As the model develops, the CMT can identify

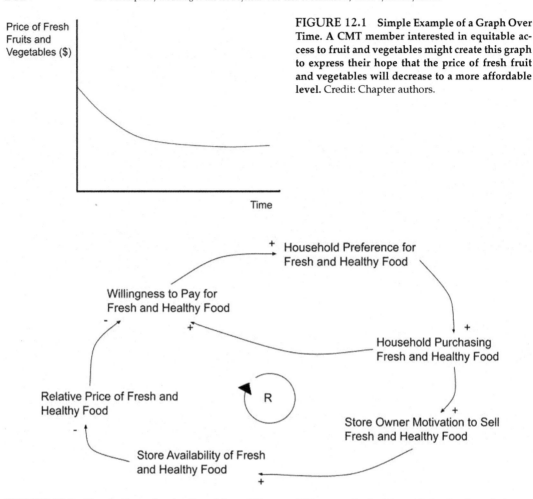

Price of Fresh
Fruits and
Vegetables ($)

Time

FIGURE 12.1 **Simple Example of a Graph Over Time. A CMT member interested in equitable access to fruit and vegetables might create this graph to express their hope that the price of fresh fruit and vegetables will decrease to a more affordable level.** Credit: Chapter authors.

FIGURE 12.2 **Simple Example of a Causal Loop Diagram. This example depicts a reinforcing loop between food pricing, food shopping, food preferences, and food availability.** Credit: Chapter authors.

full feedback loops and name them as components of the larger system. The feedback loops fall into two categories: (1) 'balancing' loops which pull the system into equilibrium–these loops have negative polarity (an odd number of negative arrows)—or, (2) "reinforcing" loops which pull the system out of homeostasis, creating exponential results–these loops have all positive arrows or an even number of negative arrows.[3] Fig. 12.2 depicts a basic example CLD that builds from the graph over time in Fig. 12.1. This CLD represents a reinforcing loop of relationships between food pricing, food shopping, food preferences, and food availability. Over time, the CMT collectively expands, complicates, and adapts their CLD.

*When is the CMT finished with its model(s)?* CBSD itself is an inherently iterative process and the CMT must expect and value adaptation over time. Yet, at some point, the CMT can agree

---

[3] See chapters 4, 7, and 10 for more information on feedback loops.

upon the main story of the model and future iterations of the model may make minor, but not substantial changes. Referring back to the guiding research questions and CMT goals can help the team identify when they have reached enough consensus on fundamental dynamics in the model to proceed to stock-and-flow models and simulations. In short, the model is done when the CMT is ready to move on.

Once consensus is achieved, the team may decide to translate the causal loop diagrams to **stock-and-flow models**. Building from the CLD structure, the team crafts a series of stock-and-flow models that build on the causal chains of the CLD. The team may also simplify the CLD to remove redundant variables and/or continue to adapt the model based on new CMT iterations. The stock-and-flow structure allows the team to run simulations to see the hypothesized effects to key outcome variables given specific changes to the system. This phase is often led by modeling experts, with less leadership and involvement by the CMT, due to the additional learning curves associated with technical skills related to data standardization and management, and the use of simulation software. However, the CMT plays a key role in determining which policy changes are important targets for simulation. Creating stock-and-flow models from CLDs requires decisions about data. This process involves the CMT examining diverse data sources ranging from publicly available prevalence data (e.g., number of grocery stores in the neighborhood from the USDA Food Atlas) to local understandings of phenomena (e.g., level of trust in the community based on perceptions among the team). Best practices for data gathering and standardization include maintaining a shared, and thorough, data dictionary, which lists each variable definition, unit of measure, and data source. In addition, writing clear coding scripts that indicate how each variable has been modified helps team members remember how raw data were transformed.[4]

Across all the aforementioned modeling activities, best practices for the planning team include the clear assignment of activity facilitator and participant roles, the intentional planning of activities through the use of scripts and recording notes throughout all activities. In addition, given the challenges inherent in collectively creating a model based on the mental models of several individuals with different backgrounds, perspectives, and life experiences, team members must take care to monitor group dynamics and provide space for reflection, deliberation, and conflict resolution. Scriptapedia includes instructions for several activities to support healthy group dynamics.

### 12.2.4 Next steps: implementation and evaluation

CBSD is designed to increase learning and capacity, and to build the trusting, long-term relationships necessary to collectively begin implementing solutions related to the multiple leverage points that the CMT identifies (Hovmand, 2014). Given the nature of systems change, it is difficult to prescribe, in advance, what will be implemented (Hovmand, 2014). Potential next steps include additional modeling, a focus on community organizing around a potential policy solution, or the implementation of a particular intervention. Goals for change may be directly from the systems dynamics models or simulation results, or may draw from what was learned in the modeling process. Low-hanging fruit might be picked first, while longer

---

[4] See chapters 4 and 9 for more information on stock-and-flow models and chapters 4, 7, 9, and 10 for more information on simulations.

cultural change is underway (Hovmand, 2014). Importantly, community members, rather than the planning team, guide the implementation phase of a CBSD project.

Teams thinking about the future of a CBSD project may benefit from insights gleaned from process and outcome evaluations. Since the CBSD process usually includes several opportunities for group reflection and feedback (such as the informal assessments of hopes and fears), it remains important that note takers record all feedback and team members demonstrate how the feedback will shape (or is already shaping) the project. A more formal process evaluation might include CMT member interviews and surveys inquiring about elements of deliberation, conflict management, and learning. Because the outcomes that result from myriad interventions inspired by a CBSD project may take several years, CBSD evaluators should consider using longitudinal outcome evaluations (Hovmand, 2014).

## 12.2.5 Potential extensions of CBSD

The sections above highlight the core components of CBSD. Since CBSD is by nature an iterative process, a CBSD project may benefit from incorporating additional methods and roles. The foodNEST 2.0 case study, discussed below, conducted a series of key informant interviews to both extend and deepen causal loop diagrams developed by the CMT. The inclusion of interviews allowed the CMT to include perspectives from stakeholders and community members unable to make the time commitment necessary to be a CMT member. In addition, as the CBSD process unfolds, the CMT may seek to recruit additional members or collaborate with external organizations that may offer additional information or resources to sustain the project in the collectively agreed upon direction. The case study below will discuss how the inclusion of these two elements added to the CBSD process. The main point of this section is to highlight that CBSD is an adaptable process that will likely change over the course of a project. In fact, this adaptability remains an important advantage of CBSD and an awareness of this flexibility and a willingness to embrace creative iteration will benefit CBSD planners and participants. The next segment of this chapter will describe the foodNEST 2.0 case study in detail, following the CBSD phases outlined above.

## 12.2.6 Useful resources

To learn more about CBSD, we recommend the following resources:

- Hovmand, P. S. (2014). *Community based system dynamics.* Springer, New York, NY
- Scriptapedia
- Washington University in Saint Louis Social System Design Lab
- International Conference of the System Dynamics Society Summer School

## 12.3 Case study: foodNEST 2.0, modeling the future of food in your neighborhood

Led by Darcy Freedman, PhD, MPH, Swetland Professor of Environmental Health Sciences at Case Western Reserve University, the foodNEST 2.0 study launched as one of five grantees of the Foundation for Food and Agriculture Research's "Tipping Points" grant program. The foodNEST 2.0 planning team sought to better understand complexities of local food systems, which strategies of change work best in different environments, as well as how parts of the food system influence one another and can be changed or combined to boost their impact on local food systems as well as overall community health and the economy. In 2018, a Core Modeling Team comprised of academic researchers and community leaders in Cleveland, Ohio, united to better understand how food systems in historically redlined neighborhoods could be leveraged in new ways to realize goals of equity and justice. Building on a long history of innovation and investment in the Greater Cleveland food system, the team decided to take a step back to examine the forces that shape the foods we are able to put on our collective Tables through a process of community-based system dynamics modeling. The foodNEST 2.0 team grounded this work by focusing on the web of forces–held together by structural racism– that shape food systems in historically redlined neighborhoods in Cleveland, Ohio (Modeling the Future of Food in Your Neighborhood Collaborative, 2020).

The following sections highlight the phases, successes, and challenges of foodNEST 2.0 and conclude with lessons learned.

### 12.3.1 Planning

foodNEST 2.0 builds on foodNEST 1.0, a quasi-experimental evaluation of a food systems intervention in Cleveland. The NIH-funded study collected rich environmental and community data, including food retailer audits, diet recalls, and surveys, to evaluate the creation of a food hub in a Cleveland neighborhood with few stores selling fresh and healthy foods. Several useful insights were gleaned from the foodNEST 1.0 project (Freedman et al., 2020). First, the food hub project was not implemented as expected. Understanding how and why the intervention changed over time drew attention to the importance of political and social dynamics often lightly considered or overlooked in food systems intervention evaluations. Second, the team realized that the idea of focusing on a single parameter, creating one new food hub, without focusing on more systematic feedback mechanisms limited the team's understanding of food system dynamics in the neighborhood and how and which interventions might affect outcomes.

In 2017, the Foundation for Food and Agriculture Research (FFAR) issued a call for proposals for its "Tipping Points" grant program. The grant program aimed to support the development of "mathematical and computational models of how factors and interventions within local food systems interact" (Foundation for Food and Agricultural, 2020). In response to FFAR's call for proposals, a research team including many of the investigators and community partners from foodNEST 1.0 as well as additional academic experts in systems dynamics, proposed a CBSD modeling project investigating local food systems dynamics in three historically redlined Cleveland neighborhoods. Having a community-university planning team allowed for transparency about goals, commitment, and compensation.

Redlining refers to the practice of state-sanctioned denial of mortgage loans to residents in particular neighborhoods based on, among other factors, a given neighborhood's ethnic and racial composition (Reece, 2021). The extension of these loans, starting as part of the New Deal in response to the Great Depression, became a main vector for wealth accumulation among the growing White middle class (Reece, 2021). Denial of these loans led to decades of neighborhood disinvestment in predominately Black and Hispanic neighborhoods and has been widely described as one of the major causes of the current racial wealth gap (Darity Jr et al., 2018). Though the practice of redlining became illegal in 1968 with the passage of the Fair Housing Act, residents of historically redlined neighborhoods continue to face low home prices, higher poverty rates, de facto racial segregation, and higher rates of health issues, including diet-related diseases (Mendez, Hogan and Culhane, 2014; Appel and Nickerson, 2016; Rothstein, 2017; McClure et al., 2019; Nardone, Thakur and Balmes, 2019; Collin et al., 2021; Ware, 2021).

The team chose to focus on four neighborhoods that had demonstrated investment (>$15 million) in food systems initiatives such as grocery store development, fruit and vegetable incentive programming, food pantries, and social marketing campaigns to promote healthy eating. Despite such investments, these neighborhoods had persistent challenges related to food insecurity, low access to fresh and healthy foods, and diet quality. The neighborhoods are located in historically redlined areas of Cleveland and remain majority African American today. In conceptualizing the study, the team viewed the food-related problems in historically redlined neighborhoods as systems issues[5] and believed that a systems dynamics approach would reveal the limitations of past interventions and highlight innovative solutions. In addition, building on the growing body of literature underscoring the racialized nature of food systems in the US, the team aimed to ground the project in a nuanced understanding of how structural racism shapes residents' experience of their local food systems.

The team chose to use CBSD, rather than generic systems dynamics modeling, to achieve the following goals: (1) cultivate trusting research relationships that are committed to power sharing, (2) democratize the research process through meaningful engagement of multiple stakeholders and collaborators, (3) support co-learning that is situated in lived experiences, and (4) generate actionable knowledge that inspires transformation of the food system to achieve justice. These goals were established to achieve the primary aim of the foodNEST 2.0 study to develop credible and relevant decision support tools that offer a systems lens to advance community driven food systems change. CBSD fit the team's research goals and aligned with team members' values, understanding of structural racism, and commitment to social equity.

Furthermore, we chose to take a deliberative systems approach to CBSD (Freedman et al., 2021). This means that our CBSD project centered on deliberation as a mechanism to uncover food system insights based on lived experiences through active listening, contestation, and reflection. The CMT paid close attention to issues of participatory inclusion, by encouraging storytelling, honoring dissent, adapting process as power differentials surfaced, taking an anti-racist stance, and focusing on current assets and hopes for the future (Holley, 2016). At 3.5 years long, foodNest 2.0 committed to a longer timeframe than most CBSD projects, a timeline enabled by the FFAR Tipping Points grant.

## 12.3.2 Core modeling team (CMT) recruitment

Recruitment of the CMT occurred by mapping out key perspectives desired to inform the modeling process—this was an evolving process as modeling efforts revealed gaps in the composition of the group. Recruitment was organized to attract collaborators from three broad

[5] See Chapter 2 for more information on defining systems problems.

groups including community residents, food retailers and providers, and regulators of the system. The study had a goal of recruiting at least one community leader from each of the four neighborhoods that were the focus of the modeling project. Food retailers and providers included people involved in both for-profit and non-profit food provision such as a chef, leaders from food bank organizations, community gardeners, and representatives of community development corporations. Regulators included people from government agencies, academia, public health, and health care. The CMT included about 20 community stakeholders and 10 academic partners. Some members remained on the CMT for three years while others rotated due to life events such as a new job, retirement, or childbirth. Several members of the CMT remained engaged throughout the process, demonstrating a deep commitment to the project and helping the project proceed on course even as the composition of CMT changed slightly over time.

The initial members of the CMT were identified when the grant was submitted, and new members were recruited through their engagement with study activities. For example, the study created a fellowship program during its second year to recruit community leaders to support dissemination of study findings at a public convening. Three of the fellows eventually became members of the CMT. All CMT members were compensated for their time related to the study either through a commitment made by their organization or by the grant funding the study. This demonstrated shared understandings of the value of the work of the modeling process. The CMT currently includes neighborhood leaders from the four study neighborhoods and representatives of the emergency food system, cooperative extension, local foundations, local government agencies, urban food enterprise, and food- and nutrition-focused nonprofits (Fig. 12.3).

In addition to the CMT, the study team members conducted 22 semi-structured interviews with individuals who represented important perspectives not covered by CMT members. In particular, interviews with younger community residents, residents facing housing insecurity, and food storeowners complemented the collaborative work of the CMT by deepening the CMTs understanding of neighborhood food systems dynamics with nuanced examples and personal stories.

### 12.3.3 Group model building activities

CMT members participated in one multi-day retreat each year, as well as three-hour, bimonthly meetings. In the early CMT meetings, the planning team (composed of principal investigators, post-doctoral scholars, doctoral students, and program staff) designed and facilitated the activities. Members of the planning team also coordinated logistics, maintained communication, and served as note-takers. Early meetings focused on two main tasks: (1) developing a sense of partnership and (2) introducing systems thinking through guided activities. The first retreat, which took place on the CWRU campus in July 2018, included the following activities: hopes and fears, principles of partnership (a collectively created agreement on the shared values guiding the team's work and relationships), problem definition using photos and Twitter tweets, graphs over time, and introductory causal loop diagrams. The retreat also included several opportunities for socializing, storytelling, and reflection. Activity leaders utilized deliberation and voting to elicit areas of consensus and areas of dissent throughout the retreat. Bimonthly meetings further developed definitions and

**FIGURE 12.3    The foodNEST 2.0 Core Modeling Team, May 2019.** Credit: Mary Ann Swetland Center for Environmental Health, Case Western Reserve University, Cleveland, OH.

feedback loops discussed at the retreat. The planning team also frequently checked in with CMT members outside of meetings.

In general, the annual retreat focused on clarifying the major goals and initiating activities tasks related to those goals, while bimonthly meetings continued to push the tasks forward. The gatherings of year one generally centered around the development of causal loop diagrams, those of year two focused on transitioning from qualitative CLDs to quantitative models and simulations, and year three's meetings concentrated on community sharing, designing decision-making tools, and planning the future of the project.

An important component of CMT gatherings was the team's commitment to a deliberative model of communication. Meetings were designed so that members, each bringing different forms of knowledge, reflected on collective understandings of how the food system works, and does not work, and its disparate effects on residents in the four neighborhoods of focus. This included encouraging the CMT to pay careful attention to words used, including specific variables in the model, and dig into the meanings behind them. Further, members were encouraged to develop narratives of how variables worked together to create a system and metaphors to convey meaning. This deliberative approach resulted in meaningful changes, such as the shift in use of "food desert" to "food apartheid." The latter term indicates that lack of access to healthy food does not arise naturally, but through a legacy of residential and retail redlining and disinvestment in racialized and low-income communities. The term "food apartheid" makes explicit the power dynamics underlying food systems, and essentially changes the problem definition.

## 12.3.4 Causal loop diagrams

The study team began the modeling process with *a priori* "tipping points" identified including use of emergency food assistance, fruit and vegetable incentives, food retail, and social marketing for healthy eating. These were selected because there was substantial investment in these four areas in the study context. Building CLDs with the CMT about these four *a priori* leverage points, however, resulted in a reconceptualization of "tipping points" for shifting the food system to achieve justice in historically redlined neighborhoods. Over three years, the CMT iterated and adapted the CLD into the current version (Fig. 12.4), which highlights three main domains of feedback: (1) Meeting basic food needs with dignity; (2) Market-based supply and demand for fresh and healthy foods; and (3) Community empowerment and food sovereignty (Table 12.1). Within these three domains, the team connected 10 feedback loops (balancing loops B1-B4 and reinforcing loops R1-R6) structuring local food systems in racialized urban neighborhoods (Table 12.1). In addition to CMT deliberation, the CLD also incorporates data from 22 key informant interviews designed to elicit points of view and lived experiences from those not on the CMT as well as information gathered from a series of public events, described in more detail in a following section.

This CLD highlights the connections between food systems issues, like food insecurity and lack of access to healthful food, and problems emerging from other systems, like housing and incarceration. Some of the most fruitful discussions and questions that emerged throughout CLD development revolved around which factors should be considered exogenous (external) to the model and which factors must be included in the system's dynamics, and how to place boundaries on the model. Boundary setting was a difficult and iterative process—the main

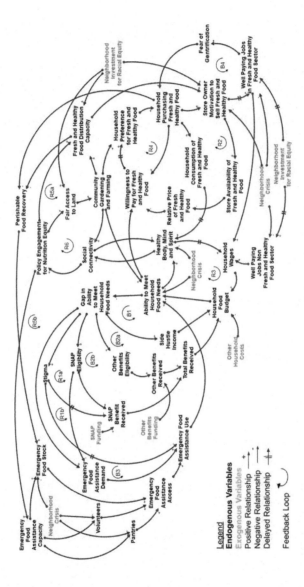

**FIGURE 12.4 foodNEST 2.0 Core Modeling Team's Complete Causal Loop Diagram, March 2021.** Credit: Mary Ann Swetland Center for Environmental Health, Case Western Reserve University, Cleveland, OH.

**TABLE 12.1** Feedback mechanisms identified within foodnest 2.0 causal loop diagram.

| Domain of feedback | Feedback mechanisms (loop label) | |
|---|---|---|
| **Domain 1** Meet Basic Food Needs with Dignity | ○ Side Hustle (B1)<br>○ Government Benefits (B2a, B2b) | ○ Emergency Food Assistance (B3)<br>○ Stigma and Stereotypes (R1a, R1b) |
| **Domain 2** Market-based Supply and Demand for Fresh and Healthy Foods | ○ Healthy Food Retail (R2)<br>○ Job Security (R3) | ○ Food Culture and Norms (R4) |
| **Domain 3** Community Empowerment and Food Sovereignty | ○ Community Power (R5a, R5b)<br>○ Urban Agriculture (R6) | ○ Risk of Gentrification (B4) |

Exogenous factors moderating the feedback mechanisms: neighborhood crises such as social (i.e., incarceration and policing, homelessness, domestic violence) and health (i.e., addiction, COVID-19) crises; (b) neighborhood investment for racial equity through provision of social, financial, material, and human capital, (c) other household costs such as expenses for transportation, child care, housing, and debt, and (d) funding for government benefits such as food assistance, disability, and health care.

philosophy was to consider broader forces outside of the neighborhoods as exogenous. For example, the concept of "neighborhood investment for racial equity" was considered critical for understanding potential change within the study neighborhoods, but was considered exogenous, while the "fear of gentrification" was deemed endogenous.

Fig. 12.5 focuses on the community empowerment and food sovereignty domain. This domain includes four feedback loops. R5a and R5b relate to community power, R6 relates to urban agriculture, and B4 relates to fear of gentrification. As an example, R5a connects the following variables in a loop: policy engagement for nutrition equity→fresh and healthy food distribution capacity→store owner motivation to sell fresh and healthy food→well-paying jobs in fresh and healthy food sector→household wages→household food budget→ability to meet household food needs→healthy body, mind, and spirit→social connectivity→policy engagement. This reinforcing loop suggests that a shift in any of these variables may stimulate a cycle of change that affects many aspects of the system. Though the CLD does not indicate the normative value of any variable because the system models change in any direction, the CMT viewed an increase in healthy body, mind, and spirit as a key desired outcome of food systems change. Many variables in R5a are also included in several other feedback loops, underscoring the interconnections between variables traditionally considered within the realm of food systems, such as emergency food system use and availability of fresh and healthy food, and those typically thought beyond the food system, such as housing policy, social capital, and civic engagement. The remaining feedback loops are described in more detail in (Freedman et al., 2021).

### 12.3.5 Community convenings

In addition to regular CMT meetings and annual CMT retreats, foodNEST 2.0 CMT and planning team members also organized annual public convenings. In 2019, about 150 people attended the day-long gathering, which included an overview of the project, an introduction to systems thinking around food systems problems, and several group activities for collectively

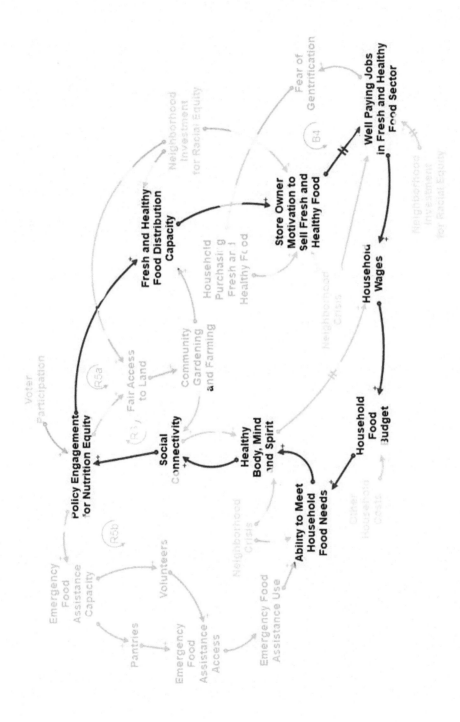

FIGURE 12.5    foodNEST 2.0 Core Modeling Team's Causal Loop Diagram of Domain 3 (Community. Empowerment and Food Sovereignty), April 2021. This domain includes feedback loops related to community power (R5a and R5b), urban agriculture (R6), and gentrification (B4). Credit: Mary Ann Swetland Center for Environmental Health, Case Western Reserve University, Cleveland, OH.

developing a vision for a shared food future (Figs. 12.6 and 12.7). The convening also included a panel of local elected officials discussing best practices for engaging policy makers to advocate for the food system solutions discussed during the convening. The convening not only spread the word about the project, but also elicited important community feedback on the emerging systems models created by the CMT. Another positive outcome of the convening included the recruitment of new CMT members.

The 2020 annual convening took place virtually, due to the COVID19 pandemic. In addition, the 2020 convening was formatted as a partnership with the Cleveland Cuyahoga County Food Policy Coalition (CCCFPC). The ninety-minute meeting formally introduced one of foodNEST 2.0's decision-making tools, the Menu of Actions, described in the following section. The CMT also held six smaller convenings in partnership with the CCCFPC to further deliberate on properties of the local food system and potential leverage points.

## 12.3.6 The Menu of Actions

Since the goals of foodNEST 2.0 include the development of a suite of decision support tools that communities can use to advocate for food systems change, in year three of the project, foodNEST 2.0 members began designing a document to highlight the potential leverage points identified by the CLDs. The CMT collectively created the document, eventually named the "Menu of Actions for Community Driven Food Systems Change" (or the Menu of Actions, for short), through a process of deliberation and voting. The CMT consulted their co-created CLDs, stakeholder interviews, and public convening feedback, resulting in a document informed by hundreds of stakeholders in the Greater Cleveland area. In addition, the team consulted with a graphic designer and a poet with expertise in metaphor to ensure that the final product would be accessible, practical, and engaging for its target audience. As the Menu of Actions describes,

> "The Menu of Actions was developed for residents, food retailers, community leaders, and elected officials working to mobilize actions that lead to community food security through efforts that advance fair access to fresh and healthy foods as well as financial strength within households. It was designed for people living and working in historically redlined urban neighborhoods. Like the process used to form this menu, it is to be used in community with people coming together to co-create a vision for the future and build the trust needed to realize change" (Modeling the Future of Food in Your Neighborhood Collaborative, 2020).

The Menu of Actions highlights five potential leverage points: (1) Fair Access to Affordable Fresh and Healthy Foods; (2) Nourished Neighborhoods; (3) Neighborhood Thriving; (4) Economic and Community Development; and (5) Social Connectivity and Policy Engagement (Modeling the Future of Food in Your Neighborhood (Modeling the Future of Food in Your Neighborhood Collaborative, 2020). These potential leverage points reflect broader concepts for change that then inspire specific actions that may be sequenced, integrated, and/or tailored in different ways for different communities. Though each leverage point might be viewed as a desired outcome in and of itself, a systems approach necessitates thinking beyond variables as discrete outcomes and underscores the interconnectedness of each component. In other words, all variables in a CLD may be viewed leverage points because changes to them will necessarily affect other parts of the system. The Menu of Action highlights the five factors that have the

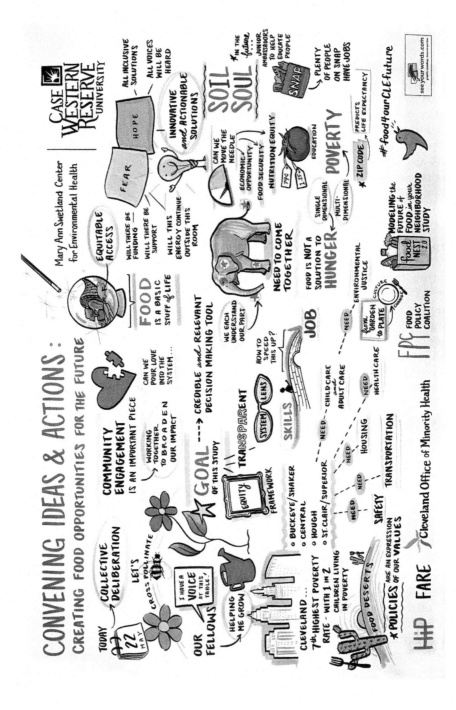

FIGURE 12.6  Graphic Recording of May 2019 Public Convening Deliberation. These images highlight several important themes gleaned from presenting foodNEST 2.0's original modeling work to a broader group of community members. Credit: Mary Ann Swetland Center for Environmental Health, Case Western Reserve University, Cleveland, OH.

FIGURE 12.7 Graphic Recording of May 2019 Public Convening Deliberation. These images highlight several important themes gleaned from presenting foodNEST 2.0's original modeling work to a broader group of community members. Credit: Mary Ann Swetland Center for Environmental Health, Case Western Reserve University, Cleveland, OH.

potential to affect significant improvements in economic opportunity, food security, and fair access to affordable, fresh, and healthy foods in historically redlined neighborhoods.

The Menu of Actions clearly defines each potential leverage point and highlights between five and eleven action items. Fig. 12.8 highlights the summary definitions for each of the five leverage points, and Fig. 12.9 provides an example of specific opportunities for transformation within the Social Connectivity and Policy Engagement leverage point.

Importantly, the Menu of Actions is not an exhaustive or prescriptive list of what is needed to transform each and every community. Instead, this menu serves as a starting point for *imagining* what it will take to ignite community driven food systems change. It is designed to begin conversations about hopes for our collective food future and provide ideas about pathways of actions that may move us there faster. Since debuting the Menu of Actions at the 2020 public convening, the team has published it on their project website to make it available to food justice advocates across the country.

## 12.3.7 Stock-and-Flow models and simulations

After the CMT reached consensus on the central dynamics of the causal loop diagrams, and concurrently with the development of the Menu of Actions, the team began to transition from building CLDs to stock-and-flow modeling. Though the CMT continued to adjust the CLD over time, the main dynamics remained intact and reflected a sense of readiness to transition to the next phase of the project. Due to the technical nature of stock-and-flow modeling, as well as quantitative simulations, this phase was guided by the team's resident experts on systems dynamics modeling.

Stock-and-flow and *simulation* modeling occurred in both smaller and full CMT group meetings. The smaller meetings were often more technical in orientation and were largely attended by academic members of the CMT. Larger meetings with the full CMT included activities to build capacity in understanding the utility of a stock-and-flow model and its capacity to simulate a "what if?" scenario of change. They also included time to practice using an online simulation tool that reflected core dynamics emerging from the broader food systems modeling project. The first stock-and-flow based simulations focused on increasing the availability of fresh and healthy foods in neighborhoods and led to discovery of system insights about levers for pulling both supply-side factors, such as establishing collective buying capacity for small retailers, and demand-demand factors, like engaging social influencers to promote fresh and healthy food products, to achieve this goal.

Collecting data for the stock-and-flow models included engagement with partners on the CMT who were able to offer organizational data about local trends, such as partners from the food bank sharing data on the number of people getting food from food pantries located in the study neighborhoods. It also included use of study-collected data such as longitudinal survey data generated initially through the foodNEST 1.0 study, which resulted in participant surveys about experiences with the food system collected among the same residents over five years. These data provided trend lines depicting behaviors resulting from food system dynamics in racialized urban neighborhoods. For instance, five years of food retail data revealed very little change in access to fresh and healthy foods despite efforts to improve food retail environments through initiatives such as corner store enhancements. These trends provide key insights about

# DEFINITIONS

### A. Fair Access to Affordable Fresh and Healthy Foods

Comprehensive, community driven approaches are in place that better link quality and affordable food supply with consumer demand so people get the foods they want.

### B. Nourished Neighborhoods

All residents have nourishment for optimal wellbeing—in body, mind, and spirit—to achieve fully individual, family, and neighborhood capabilities.

### C. Neighborhood Thriving

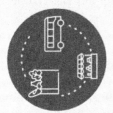

Connections within the neighborhood feed the soul of the community while growing local wealth and ownership.

### D. Economic and Community Development

Investments are made to advance neighborhood sovereignty and community driven development so people are empowered to thrive.

### E. Social Connectivity and Policy Engagement

Collective power is cultivated to transform political, social, and economic forces shaping community capacity to nurture dignified and flourishing lives.

**FIGURE 12.8** Definition of Five Leverage Points from *The Menu of Actions*, September 2020. Credit: Mary Ann Swetland Center for Environmental Health, Case Western Reserve University, Cleveland, OH.

## E. SOCIAL CONNECTIVITY AND POLICY ENGAGEMENT

Collective power is cultivated to transform political, social, and economic forces shaping community capacity to nurture dignified and flourishing lives.

**E1.** Get neighbors out to vote and ensure voters are aware of issues that directly and indirectly impact fair access to fresh and healthy foods through voter registration, voter education, and reducing barriers on voting day.

**E2.** Empower community champions who are committed to promoting fair access to fresh and healthy foods to take on positions in elected offices (i.e., organizational boards, city council, school board, local government, state legislature, US congress).

**E3.** Develop new or coordinate with existing grassroots coalition(s) to mobilize policy changes that have a direct impact on fair access to fresh and healthy foods.

**E4.** Train neighbors in advocacy skills to ensure messages about fair access to fresh and healthy foods are conveyed in a manner that ignites policy actions (i.e., testimony, op-ed, social media).

**E5.** Establishing a full-time position for a food policy coalition coordinator.

**E6.** Identify and increase utilization of avenues of input for community members to collectively review and authorize neighborhood changes that directly and indirectly impact fair access to fresh and healthy foods (i.e., zoning commissions, food licensing, purchasing processes, etc.).

FIGURE 12.9   **Specific Recommendations for the Social Connectivity and Policy Engagement Leverage Point, September 2020.** Credit: Mary Ann Swetland Center for Environmental Health, Case Western Reserve University, Cleveland, OH.

the status quo of the system and the type of leverage that will be needed to achieve hopes for the future.

After deciding on appropriate measures for CLD variables, the team began building stock-and-flow models using STELLA and Vensim simulation software. As of this chapter's writing, running simulations remains a current area of work for the foodNEST 2.0; future research papers will disseminate the simulation results.

### 12.3.8 Moving forward

The foodNEST 2.0 planning team initially envisioned the priorities of the third year of the project as tailoring simulation models to each of the study neighborhoods and designing a community sharing process that would transition the project firmly into the hands of the community. Originally designed to roll out between September 2020 and September 2021, the community sharing plan was delayed due to the COVID-19 pandemic. Facing a potential shift to a virtual community sharing process, the planning team further consulted with community members on the CMT. As an example of the iterative and adaptive nature of CBSD, the resulting conversations led to a shift in the design of the community sharing process as well as the nature of the project.

One theme that emerged from the conversations was a need to rethink the organizing structure of the project as it pivoted to the broader community. To help with this transition, the team sought the help of Renee Wallace, an experienced facilitator and CMT member for another CBSD project based in Flint, Michigan. Whereas the planning team had previously convened separate planning meetings, future meetings would be open to all interested. CMT members took on research-related responsibilities and planning roles.

The team then re-conceptualized the community sharing process as a food justice fellowship. The fellowship serves as a "test kitchen" for incubating community fellows' ideas related to potential leverage points identified in the Menu of Actions. The driving philosophy of the fellowship is that change is community-centered and informed by experimentation, co-learning and systems analysis. The fellowship kicked off in early 2021. After several years of learning, deliberating, and collectively visioning, the fellowship marks a transition to action.

Lessons Learned

1. **Shifting power from research institutions to communities (or, action matters).** Since CBSD remains a relatively new research practice (developed out of group model building in the late aughts), CBSD practitioners understandably desire to understand (and utilize or improve upon) best practices and highlight model co-creation as a major output of the process. The models created by the foodNEST 2.0 remain an important contribution to the literature on urban food systems and food justice. However, the prioritization (in terms of time and funding) of model development became a point of frustration for many CMT members who sought to put the project's findings into action. As described above, this led to a reconceptualization of the project's future with an emphasis on funding direct action projects led by program fellows and applying for grants to support community initiatives in addition to continued research efforts.

2. **A deliberation-based approach meaningfully extends CBSD.** Anchoring a multi-year CBSD process in listening, contestation, reflection and attention enhances participatory inclusion of lived experience.

3. **Language matters and storytelling builds mutual understanding.** Words chosen and an intent to understand their meaning resulted in several critical shifts in the project. For example, instead of equity, the project now focuses on community empowerment and sovereignty, making our work more relevant and meaningful to CMT members. The team also found it critical to allow time and space to provide personal narratives during the

activities. This helped to ground us in mutual understanding of the problem, and the broader system, and to develop trust, respect, and understanding between CMT members.

4. Qualitative research complements CBSD. The stakeholder interviews conducted in the first year of the project enriched team members' understandings of the CLDs. Since the CMT co-created CLDs based on a combination of lived experience, professional expertise, and implicit mental models, conducting formal qualitative research offered an opportunity to validate and/or critique important components of the CLDs.

5. Openness to change and adaptation is essential. As an inherently iterative and collaborative process, CBSD invites continuous interrogation of group norms and processes. Though careful planning remains important, responsiveness to CMT member feedback means that sometimes plans change. Often these changes help the team better align with its core values, thus improving both the process and outcomes of the project. Openness to expanding the CMT and consistent public outreach also allowed participants with new and important perspectives to join the project over time.

6. Boundaries help. One benefit of CBSM remains the modeling of complex problems as embedded within a larger system that functions with several feedback loops. Understanding a system often means making connections between various factors usually considered separately. However, broadening the understanding of a complex problem runs the risk of expanding the problem to include numerous interrelated systems, leading to an overwhelming model with unclear leverage points. At several points throughout the project, the team found it helpful to review project goals and use those goals to make decisions about what to include in the model and what to consider exogenous to the model. Having clear project goals can facilitate the negotiation of where to set model boundaries, a necessary aspect of systems dynamics modeling.

7. Trust in the collaborative. Trust, rooted in deep respect and appreciation for the contributions of each team member, was an essential ingredient to the entire process. It required work to build and maintain. When conflict emerges and challenges trust, it is important for the team to prioritize time for healing and reflection and to codify emergent lessons into formal structures within the team.

## 12.4 Conclusion

Community-based systems dynamics centers solution building with those most affected by a systems' problems. In the case of urban food systems, CBSD illuminates root causes of food insecurity, inequitable access to healthy foods, and economic instability. The collective insight of CMT members and other community stakeholders underscores the importance of understanding structural racism as a key factor shaping food systems in historically redlined neighborhoods and points to intersecting systems of disadvantage, such as mass incarceration and gentrification, as fundamentally intertwined with food systems. In the case of foodNEST 2.0, CBSD has been a helpful tool to reckon with the complexity of systems while envisioning potential leverage points for change.

Though it remains desirable and necessary to improve the lives of those living in historically redlined neighborhoods, systemic change can be difficult in a political system built to favor incrementalism. Systems dynamics modeling helps participants understand that systems often function as designed. In other words, the status quo currently benefits some who may be resistant to structural change. Facing this reality, rather than focusing on politically feasible surface-level solutions, can sometimes feel overwhelming and potentially disempowering. CBSD practitioners must reckon with the fact that CBSD can lead to transformative insight and that acting upon those insights remains complex and resource intensive. Conducting CBSD as a type of formal public participation, convened by a government agency rather than an external research institution, may help build capacity and political power around systems level change.

Finally, CBSD offers a tool to build long-term relationships, modeling expertise, and stakeholder connections within communities. As an ongoing iterative process, CBSD strives not only to elucidate community-based knowledge to improve systems models (i.e. epistemological justice), but to empower community members to take ownership of the research process and build political will to act upon identified leverage points. Future CBSD projects must build upon CBSDs existing strengths and work towards translating research into action.

# References

Appel, I., & Nickerson, J. 2016. Pockets of poverty: the long-term effects of redlining. *Available at SSRN 2852856*.

Bonilla-Silva, E., 1997. Rethinking racism: toward a structural interpretation. American Sociological Review 465–480.

Clark, J.K., Freedgood, J., Irish, A., Hodgson, K., Raja, S., 2017. Fail to include, plan to exclude: reflections on local governments' readiness for building equitable community food systems. Built Environ 43 (3), 315–327.

Collin, L.J., Gaglioti, A.H., Beyer, K.M., Zhou, Y., Moore, M.A., Nash, R., et al., 2021. Neighborhood-level redlining and lending bias are associated with breast cancer mortality in a large and diverse metropolitan area. Cancer Epidemiology, Biomarkers & Prevention 30 (1), 53–60.

Cunliffe, A.L., 2016. On becoming a critically reflexive practitioner" redux: what does it mean to be reflexive? Journal of Management Education 40 (6), 740–746.

Darity Jr,, W., Hamilton, D., Paul, M., Aja, A., Price, A., Moore, A., et al., 2018. What we get wrong about closing the racial wealth gap. Samuel DuBois Cook Center on Social Equity and Insight Center for Community Economic Development 1–67.

Freedman, D.A., Bell, B.A., Clark, J., Ngendahimana, D., Borawski, E., Trapl, E., et al., 2020. Small improvements in an urban food environment resulted in no changes in diet among residents. Journal of Community Health 1–12.

Foundation for Food and Agricultural Research. (2020). "Tipping Points." Accessed via https://foundationfar.org/programs/tipping-points/#:~:text=The%20Tipping%20Points%20Program%20explores,health%20and%20enhance%20economic%20outcomes.

Freedman, D.A., Clark, J., Lounsbury, D.W., Boswell, L., Burns, M., Jackson, M.B. et al., 2021. (In Review) Deliberative and Situated Systems Research to Advance Nutrition Equity in Racialized Urban Neighborhoods.

Gee, G.C., Ford, C.L., 2011. Structural racism and health inequities: old issues, New Directions. Du Bois review: social science research on race 8 (1), 115.

Griffin, K.A., 2019. Institutional barriers, strategies, and benefits to increasing the representation of Women and Men of Color in the professoriate: looking beyond the pipeline. Higher Education: Handbook of Theory and Research, 35, pp. 1–73.

Holley, K. (2016). The Principles for Equitable and Inclusive Civic Engagement: a Guide to Transformative Change.

Hovmand, P.S., 2014. Community based system dynamics. Springer, New York, NY.

Jones, C.P., 2000. Levels of racism: a theoretic framework and a gardener's tale. American Journal of Public Health 90 (8), 1212.

Link, B.G., Phelan, J., 1995. Social conditions as fundamental causes of disease. Journal of Health and Social Behavior 80–94.

McClure, E., Feinstein, L., Cordoba, E., Douglas, C., Emch, M., Robinson, W., et al., 2019. The legacy of redlining in the effect of foreclosures on Detroit residents' self-rated health. Health & Place 55, 9–19.

Mendez, D.D., Hogan, V.K., Culhane, J.F., 2014. Institutional racism, neighborhood factors, stress, and preterm birth. Ethnicity & Health 19 (5), 479–499.

Modeling the Future of Food in Your Neighborhood Collaborative. (2020). Menu of Actions for Community Driven Food Systems Change. Cleveland, OH: Mary Ann Swetland Center for Environmental Health, Case Western Reserve University.

Nardone, A., Thakur, N., Balmes, J.R., 2019. Historic redlining and asthma exacerbations across eight cities of California: a foray into how historic maps are associated with asthma risk. D96. Environmental Asthma Epidemiology. American Thoracic Society, p. A7054.

Powell, J.A., 2008. Structural racism: building upon the insights of John Calmore. NCL Rev 86, 791.

Reece, J., 2021. Confronting the Legacy of "Separate but Equal": can the History of Race, Real Estate, and Discrimination Engage and Inform Contemporary Policy?. RSF: the Russell Sage Foundation. Journal of the Social Sciences 7 (1), 110–133.

Richardson, G.P., 2011. Reflections on the foundations of system dynamics. Syst Dyn Rev 27 (3), 219–243.

Rothstein, R., 2017. The color of law: A forgotten history of how our government segregated America. Liveright Publishing.

Rouwette, E., Vennix, J.A.M., van Mullekom, T., 2006. Group model building effectiveness: a review of assessment studies. Syst Dyn Rev 18 (1), 5–45. U.S. Office of Personnel Management . (2018). FedScope employment cubes. Federal human resources data. Retrieved from http://www.fedscope.opm.gov/employment.asp.

Vivero-Pol, J. L., Ferrando, T., De Schutter, O., & Mattei, U. (Eds.). (2018). *Routledge handbook of food as a commons*. Routledge.

Ware, L., 2021. Plessy's Legacy: the Government's Role in the Development and Perpetuation of Segregated Neighborhoods. RSF: The Russell Sage Foundation Journal of the Social Sciences 7 (1), 92–109.

# Using models to understand community interventions for improving public health and food systems

*Erin Hennessy*[a,b], *Larissa Calancie*[a,b] *and*
*Christina Economos (D.)*[a,b]

[a]Friedman School of Nutrition Science and Policy, Tufts University, Boston, MA,
United States [b]ChildObesity180

## 13.1 Introduction

This book presents a comprehensive overview of various conceptual and computational modeling approaches to understand food systems. Each approach or tool provides a unique lens to understanding a specific facet of the system. However, modeling approaches do not need to be used in isolation from one another. In fact, an integrated approach – utilizing multiple qualitative and/or quantitative methods simultaneously – may offer novel insights above and beyond what a single modeling tool can offer.

This chapter builds from concepts that underlie food systems modeling (Chapter 2) such as systems thinking and complex systems modeling, and concepts related to a specific modeling approach like group modeling building (Chapter 7), social network analysis (Chapter 11) and community-based system dynamics (Chapter 12). In doing so this chapter is meant to provide an illustration of how modeling techniques can be integrated or blended to study complex issues in food systems. This illustration shows how modeling can be used to help researchers understand how community stakeholders perceive food-related issues in their community and how such stakeholders can create and diffuse actionable strategies to drive policy, system, and environmental change. Table 13.1 provides examples of the modeling approaches that are mentioned throughout this chapter.

To begin, the chapter presents systems mapping, which we used to understand the dynamics of community change that occurred as part of the Shape Up Somerville intervention. As

TABLE 13.1 Examples of complex systems approaches applied to food systems either independently or in combination with one another as presented in this chapter.

| Modeling Approach | Description | Illustrative Example(s) |
|---|---|---|
| Agent-based modeling | Agent-based modeling is a computational tool that models the actions or interaction of "agents" and allows for the exploration of dynamic mechanisms that link individual behavior to overall outcomes among populations (Auchincloss and Diez Roux, 2008). | Chapter 13 |
| Community-based system dynamics | Community-based system dynamics (CBSD) is a participatory method for involving communities in the process of understanding and changing systems. The method emphasizes capability development where community members develop skills to create, understand, and use models to promote systems thinking and guide action (Hovmand, 2014). | Chapter 12, 13 |
| Group model building | Group model building is a participatory approach widely used to create a shared understanding of systems that drive problematic trends (Vennix, 1996). Group model building improves understanding of the problem, increases engagement in systems thinking, builds confidence in the use of systems ideas, and creates consensus for action among diverse stakeholders (Vennix, 1996). It is often used in the broader approaches of community-based system dynamics and system dynamics in general. | Chapter 7, 13 |
| Social network analysis | Social network analysis is a way to map and measure the relationships between people (or groups, organizations, etc.) (Scott, 1988). This analytic tool has recently been applied to the study of interventions, including obesity prevention interventions. | Chapter 11, 13, 15 |
| System dynamics | System dynamics is an approach to understanding the nonlinear behavior of complex systems over time using informal maps and formal models with computer simulation to understand endogenous sources of system behavior (National Cancer Institute, 2007). | Chapter 4, 7, 9 |
| Systems mapping | Systems mapping is a qualitative systems modeling approach that looks at how variables interact over time and form patterns of behaviors across the system. It is a visual tool to help share a simplified conceptual understanding of a complex system. Causal loop diagrams, a causal diagram that aids in visualizing how different variables (or elements) in a system are interested, are a type of systems mapping (Forrester, 1975). | Chapter 13 |

part of Shape Up Somerville, Dr Economos co-led with a community coalition the design and implementation of community-wide policy and environmental change strategies including a focus on local food system change. Next, we describe the foundation and development of our Stakeholder-driven Community Diffusion theory, informed by Shape Up Somerville and related efforts, and our efforts to test the theory. We then present case study examples of ongoing interventions and engagement with coalitions in several communities nationwide, all informed by the Stakeholder-driven Community Diffusion theory. These case studies are intended to illustrate the application of our blended modeling approach. The chapter concludes with a synthesis of how the various modeling approaches presented can inform future food systems modeling work more broadly.

## 13.2 Food systems change at the local level: Shape up Somerville intervention

The Shape Up Somerville (SUS) study (2002–05) was a landmark study: it was the first to demonstrate that, compared to control communities, children living in the intervention community had a significant decrease in body mass index (BMI) z-scores (the number of standard deviations from the mean) one year and two years after the intervention began (Economos et al., 2007, 2013; Economos and Curtatone, 2010). This study involved a community coalition and utilized a participatory, whole-of-community approach to disseminate and implement evidence-based interventions, including those targeting the community-level food system. Following the research period, SUS became ingrained into city and public health programming in Somerville. To this day, SUS continues as a city-wide program that seeks to build healthy, equitable communities in Somerville through interdisciplinary partnerships, programming, and policies related to food systems and active living (City of Somerville, 2021).

Much was written about SUS in the years immediately following the intervention, but even those involved had a limited understanding of the systems processes that led to its success. To fill this gap, a small group of researchers and practitioners familiar with the study used systems mapping to retrospectively describe and visualize the key mechanisms of how SUS was adopted, disseminated, implemented, and sustained over time. Systems mapping is a qualitative modeling approach (National Research Council, 2012; Foresight 2007; Forrester, 1975) that looks at how variables interact over time and form patterns of behaviors across a system. Systems maps have been used for decades for a range of undertakings, including as the starting point for simulation modelling (National Cancer Institute, 2007). However, their utilization as a public health tool has increased dramatically over the past decade.

The SUS map (as shown in Fig. 13.1) helped to move beyond communicating fragments or parts of the SUS community intervention to communicating the whole of SUS, and thus, more fully conveying the system structure and interrelationships that comprised SUS. Numerous policy, environmental, and practice strategies to increase children's healthy eating and physical activity were implemented throughout the intervention period. For example, increased school food service capacity enabled child nutrition staff to source more nutritious food and thereby increase the quality and variety of offerings on the school menu, ultimately leading to student adoption and acceptance of new, healthier foods (Fig. 13.1, R7) (Folta et al., 2013; Goldberg et al., 2009).

In brief, the systems map helped to illustrate the role of the SUS coalition in diffusing, disseminating, and/or implementing intervention strategies throughout the community that drove the success of SUS (Hennessy, Economos, and Hammond, 2020). Around a similar time as the original SUS study, another whole-of-community intervention was conducted in Australia. Like SUS, it was a successful intervention that involved a community coalition to guide the adoption, dissemination, and implementation of intervention strategies focused on policy and environmental change (de Groot, Robertson, Swinburn, and de Silva-Sanigorski, 2010; de Silva-Sanigorski et al., 2010, 2012). A systems map was also created to retrospectively

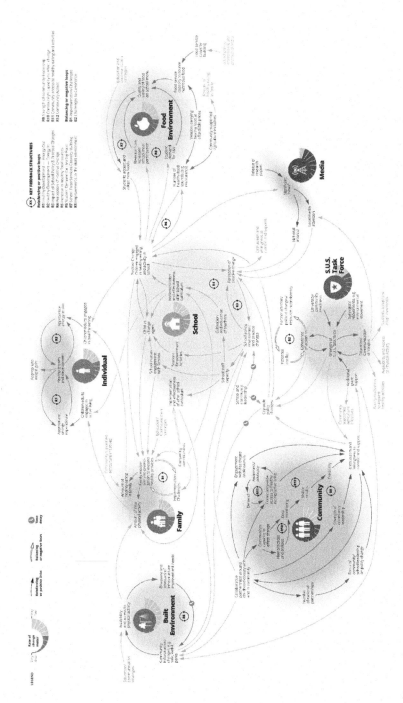

**FIGURE 13.1   Shape Up Somerville Systems Map (reprinted with permission from (Hennessy et al., 2020)).** The Shape Up Somerville (SUS) Systems Map was developed using a systems mapping approach and depicts 8 subsystems: Individual, Family, School, Community, Built Environment, Food Environment, Media and the SUS Task Force that were focal elements of the SUS intervention. The SUS Task Force (community coalition) was tasked with diffusing, disseminating and/or implementing education/communication strategies throughout the community of Somerville, MA, which ultimately helped reduce unnecessary weight gain among children in the community (Economos et al., 2007).

## BOX 13.1

### The importance of community coalitions

The scientific literature on community coalitions underscores the critical role they can play in creating change in their communities. They do so by creating and sustaining relationships or social networks, improving stakeholders' capacity to create change in their communities, and addressing challenges synergistically where the group is able to do more together than organizations could do working independently (Butterfoss and Kegler, 2012; Calancie et al., 2021). Community coalitions, defined broadly as multisector partnerships working toward shared goals, have been effective in addressing complex public health problems including those related to food systems (Butterfoss, Goodman, and Wandersman, 1996; Wolff and Maurana, 2001). Community coalitions often consist of groups of leaders and stakeholders from diverse organizations, settings, and sectors working collectively, and they have been identified by the scientific community and various organizations and agencies as strategic entities to amplify public health prevention efforts (Butterfoss et al., 1996; Butterfoss, Goodman, and Wandersman, 1993; Cormier, Wargo, and Winslow, 2015; de Silva-Sanigorski et al., 2010; Khan et al., 2009; Korn et al., 2018). Community coalitions bridge traditionally siloed stakeholders and organizations to (a) generate broad and diverse community representation and (b) increase stakeholder capacity to effect change in local communities (Bess, 2015; Foster-Fishman, Berkowitz, Lounsbury, Jacobson, and Allen, 2001). Community coalitions involved in health promotion have already been identified as effective in reducing health disparities among racial and ethnic minorities (Anderson et al., 2015; Korn et al., 2018); reducing food insecurity in an Indigenous reservation (Blue Bird Jernigan, Salvatore, Styne, and Winkleby, 2012); reducing childhood obesity through policy and environmental change strategies (Coffield, Nihiser, Sherry, and Economos, 2015; Economos and Curtatone, 2010; Economos et al., 2007, 2013, 2009; Folta et al., 2013; Goldberg et al., 2009), and many other examples related to coalition action for community health improvement (Bailey et al., 2011). In other words, gathering decision-maker stakeholders into community coalitions can make change happen in their respective localities. While evidence for community coalitions effectiveness continues to grow, a dearth of research examines the mechanisms and processes by which coalitions exert their influence. Understanding the processes by which coalitions conceive, design, diffuse, disseminate and/or implement interventions is a critical step to inform public prevention efforts and impact research outcomes (Gillman and Hammond, 2016; Sandoval et al., 2012). The drive to understand these processes is what led to the development of the Stakeholder-Driven Community Diffusion theory.

understand the dynamics of project implementation, which highlighted the important role of collaboration in achieving intervention outcomes (Owen et al., 2018). Collectively this work and that of others (see Box 13.1) helped inform the development of a new theory and dynamic hypothesis focused on identifying a mechanism by which community coalitions exert their influence (Belone et al., 2016; Boelsen-Robinson et al., 2015; Cacari-Stone, Wallerstein, Garcia, and Minkler, 2014; Ewart-Pierce, Mejia Ruiz, and Gittelsohn, 2016; Flego, Keating, and Moodie, 2014; Hennessy, Economos and Hammond, 2020; Jenkins, Lowe, Allender, & Bolton, 2020; Korn et al., 2018; Lucero et al., 2018; Oetzel et al., 2018, 2015).

## 13.3   Understanding how coalitions achieve change: the stakeholder-driven community diffusion theory

The Stakeholder-Driven Community Diffusion (SDCD) theory is aligned with recent reports that emphasize a focus on 'change agents' who can catalyze and harness bottom-up demand and readiness for change (Swinburn et al., 2019) and is grounded in several prior studies (Appel et al., 2019; Economos et al., 2013; Korn et al., 2018). Specifically, SDCD describes an empirically testable mechanism whereby stakeholder knowledge of and engagement with a complex issue, such as the food system, can diffuse through stakeholder social networks (Kasman et al., 2019). This diffusion is, in turn, expected to facilitate evidence-based policy, system, and environmental change via stakeholder and stakeholder-led change in multisector settings (Jou et al., 2018). SDCD is informed by existing theories and frameworks from community psychology, implementation science, systems thinking, and social network theory (Belone et al., 2016; Butterfoss and Kegler, 2012).

Systems thinking is a way of approaching problems that asks how various parts within a system are arranged and interact to yield a specific function or outcomes over time (Meadows, 2008). Systems thinking and related approaches are well suited to address the complexity inherent in public health problems such as poor nutrition. Such problems are often influenced by many actors operating at multiple levels, and influenced by feedback mechanisms throughout and across levels (Glass and McAtee, 2006). While systems thinking considers the full system, social network theory (Rogers, 2002; Valente and Pitts, 2017) focuses on the relationships between actors (e.g., individuals, organizations, etc.) and the web of relationships in which actors are embedded.

Conceptualizing, designing, and testing the SDCD theory was possible through a groundbreaking five-year international collaboration to apply the principles of systems science to community-based childhood obesity interventions (NIH R01HL115485). This was a multiphased research endeavor that led to the creation of new tools to measure stakeholder engagement and techniques to understand the processes by which community coalition stakeholder groups influence outcomes. The study drew from our successful studies described above and included a prospective phase to test the theory in a new pilot study modeled after SUS targeting a different population: children ages 0–5 years ("Shape Up Under-5").

In reviewing existing survey tools to develop the knowledge and engagement constructs, we noted numerous weaknesses that limited their utility, such as lack of detail on reliability and validity, narrow or limited domain areas, or lack of context specificity (Korn et al., 2018). Sensitive, reliable, and valid tools to measure longitudinal information on context, including differences in stakeholder social networks and diffusion of information, were needed to shift how investigators approach, understand, and work with community partners. To overcome this gap, we first created a new survey tool to measure the knowledge, engagement and social

## BOX 13.2

## Key Constructs of the Stakeholder-Driven Community Diffusion Theory

Knowledge
- Stakeholders' understanding of community-wide efforts to prevent a complex problem (e.g., childhood obesity) (1).[p.2]
- Consists of five attributes drawn from existing measures, frameworks (Belone et al., 2016; de Groot et al., 2010; Edwards, Jumper-Thurman, Plested, Oetting, and Swanson, 2000; Lucero et al., 2018; Oetzel et al., 2011; Sandoval et al., 2012), and literature (Economos et al., 2007)
- Five attributes: Knowledge of... (a) the problem; (b) modifiable determinants; (c) stakeholders' roles related to intervening; (d) sustainable intervention approaches; and (e) available resources.

Engagement
- Stakeholders' enthusiasm for and commitment to preventing the complex problem (e.g., childhood obesity) in their community (Korn et al., 2018)[p.3]
- Motivates stakeholders to share their knowledge with others and represents stakeholders' desires and ability to translate their knowledge into effective action for community interventions. (Korn et al., 2018)

- Five attributes: (a) exchange of skills and understanding ("Dialogue and mutual learning"); (b) willingness to compromise and adapt ("Flexibility"); (c) ability or capacity to affect the course of events, others' thinking, and behavior ("Influence and power"); (d) action of directing and being responsible for a group of people or course of events ("Leadership and stewardship"); and (e) belief and confidence in others ("Trust").

Social Networks
- The network structure of stakeholder's professional relationships related to focal problem.
- Networks of social interactions and personal relationships
- The number, strength, and influence of stakeholder social network connections represent the pathways for knowledge and engagement diffusion.
- Two properties are of main importance: the number of undirected (bi-directional) ties an individual has with others in the coalition and community and the propensity of individuals to be socially connected with other members of their own coalition.

networks of community coalition members that then serve as the data inputs into our agent-based model described below (see Box 13.2 for definitions of these constructs). Complete details on the development and testing of this survey are provided in Korn et al. (Korn et al., 2018).

To test the SDCD theory and understand whether and how knowledge and engagement of community coalition stakeholders diffuse through stakeholder social networks to the broader community, we turned to agent-based modeling (ABM). ABM is a computational tool that models the actions or interaction of 'agents' and allows for the exploration of dynamic mechanisms that link individual-level behavior to population-level outcomes (Auchincloss and Diez Roux, 2008; Shoham, Hammond, Rahmandad, Wang, and Hovmand, 2015). Several papers have addressed the rationale and advantages of applying ABM to the study of complex public health problems (El-Sayed, Seemann, Scarborough, and Galea, 2013; Hammond, 2009; Rahmandad and Sterman, 2008). ABM also focuses on individual-level interaction and can include social network structures, allowing the model to better capture real-world dynamics that are influenced by social interactions.

In brief, the 'agents' in our ABM are community coalition stakeholders with attributes like knowledge, engagement, and social network position, among others (Kasman et al., 2019). The ABM simulates the interaction of agents within community coalition committee meetings and their interactions within the broader community. During each simulated day, there is a probability that agents influence the knowledge and engagement of those with whom they are socially connected (Kasman et al., 2019). The probability is allowed to change as a function of change in mean community engagement. This reflects the theory that highly engaged communities more frequently discuss the focal problem. In the case of our ABM, the focal problem was defined as childhood obesity prevention. The model generates knowledge and engagement values on every simulated day of each run.

To test hypotheses about SDCD, we relied on retrospective data from completed intervention trials (de Silva-Sanigorski et al., 2010). Model outputs were compared to criteria derived from empirical data and experts' observations. Our work has shown that the ABM can estimate trends in community knowledge about and engagement with childhood obesity prevention given plausible model inputs and effect pathways, and could be used to compare the intervention to a "virtual counterfactual" where it had not taken place (Kasman et al., 2019). Simulating the presence of a community coalition committee meeting amplified the diffusion of knowledge and engagement in the community, providing strong suggestive evidence in support of the Stakeholder-driven Community Diffusion (SDCD) hypothesis (Kasman et al., 2019). The next step was then to test the theory with prospective data.

## 13.4 The SDCD theory in action

### 13.4.1 Shape up under 5 case study

To prospectively test the SDCD Theory, we convened a new community coalition, Shape Up Under 5, to prevent childhood obesity in children ages 0 to 5 years (Appel et al., 2019). Over a two-year period (2015–17) the coalition interacted during planned meetings held every four to six weeks throughout the study period. To structure these meetings, we relied on another modeling technique known as Group Modeling Building (GMB). GMB is rooted in system dynamics and is a participatory method led by trained facilitators who follow scripted group exercises (Vennix, 1996). The process helps diverse stakeholders share their mental models (meaning their beliefs, assumptions, and viewpoints of a problem) to visualize a complex and dynamic system, develop and prioritize action steps, and view connections across time and scale (Carey et al., 2015). It can be useful for stakeholders working together over time in a specific community and has been cited as a promising approach for designing and adapting intervention strategies that take the inherent complexities into account (Burke et al., 2015).

The logic behind the SDCD-informed intervention work is illustrated in Fig. 14.2. As stated above, a convened group of community stakeholders – a community coalition – can diffuse their knowledge and engagement through their network that will then lead to changes in policy, practices, and environments, and eventually health outcomes. To accelerate systems change, GMB is used as an intervention to guide several meetings during the study period (Korn et al., 2016). Complementing this strategy, the research team also provides evidence-based strategies that align with coalition priorities, and supplies seed funding and technical support as the coalition works to implement selected evidence-based interventions and evaluate action(s) as illustrated in Fig. 13.2.

In brief, the research team utilized GMB facilitation guides, or 'scripts', with the Shape Up Under 5 coalition. These scripts are freely available online (https://en.wikibooks.org/wiki/Scriptapedia). GMB generally flows through a sequence of scripts typically beginning with those that build trust and open communication among group members, then moving to divergent scripts where groups explore factors that influence trends over time and connections between those factors, and finally convergent scripts that help groups build a collective understanding of a complex problem and then prioritize and commit to actions to address that problem (Calancie et al., 2020). Fig. 13.3 illustrates GMB sessions and outputs: the 'Hopes and Fears' trust-building activity (A), the Causal Loop Diagram (CLD) created by the coalition (B), and a convergent activity, 'Action Priorities Grid' to drive action in areas depicted in the CLD (C).

The outcome of the GMB process was the creation and dissemination of a health messaging campaign. The goal of the campaign was to provide evidence-based messages to reach the entire community, including special efforts to reach immigrants and non-English speaking community members. GMB allowed us to engage with the coalition and facilitate coalition meetings, explicating the mental models of stakeholders of the drivers and solutions for obesity in their community. This purposeful, participatory technique helped members to better understand complex systems that influence problems of interest, take effective actions to address those problems, and aimed to equalize power among group members, enabling input from all participants and promoting group cohesion.

Additionally, as reported in Korn et al. (Korn et al., 2021), we analyzed social network data from community stakeholders involved in the Shape Up Under 5 (SUU5) coalition. As described in Table 13.1, social network analysis (SNA) is a way to map and measure the relationships between people (or groups, organizations, etc.) (Valente and Pitts, 2017). This analytic tool has recently been applied to the study of interventions and offers a framework and toolkit to understand coalition structure (Allender et al., 2016; Marks et al., 2018; McGlashan et al., 2018; McGlashan, Johnstone, Creighton, de la Haye, and Allender, 2016). SNA allowed us to observe change in the structure of the social network over time, as shown in Fig. 13.4.

While details of the methodology for the SNA is described in Korn et al. (Korn et al., 2021), in brief, committee members of the Shape Up Under 5 coalition ($n = 16$) periodically completed a web-based network survey asking them to nominate up to 20 individuals with whom they 'discuss issues related to early childhood obesity prevention'. These are referred to as discussion ties. Survey respondents then were asked to characterize the strength of the relationship (the 'tie') by two qualities: interaction frequency (5-point Likert scale, yearly (Korn et al., 2018) to daily (Sandoval et al., 2012)) about childhood obesity prevention and perceived influence (5-point Likert scale, not influential (Korn et al., 2018) to very influential

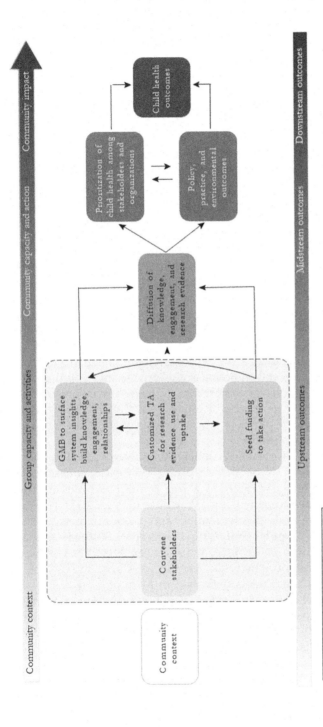

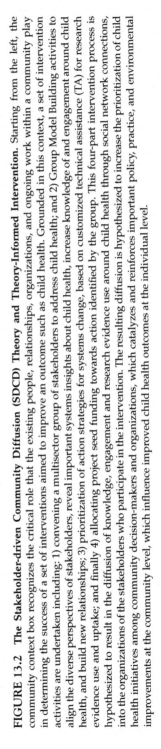

**FIGURE 13.2   The Stakeholder-driven Community Diffusion (SDCD) Theory and Theory-Informed Intervention.** Starting from the left, the community context box recognizes the critical role that the existing people, relationships, organizations, and ongoing work within a community play in determining the success of a set of interventions aimed to improve an outcome such as child health. Grounded in this context, a set of intervention activities are undertaken including: 1) convening a multisector group of stakeholders to address child health; and 2) Group Model Building activities to align the diverse perspectives of stakeholders, reveal important systems insights about child health, increase knowledge of and engagement around child health, and build new relationships; 3) prioritization of action strategies for systems change, based on customized technical assistance (TA) for research evidence use and uptake; and finally 4) allocating project seed funding towards action identified by the group. This four-part intervention process is hypothesized to result in the diffusion of knowledge, engagement and research evidence use around child health through social network connections, into the organizations of the stakeholders who participate in the intervention. The resulting diffusion is hypothesized to increase the prioritization of child health initiatives among community decision-makers and organizations, which catalyzes and reinforces important policy, practice, and environmental improvements at the community level, which influence improved child health outcomes at the individual level.

FIGURE 13.3 (A-C). Group Model Building GMB Activities with the Shape Up Under 5 Community Coalition. A) 'Hopes and Fears' GMB session to build trust and open communication among coalition members. Hopes were written on yellow paper while Fears were written on blue paper. Each committee member shared their Hopes and Fears in response to a prompt tailored to the project. B) Final causal loop diagram generated by the coalition that formed the basis and input for convergent GMB sessions. C) The Fig. shows coalition members discussing the action priority grid as part of a convergent GMB session to commit to actions to address the problem of early childhood obesity in the community.

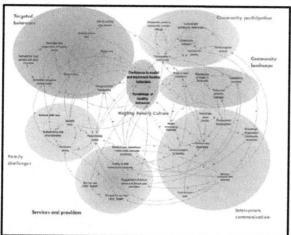

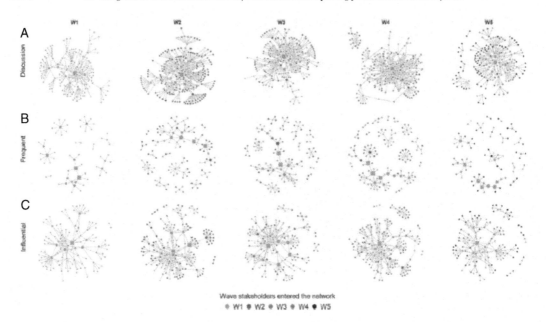

FIGURE 13.4  (A–C). Shape Up Under 5 early childhood obesity prevention network by survey wave and tie type (discussion, frequent, influential) adapted from Korn et al. (Korn et al., 2021). Square nodes represent Shape Up Under 5 Committee members and circular nodes represent their first- and second-degree alters. Node color indicates when stakeholders entered the network. Stakeholders observed at W1 (orange) were prevalent in later waves. We observed the largest proportion of new stakeholders at W2 during intervention planning (see blue nodes, which make up 63 percent, 45 percent, and 55 percent of the discussion, frequent, and influential networks, respectively). Approximately one-third of stakeholders observed at W5 follow-up were new to the discussion, frequent, and influential networks (black nodes). Node size is scaled to represent betweenness centrality. Stakeholders with high betweenness centrality are involved with many communication paths in the network and may have greater control over information flows related to early childhood obesity prevention. Stakeholders with high betweenness centrality over time were often Committee members (larger square nodes). Among alters with higher relative betweenness centrality over time, the majority entered the network at W1 (larger circular orange nodes). Network ties perceived as influential related to early childhood obesity prevention were more prevalent than frequent ties with daily or weekly communication. We observed the largest proportions of stakeholders communicating frequently about early childhood obesity prevention at W3 and W4 during the peak intervention period.

(Sandoval et al., 2012)) in shaping one's understanding about childhood obesity efforts. Therefore, ties could be described in three ways: discussion, frequent, influential.

From a core group of 16 committed coalition members, the total network grew to 558 stakeholders. Thus, the network included Committee members ($n = 16$), nominees of Committee members ('first-degree alters', $n = 201$), and nominees of first-degree alters ('second-degree alters', $n = 341$). This network represented community-based organizations, parents, health care, childcare, and universities, among others (Korn et al., 2021). While size and membership varied over the two-year period, the network grew most significantly during the planning stages and had the most frequent communication between stakeholders (daily/weekly) at that time as well. Over time though, social network ties were increasingly perceived as influential yet siloed within community groups (Korn et al., 2021). The network's extensive evolving membership may indicate access to a wide range of resources, ideas, and an ability to broadly disseminate intervention messages. The change in network structure over time may have supported more equal participation and control over intervention efforts.

Shape Up Under 5 provides another comprehensive example of how various modeling techniques can be used independently and together to address a complex public health problem. ABM and SNA allow us to test our SDCD theory, while GMB has been utilized as an intervention strategy with a community coalition. Moreover, SNA can be used as an independent tool to understand change in the social network structure of the coalition and their interaction with various stakeholders in the community throughout the intervention period. Building from these early insights, we are now using our tools and approaches with other community coalitions nationwide, expanding beyond GMB techniques alone to draw more from the broader field of Community-based System Dynamics (CBSD) to design and implement our SDCD-informed intervention (Hovmand, 2014). Details on CBSD can be found in Chapter 12.

## 13.5 The SDCD theory in action – community expansion

From Somerville, our work was expanded to more communities across the U.S. This new body of work is called Catalyzing Communities (https://catalyzingcommunities.org/). While each community is different and unique, the process and our methods are replicable. To begin, our team identifies 2–3 well-connected and impassioned community changemakers who are knowledgeable about complex drivers of health in their community, and often represent or lead an existing coalition. We work with the changemakers to identify and convene a multisector stakeholder group to engage in participatory, structured group activities, including GMB. The stakeholder group is selected from different areas that impact children's health (e.g., education, food security, healthcare, public health, childcare, public policy, transportation) and often represent a smaller subset of an existing coalition. As the stakeholder group participates in structured group activities, stakeholders explore the complex relationships between drivers of health, and gain insights about how to intervene in their community system to address those drivers. It is an iterative process; every subsequent session builds on outputs from the previous one. We share back the knowledge, engagement, and network data collected with the stakeholder group to guide their work and generate buy-in for sustained action and impact.

An important outcome of the intervention, particularly within the stakeholder group, is the enhanced networking and capacity building that arises from the regular meetings. While our social network data document the formation of relationship ties between stakeholders and community members in different sectors, we are also developing additional tools to capture the extent to which these ties form within the stakeholder group as a direct result of the meetings themselves. Our research team supports the coalition's transition from exploration and understanding to action by sharing relevant evidence to inform their work, facilitating new partnerships to address the myriad social determinants that surface, and providing seed funding and technical assistance to guide successful implementation of evidence-based strategies.

Each of the coalitions worked together to build a shared understanding of the system that shaped the trends each group was focused on. The groups explored topics such as i) the role of federal nutrition programs in helping families obtain healthy foods and ii) coordinating hunger relief efforts (Greenville, SC); i) the relationship between funding for education and extracurricular activities that promote physical activity and nutrition and ii) the balance of power between corporations and citizens (Tucson, AZ); i) parent-child dynamics that support or challenge the goal of promoting healthy eating at home and ii) the role of community gardens as gathering places in neighborhoods (Milwaukee, WI); and finally, i) the relationship between immigration policies and access to work opportunities and ii) the effect of house sharing to lower housing costs and how house sharing affects the home food environment (East Boston, MA). Each of these topics are embedded in the groups' large system maps that show the dynamic complexity of feedback mechanisms operating throughout the system.

### 13.5.1  Case study: Tucson, AZ

Stakeholders in Tucson, AZ formed a group called the Tucson Child Health Working Group that is affiliated with an existing coalition called Activate Tucson.[1] The group focused on the drivers and consequences of childhood obesity and poverty in their community. As part of one of the initial GMB sessions, stakeholders developed 'behavior-over-time-graphs', which are line graphs that show a pattern of change over time -how something increases or decreases as time passes. It is a common GMB activity for modeling and understanding systems. The group ultimately created a CLD, which identified many of the social determinants of health, including housing, transportation, access to health care, and access to healthy foods (Braveman and Gottlieb, 2014). The CLD represents the group's shared understanding of systems that drive and are influenced by childhood obesity. See Fig. 13.5 for these illustrations. (Note: due to COVID-19, GMB activities were conducted virtually).

One section of a larger CLD created by stakeholders in Tuscon focuses on the relationships between power in the food industry and the power of citizens to shape policies that govern the industry through advocacy. Reinforcing loop one (R1, 'marketing') shows that as advertising for unhealthy food goes up, promotion of unhealthy options and purchases and consumption of unhealthy foods goes up. Purchases lead to profits for food companies, which provide resources for even more advertising. In the feedback loop labeled 'resources for opposition' (R2), company profits increase wealth of corporations and resources available to oppose advocacy efforts aiming to curb unhealthy food purchasing and consumption. Profits also drive R3 ('powerful stay powerful') by increasing resources for corporate lobbying. Such lobbying targets policies that favor corporations, leading to more power for companies over time. As power accumulates within a company, the company may become even more effective in their lobbying. Citizens can act as a check on corporate power by becoming engaged and developing their own power (R4, 'people power' and R5, 'citizen power and engagement'). High citizen engagement may take the form of advocacy against unhealthy foods, which could lead to less profits and power for companies over time (R6, 'advocacy and opposition'). Alternatively, R6 also operates such that low citizen power and engagement leads to less advocacy, more purchases, more profits, more corporate power, and even lower citizen power. Additional factors and systems outside of the model certainly influence corporate and citizen power, but those were beyond the scope of this model.

After creating a CLD that explored system structure related to unhealthy food purchasing and how that is shaped by corporate and citizen power as well as a variety of other topics,

---

[1] More information about the working group is available on the Catalyzing Communities website (https://catalyzingcommunities.org/communities-tucson-az/).

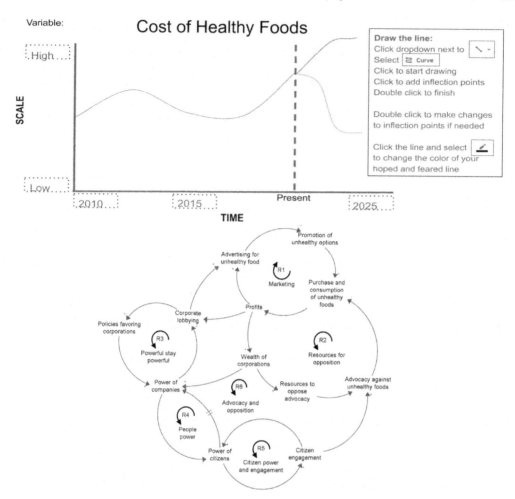

**FIGURE 13.5** (A-B). Tucson, AZ community coalition depiction of a behavior-over-time-graph (BOTG) showing the change in the cost of healthy foods from 2010 to a predicted level in 2025 (A) and a subset of a causal loop diagram exploring relationships between food system industry power, food purchasing patterns, and citizen advocacy (B).

the group prioritized where within the system they wanted to learn from the evidence, and then where to act. One action area the group is exploring now is to work with advocacy organizations to create opportunities for their coalition members and the broader community to develop advocacy skills that can be applied to promote healthy eating and physical activity policies, systems, and environments.

## 13.5.2 Case study: Milwaukee, WI

The Milwaukee County Organizations Promoting Prevention (MCOPP) is a strong coalition that brings together individuals from across the city with the goal of supporting a healthy community and eliminating health disparities. The Healthy & Fit Kids MCOPP Strategic

Planning Committee was convened through the SDCD intervention and included MCOPP members and community leaders focused on promoting healthy weights among children ages 0-5 years old. The Healthy & Fit Kids MCOPP Strategic Planning Committee included several individuals with strong ties to community gardens.[2] They shared how gardens are spaces where people feel comfortable gathering and having conversations with their neighbors. Urban agriculture is thriving in Milwaukee and is helping to revitalize the area. The midwestern city of 600,000 residents hosts 177 community gardens, 30 farms, and 26 farmers' markets (Mazur, 2017). Two gardens stood out to the predominately African American stakeholders: Alice's Garden Urban Farm (https://www.alicesgardenmke.com/) and We Got This (https://wegotthismke.com/). The purpose of Alice's Garden is to provide "models of regenerative farming, community cultural development, and economic agricultural enterprises for the global landscape." The garden hosts events, workshops, weekly walks, and other opportunities to learn, reflect, and build community. We Got This provides opportunities for boys and young men to garden, clean up their neighborhood, earn a stipend, and develop relationships with positive role models in their communities. Started in 2013, the garden has reached hundreds of youth in the community (McCoy, 2019). Community gardens are an integral piece of Milwaukee's social fabric and an important component of the city's growing urban agriculture sector.

Community gardens are a microcosm of interdependencies between social and natural systems. Stakeholders described how opportunities to gather in community gardens increases conversations between neighbors, creating relationships and a desire to gather, leading to more gatherings in community gardens (R1, 'growing community'). Stakeholders reported that neighbors shared information about cooking and eating produce, which can lead to increased knowledge of healthy foods. Coupled with this social feedback loop is another loop where gatherings in community gardens bring in volunteers who can help tend to gardens, helping gardens to thrive and serve as an appealing space to gather (R2, 'growing food'). When gardens thrive, produce becomes available to those working in the gardens and to community members through donations, farm stands and farmers' markets. Without people tending to gardens, they become overgrown, unappealing places to gather and no longer serve the purpose of a community garden.

The Healthy & Fit Kids MCOPP Strategic Planning Committee developed a systems map that included community garden dynamics, poverty, structural racism, access to and trust in the healthcare system, resource sharing between healthcare providers and patients, stress and trauma, the home food environments, and parent-child dynamics related to healthy eating (Healthy and Fit Milwaukee, 2021). The group narrowed down topic areas, requested more evidence to inform their choices, and then prioritized action areas. One action they are currently working on is coordinating a health and wellness event in multiple community gardens across the city. The goals are to attract new people to gardens and to provide opportunities to learn about and practice holistic wellness to promote physical and mental health. By attracting new people to gardens, the group may be able to invigorate virtuous cycles of community gathering and tending to gardens shown in Fig. 13.6. Additional actions the group are taking include investigating technological solutions, such as NowPow (https://nowpow.com/),

---

[2] More information about the working group is available on the Catalyzing Communities website (https://catalyzingcommunities.org/communities-milwaukee-wi/).

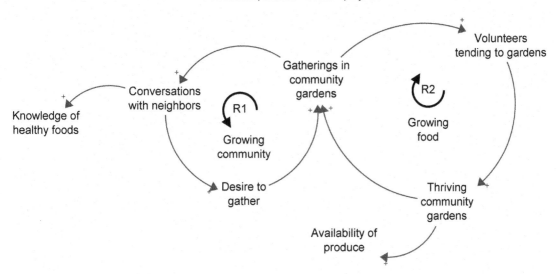

**FIGURE 13.6** A causal loop diagram extracted from a larger systems map created by community stakeholders in Milwaukee, WI showing how community gardens are spaces for growing food and community.

https://www.streetwyze.com/for connecting patients to community resources and researching strategies for reaching out to young families to get them connected to the healthcare system and community supports early. Overall, the group discussed a wide range of topics during the SDCD-theory informed intervention, they developed new relationships within and beyond the group, and are now working with new partners to put their ideas into action by writing grants, coordinating programs, and developing an advocacy agenda that is grounded in their systems map.

### 13.5.3 Case study: Greenville, SC

LiveWell Greenville is a coalition of organizations partnering to ensure access to healthy food and physical activity for Greenville County, SC residents. In 2019, we partnered with LiveWell Greenville to engage a group of 20 influential leaders, representing the many social determinants of health, to explore a broader, bolder strategic plan for the organization (see Fig. 13.7). Through GMB activities, the group explored the systems in place that impact childhood obesity and considered evidence-based opportunities and prioritized the opportunities for change. As shown in the shaded area of Fig. 13.7, the group discussed how many hunger relief agencies are located in Greenville's city center and help meet household needs (B-'Hunger Relief'). However, many low-income families are moving out of the city center to more remote locations throughout the County (R-'Effects of Moving'). The Balancing Loop Transportation (B-'Transportation') highlights how reliable transportation becomes a larger driver of access to hunger relief and food security as a result of families moving.

To further this work, LiveWell Greenville is partnering with the Institute for Advancement of Community Health at Furman University to better define the factors that contribute to food insecurity through the development of a Food Insecurity Index. This visual tool considers

FIGURE 13.7    (A-B) LiveWell Greenville coalition engaged in a GMB session (A) and their CLD (B).

more than proximity to a grocery store and also includes housing/rent burden, educational attainment, access to transportation, and other census data to determine density of populations experiencing food insecurity. This data will be used to determine placement of future infrastructure investments; opportunities to increase food equity; and better coordination between and among food security partner agencies.

## 13.6  Implications for food systems modeling

Integrating different modeling approaches can help generate new ideas and hypotheses and offer greater analytic capacity than any single approach can achieve on its own. In our work with whole-of-community interventions, systems mapping helped formulate our initial hypothesis, which was expanded into a full mechanistic theory and operationalized in an ABM that contains social network structure (Hennessy et al., 2020). Systems mapping also elucidated the important contributions of the coalition member's social networks. Ultimately, GMB and CBSD are now used prospectively with new communities to facilitate coalition meetings and build capabilities among community stakeholders in systems thinking and methods. This work generates community-specific systems maps that are the starting point for the coalition to prioritize action. Ongoing evaluation efforts draw from the various methods to provide a comprehensive assessment of the predictive validity of the SDCD theory, monitoring change in community social network structures, and providing the basis for evaluating mid- and downstream outcomes like policy, practice and environmental change and health outcomes.

Recently, there has been a call for 'whole-of-systems interventions to address the global burden of diet-related disease and re-consider the way we grow, process, distribute and commercialize our food. However, there is little evidence or guidance on how best to achieve this goal'. (Waterlander et al., 2018).[p.124] Integrating systems methods is one potential approach. As illustrated in this book, modeling a food system is challenging. Specifically, the

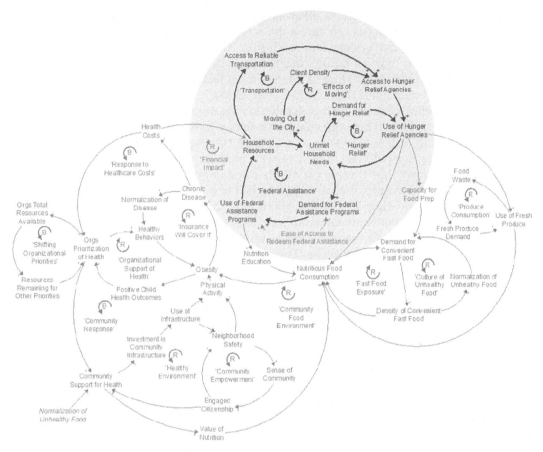

FIGURE 13.7, cont'd.

U.S. food and agriculture system has been shown to demonstrate many of the characteristics of a complex adaptive system – there are diverse actors who interact with each other and with the social and natural environment, and it includes both spatial and temporal heterogeneity (Committee on a Framework for Assessing the Health E, and Social Effects of the Food System; Food and Nutrition Board; Board on Agriculture and Natural Resources; Institute of Medicine; National Research Council. A Framework for Assessing Effects of the Food System 2015). Although no single method or approach is likely able to capture all elements of the system at once, the methods presented in this book and in this chapter can be used as a guide for food system modeling options.

A systems mapping approach to analyze food systems interventions may yield insights into common structures or archetypes of how these interventions are adopted, disseminated, implemented, and sustained. The process of system mapping often forces modelers to make model boundaries explicit as they decide what to include, and what not to include, in a model. A priori decisions like focusing the system boundary at the local level can help to increase the transferability of the findings to other similar contexts (Foster-Fishman and Behrens, 2007). A social network analysis can assess the actors in the system and the nature of their interactions while a broader systems approach may help to expose conflicts of interest and power dimensions that may be influencing the food system. This work may also help to identify necessary shifts in the nature of interventions focusing deeper on the relationships, feedback loops, or system goals (Waterlander et al., 2018).

Of all the approaches discussed in this chapter, to date, systems mapping (Jacobi et al., 2019; Savary et al., 2020) and social network analysis (Brinkley, Manser, and Pesci, 2021; Brown and Brewster, 2015; McIntyre, Jessiman-Perreault, Mah, and Godley, 2018; Wu and Guclu, 2013) are two that appear to be most utilized in food systems modeling; however, even their utilization is limited. To our knowledge, there is very limited use of agent-based modeling despite this approach having many features that are well matched to the complexity of modeling the food system (Auchincloss and Diez Roux, 2008; Hammond, 2009). More studies are beginning to utilize this approach to better understand policies (Widener, Metcalf, and Bar-Yam, 2013), decision-making (Namany, Govindan, Alfagih, McKay, and Al-Ansari, 2020) and dynamics within a food system hub (Krejci, Stone, Dorneich, and Gilbert, 2016) among other topics. However, the next frontier for food systems modeling may lie within the integration of these approaches.

## 13.7  Conclusion

Over a decade ago there was a call to apply systems approaches to complex public health issues, and we are now shepherding a new era of integrating complex systems approaches to study food systems. This work is part of a growing movement not only to apply systems thinking, theory, and methods to the study of complex public health problems but also to apply it to the study of interventions to solve those problem (Waterlander et al., 2018). This chapter demonstrates the value of blending multiple methods from the complex systems toolkit to suit the research questions and topic domain (rather than driving research with a single tool as the lens). The specific multi-method research process described in this chapter—qualitative domain-specific systems mapping, theory building by an interdisciplinary team informed by multiple fields of science, and quantitative tools such as SNA and ABM—may provide a roadmap for future applications in food systems modeling.

## Acknowledgement

We would like to thank the communities and coalition members who have made this work possible. Portion of this work have been funding by the National Institutes of Health (5R01HL115485) and the JPB Foundation. The views expressed in this article do not necessarily represent the views of the US Government, the Department of Health and Human Services or the National Institutes of Health.

## References

Allender, S., Millar, L., Hovmand, P., Bell, C., Moodie, M., Carter, R., Morgan, S., et al., 2016. Whole of Systems Trial of Prevention Strategies for Childhood Obesity: WHO STOPS Childhood Obesity. International Journal of Environmental Research and Public Health 13 (11).
Anderson, L.M., Adeney, K.L., Shinn, C., Safranek, S., Buckner-Brown, J., Krause, L.K., 2015. Community coalition-driven interventions to reduce health disparities among racial and ethnic minority populations. Cochrane Database of Systematic Reviews (Online) (6), CD009905.

Appel, J.M., Fullerton, K., Hennessy, E., Korn, A.R., Tovar, A., Allender, S., Economos, C.D., et al., 2019. Design and methods of Shape Up Under 5: integration of systems science and community-engaged research to prevent early childhood obesity. Plos One 14 (8) DOI at: https://doi.org/10.1371/journal.pone.0220169.

Auchincloss, A.H., Diez Roux, A.V., 2008. A new tool for epidemiology: the usefulness of dynamic-agent models in understanding place effects on health. American Journal of Epidemiology 168 (1), 1–8.

Bailey, M.B., Shiau, R., Zola, J., Fernyak, S.E., Fang, T., So, S.K., Chang, E.T., et al., 2011. San Francisco hep B free: a grassroots community coalition to prevent hepatitis B and liver cancer. Journal of Community Health 36 (4), 538–551.

Belone, L., Lucero, J.E., Duran, B., Tafoya, G., Baker, E.A., Chan, D., Wallerstein, N., et al., 2016. Community-Based Participatory Research Conceptual Model: community Partner Consultation and Face Validity. Qualitative Health Research 26 (1), 117–135.

Bess, K.D., 2015. Reframing Coalitions as Systems Interventions: a Network Study Exploring the Contribution of a Youth Violence Prevention Coalition to Broader System Capacity. American Journal of Community Psychology 55, 381–395.

Blue Bird Jernigan, V., Salvatore, A.L., Styne, D.M., Winkleby, M., 2012. Addressing food insecurity in a Native American reservation using community-based participatory research. Health Education Research 27, 645–655.

Boelsen-Robinson, T., Peeters, A., Beauchamp, A., Chung, A., Gearon, E., Backholer, K., 2015. A systematic review of the effectiveness of whole-of-community interventions by socioeconomic position. Obesity Reviews 16 (9), 806–816.

Braveman, P., Gottlieb, L., 2014. The social determinants of health: it's time to consider the causes of the causes. Public Health Reports 129 (Suppl 2), 19–31.

Brinkley, C., Manser, G.M., Pesci, S., 2021. Growing pains in local food systems: a longitudinal social network analysis on local food marketing in Baltimore County, Maryland and Chester County, Pennsylvania. Agric Human Values 38, 911–927.

Brown, D.R., Brewster, L.G., 2015. The food environment is a complex social network. Social Science & Medicine 133, 202–204.

Burke, J.G., Lich, K.H., Neal, J.W., Meissner, H.I., Yonas, M., Mabry, P.L., 2015. Enhancing dissemination and implementation research using systems science methods. International Journal of Behavioral Medicine 22 (3), 283–291.

Butterfoss, F.D., Goodman, R.M., Wandersman, A., 1993. Community coalitions for prevention and health promotion. Health Education Research 8 (3), 315–330.

Butterfoss, F.D., Goodman, R.M., Wandersman, A., 1996. Community coalitions for prevention and health promotion: factors predicting satisfaction, participation, and planning. Health Education Quarterly 23 (1), 65–79.

Butterfoss, F.D., & Kegler, M.C. (.2012).A Coalition Model for Community Action. Minkler M. New Brunswick, NJ: Rutgers University Press;

Cacari-Stone, L., Wallerstein, N., Garcia, A.P., Minkler, M., 2014. The promise of community-based participatory research for health equity: a conceptual model for bridging evidence with policy. American Journal of Public Health 104 (9), 1615–1623.

Calancie, L., Frerichs, L., Davis, M.M., Sullivan, E., White, A.M., Cilenti, D., Lich, K.H., et al., 2021. Consolidated Framework for Collaboration Research derived from a systematic review of theories, models, frameworks and principles for cross-sector collaboration. Plos One 16 (1) e0244501, DOI: https://doi.org/10.1371/journal.pone.0244501.

Calancie, L., Fullerton, K., Appel, J.M., Korn, A.R., Hennessy, E., Hovmand, P., Economos, C.E., et al., 2020. Implementing Group Model Building With the Shape Up Under 5 Community Committee Working to Prevent Early Childhood Obesity in Somerville, Massachusetts. Journal of Public Health Management and Practice Doi: 10.1097/PHH.0000000000001213.

Carey, G., Malbon, E., Carey, N., Joyce, A., Crammond, B., Carey, A., 2015. Systems science and systems thinking for public health: a systematic review of the field. BMJ Open 5 (12), e009002 Doi: 10.1136/bmjopen-2015-009002.

Coffield, E., Nihiser, A.J., Sherry, B., Economos, C.D., 2015. Shape Up Somerville: change in parent body mass indexes during a child-targeted, community-based environmental change intervention. American Journal of Public Health 105 (2), e83–e89.

City of Somerville, MA. Shape Up Somerville (2021). https://www.somervillema.gov/departments/health-and-human-services/shape-somerville

Committee on a Framework for Assessing the Health E, and Social Effects of the Food System; Food and Nutrition Board; Board on Agriculture and Natural Resources; Institute of Medicine; National Research Council. A Framework for Assessing Effects of the Food System. Washington (DC) National Academies Press; (2015).

Cormier, S., Wargo, M., Winslow, W., 2015. Transforming Health Care Coalitions From Hospitals to Whole of Community: lessons Learned From Two Large Health Care Organizations. Disaster Med Public Health Prep 9 (6), 712–716.

de Groot, F.P., Robertson, N.M., Swinburn, B.A., de Silva-Sanigorski, A.M., 2010. Increasing community capacity to prevent childhood obesity: challenges, lessons learned and results from the Romp & Chomp intervention. Bmc Public Health [Electronic Resource] 10, 522.

de Silva-Sanigorski, A.M., Bell, A.C., Kremer, P., Nichols, M., Crellin, M., Smith, M., Swinburn, B.A., et al., 2010. Reducing obesity in early childhood: results from Romp & Chomp, an Australian community-wide intervention program. American Journal of Clinical Nutrition 91 (4), 831–840.

de Silva-Sanigorski, A.M., Bell, A.C., Kremer, P., Park, J., Demajo, L., Smith, M., Swinburn, B., et al., 2012. Process and impact evaluation of the Romp & Chomp obesity prevention intervention in early childhood settings: lessons learned from implementation in preschools and long day care settings. Child Obes 8 (3), 205–215.

Economos, C.D., Curtatone, J.A., 2010. Shaping up Somerville: a community initiative in Massachusetts. Preventive Medicine 50 (Suppl 1), S97–S98.

Economos, C.D., Folta, S.C., Goldberg, J., Hudson, D., Collins, J., Baker, Z., Nelson, M., et al., 2009. A community-based restaurant initiative to increase availability of healthy menu options in Somerville, Massachusetts: shape Up Somerville. Preventing Chronic Disease [Electronic Resource] 6 (3), A102.

Economos, C.D., Hyatt, R.R., Goldberg, J.P., Must, A., Naumova, E.N., Collins, J.J., Nelson, M.E., et al., 2007. A community intervention reduces BMI z-score in children: shape Up Somerville first year results. Obesity (Silver Spring) 15 (5), 1325–1336.

Economos, C.D., Hyatt, R.R., Must, A., Goldberg, J.P., Kuder, J., Naumova, E.N., Nelson, M.E., et al., 2013. Shape Up Somerville two-year results: a community-based environmental change intervention sustains weight reduction in children. Preventive Medicine 57 (4), 322–327.

Edwards, R.W., Jumper-Thurman, P., Plested, B.A., Oetting, E.R., Swanson, L., 2000. Community readiness: research to practice. Journal of Community Psychology 28 (3), 291–307.

El-Sayed, A.M., Seemann, L., Scarborough, P., Galea, S., 2013. Are network-based interventions a useful antiobesity strategy? An application of simulation models for causal inference in epidemiology. American Journal of Epidemiology 178 (2), 287–295.

Ewart-Pierce, E., Mejia Ruiz, M.J., Gittelsohn, J., 2016. Whole-of-Community" Obesity Prevention: a Review of Challenges and Opportunities in Multilevel, Multicomponent Interventions. Curr Obes Rep 5 (3), 361–374.

Flego, A., Keating, C., Moodie, M., 2014. Cost-effectiveness of whole-of-community obesity prevention programs: an overview of the evidence. Expert Rev Pharmacoecon Outcomes Res 14 (5), 719–727.

Folta, S.C., Kuder, J.F., Goldberg, J.P., Hyatt, R.R., Must, A., Naumova, E.N., Economos, C.D., et al., 2013. Changes in diet and physical activity resulting from the Shape Up Somerville community intervention. Bmc Pediatrics [Electronic Resource] 13, 157.

Foresight. Building the Obesity System Map. . (2007).

Forrester, J.W., 1975. Common Foundations Underlying Engineering and Management Collected Papers of Jay W Forrester. Wright-Allen Press Inc., Cambridge, MA, pp. 61–80.

Foster-Fishman, P.G., Behrens, T.R., 2007. Systems change reborn: rethinking our theories, methods, and efforts in human services reform and community-based change. American Journal of Community Psychology 39 (3–4), 191–196.

Foster-Fishman, P.G., Berkowitz, S.L., Lounsbury, D.W., Jacobson, S., Allen, N.A., 2001. Building collaborative capacity in community coalitions: a review and integrative framework. American Journal of Community Psychology 29 (2), 241–261.

Gillman, M.W., Hammond, R.A., 2016. Precision Treatment and Precision Prevention: integrating "Below and Above the Skin". JAMA Pediatr 170 (1), 9–10.

Glass, T.A., McAtee, M.J., 2006. Behavioral science at the crossroads in public health: extending horizons, envisioning the future. Social Science & Medicine 62 (7), 1650–1671.

Goldberg, J.P., Collins, J.J., Folta, S.C., McLarney, M.J., Kozower, C., Kuder, J., Economos, C.D., et al., 2009. Retooling food service for early elementary school students in Somerville, Massachusetts: the Shape Up Somerville experience. Preventing Chronic Disease [Electronic Resource] 6 (3), A103.

Hammond, R.A., 2009. Complex systems modeling for obesity research. Preventing Chronic Disease [Electronic Resource] 6 (3), A97.

Healthy and Fit Milwaukee County Organizations Promoting Prevention Strategic Planning Committee. (2021). System changes to support child health in Milwaukee. https://catalyzingcommunities.org/communities-milwaukee-wi/

Hennessy, E., Economos, C.D., Hammond, R.A., 2020. Team SUSSM, the CT. Integrating Complex Systems Methods to Advance Obesity Prevention Intervention Research. Health Education & Behavior 47 (2), 213–223.

Hovmand, P.S., 2014. Community Based System Dynamics. Springer, New York.

Jacobi, J., Wambugu, G., Ngutu, M., Augstburger, H., Mwangi, V., Zonta, A.L., Rist, S., et al., 2019. Mapping Food Systems: a Participatory Research Tool Tested in Kenya and Bolivia. Mt Res Dev 39 (1), R1–R11.

Jenkins, E., Lowe, J., Allender, S., Bolton, K.A., 2020. Process evaluation of a whole-of-community systems approach to address childhood obesity in western Victoria, Australia. BMC Public Health 20 (1), 450.

Jou, J., Nanney, M.S., Walker, E., Callanan, R., Weisman, S., Gollust, S.E., 2018. Using Obesity Research to Shape Obesity Policy in Minnesota: stakeholder Insights and Feasibility of Recommendations. Journal of Public Health Management and Practice 24 (3), 195–203.

Kasman, M., Hammond, R.A., Heuberger, B., Mack-Crane, A., Purcell, R., Economos, C., Nichols, M., et al., 2019. Activating a Community: an Agent-Based Model of Romp & Chomp, a Whole-of-Community Childhood Obesity Intervention. Obesity (Silver Spring) 27 (9), 1494–1502.

Khan, L.K., Sobush, K., Keener, D., Goodman, K., Lowry, A., Kakietek, J., Zaro, S., et al., 2009. Recommended community strategies and measurements to prevent obesity in the United States. Mmwr Recommendations and Reports 58 (RR–7), 1–26.

Korn, A., Hovmand, P.S., Fullerton, K., Zoellner, N., Hennessy, E., Tovar, A., et al., 2016. Use of group model building to develop implementation strategies for early childhood obesity prevention. 9th Annual Conference on the Science of Dissemination and Implementation in Health: Mapping the Complexity and Dynamism of the Field. Washington, DC.

Korn, A.R., Hammond, R.A., Hennessy, E., Must, A., Pachucki, M.C., Economos, C.D., 2021. Evolution of a Coalition Network during a Whole-of-Community Intervention to Prevent Early Childhood Obesity. Child Obes 17 (6), 379–390.

Korn, A.R., Hennessy, E., Hammond, R.A., Allender, S., Gillman, M.W., Kasman, M., Economos, C.D., et al., 2018. Development and testing of a novel survey to assess Stakeholder-driven Community Diffusion of childhood obesity prevention efforts. Bmc Public Health [Electronic Resource] 18 (1), 681.

Korn, A.R., Hennessy, E., Tovar, A., Finn, C., Hammond, R.A., Economos, C.D., 2018. Engaging Coalitions in Community-Based Childhood Obesity Prevention Interventions: a Mixed Methods Assessment. Child Obes 14 (8), 537–552.

Korn, A.R., Hennessy, E., Tovar, A., Finn, C., Hammond, R.A., Economos, C.D., 2018. Engaging Coalitions in Community-Based Childhood Obesity Prevention Interventions: a Mixed Methods Assessment. Child Obes 14 (8), 537–552.

Krejci, C.C., Stone, R.T., Dorneich, M.C., Gilbert, S.B., 2016. Analysis of Food Hub Commerce and Participation Using Agent-Based Modeling: integrating Financial and Social Drivers. Human Factors 58 (1), 58–79.

Lucero, J., Wallerstein, N., Duran, B., Alegria, M., Greene-Moton, E., Israel, B., Hat, E.R.W., et al., 2018. Development of a Mixed Methods Investigation of Process and Outcomes of Community-Based Participatory Research. J Mix Methods Res 12 (1), 55–74.

Marks, J., Sanigorski, A., Owen, B., McGlashan, J., Millar, L., Nichols, M., Allender, S., et al., 2018. Networks for prevention in 19 communities at the start of a large-scale community-based obesity prevention initiative. Transl Behav Med 8 (4), 575–584.

Mazur, L. (2017).Milwaukee is Showing How Urban Gardening Can Heal a City: civil Eats; https://civileats.com/2017/10/04/milwaukee-is-showing-how-urban-gardening-can-heal-a-city/

McCoy, M.K. (.2019).Milwaukee's 'We Got This' Garden Offers Work, Life Skills To Young Men: wisconsin Public Radio https://www.wpr.org/milwaukees-we-got-garden-offers-work-life-skills-young-men]

McGlashan, J., Johnstone, M., Creighton, D., de la Haye, K., Allender, S., 2016. Quantifying a Systems Map: network Analysis of a Childhood Obesity Causal Loop Diagram. Plos One 11 (10), e0165459.

McGlashan, J., Nichols, M., Korn, A., Millar, L., Marks, J., Sanigorski, A., Economos, C., et al., 2018. Social network analysis of stakeholder networks from two community-based obesity prevention interventions. Plos One 13 (4), e0196211.

McIntyre, L., Jessiman-Perreault, G., Mah, C.L., Godley, J., 2018. A social network analysis of Canadian food insecurity policy actors. Canadian Journal of Dietetic Practice and Research 79 (2), 60–66.

Meadows, D.H., 2008. Thinking in Systems: A Primer. Chelsea Green Publishing, White River Junction.

Namany, S., Govindan, R., Alfagih, L., McKay, G., Al-Ansari, T., 2020. Sustainable food security decision-making: an agent-based modelling approach. J Clean Prod 225 (120296).

National Cancer Institute, 2007. Greater than the sum: systems thinking in tobacco control. NCI Tobacco Control Monographs Series. Department of Health and Human Services, National Institutes of Health, Bethesda, MD.

National Research Council, 2012. Accelerating Progress in Obesity Prevention: Solving the Weight of the Nation. The National Academies Press, Washington, DC.

Oetzel, J., Wallerstein, N., Solimon, A., Garcia, B., Siemon, M., Adeky, S., Tafoya, G., et al., 2011. Creating an instrument to measure people's perception of community capacity in American Indian communities. Health Education & Behavior 38 (3), 301–310.

Oetzel, J.G., Wallerstein, N., Duran, B., Sanchez-Youngman, S., Nguyen, T., Woo, K., Alegria, M., et al., 2018. Impact of Participatory Health Research: a Test of the Community-Based Participatory Research Conceptual Model. BioMed research international 2018, 7281405.

Oetzel, J.G., Zhou, C., Duran, B., Pearson, C., Magarati, M., Lucero, J., Villegas, M., et al., 2015. Establishing the psychometric properties of constructs in a community-based participatory research conceptual model. American Journal of Health Promotion 29 (5), e188–e202.

Owen, B., Brown, A.D., Kuhlberg, J., Millar, L., Nichols, M., Economos, C., Allender, S., et al., 2018. Understanding a successful obesity prevention initiative in children under 5 from a systems perspective. Plos One 13 (3), e0195141.

Rahmandad, H., Sterman, J.D., 2008. Heterogeneity and network structure in the dynamics of diffusion: comparing agent-based and differential Eq. models. Manage Sci 54, 998–1014.

Rogers, E.M., 2002. Diffusion of preventive innovations. Addictive Behaviors 27 (6), 989–993.

Sandoval, J.A., Lucero, J., Oetzel, J., Avila, M., Belone, L., Mau, M., Wallerstein, N., et al., 2012. Process and outcome constructs for evaluating community-based participatory research projects: a matrix of existing measures. Health Education Research 27 (4), 680–690.

Savary, S., Akter, S., Almekinders, C., Harris, J., Korsten, L., Rotter, R., Watson, D., et al., 2020. Mapping disruption and resilience mechanisms in food systems. Food Secur 1–23.

Scott, J., 1988. Social Network Analysis. Sociology 22 (1), 109–127.

Shoham, D.A., Hammond, R., Rahmandad, H., Wang, Y., Hovmand, P., 2015. Modeling social norms and social influence in obesity. Curr Epidemiol Rep 2 (1), 71–79.

Swinburn, B.A., Kraak, V.I., Allender, S., Atkins, V.J., Baker, P.I., Bogard, J.R., Dietz, W.H., et al., 2019. The Global Syndemic of Obesity, Undernutrition, and Climate Change: the Lancet Commission report. Lancet 393 (10173), 791–846.

Valente, T.W., Pitts, S.R., 2017. An Appraisal of Social Network Theory and Analysis as Applied to Public Health: challenges and Opportunities. Annual Review of Public Health 38, 103–118.

Vennix, J., 1996. Group model building: Facilitating team learning in system dynamics. John Wiley and Sons Ltd, Hoboken, NJ.

Waterlander, W.E., Ni Mhurchu, C., Eyles, H., Vandevijvere, S., Cleghorn, C., Scarborough, P., Seidell, J., et al., 2018. Food Futures: developing effective food systems interventions to improve public health nutrition. Agricultural Systems 160, 124–131.

Widener, M.J., Metcalf, S., Bar-Yam, Y., 2013. Agent-based modeling of policies to improve urban food access for low-income populations. Applied Geography 40, 1–10.

Wolff, M., Maurana, C., 2001. Building Effective Community—Academic Partnerships to Improve Health: a Qualitative Study of Perspectives from Communities. Academic Medicine 76, 166–172.

Wu, F., Guclu, H., 2013. Global maize trade and food security: implications from a social network model. Risk Analysis 33 (12), 2168–2178.

# Applying environmental models in the food business context

*Christina Skonberg[a] and Mariko Thorbecke[b]*

[a] Simple Mills, Sustainability & Strategic Sourcing, Chicago, IL, United States
[b] Anthesis Group, Analytics, Boulder, CO, United States

## 14.1 Motivations for food companies to leverage environmental models

Mounting concern about the environmental impacts of food production has led investors, consumers, and other stakeholders in the food industry to scrutinize how food companies address environmental impacts across their value chains (Röös et al., 2014). It is estimated that more than one quarter of global greenhouse gas emissions stems from the food system (Crippa et al., 2021; Vermeulen 2012), with agriculture driving an estimated 70 percent of global freshwater withdrawal (Huang et al., 2019) and more than 70 percent of global deforestation (Hosonuma et al., 2012). As awareness builds around the environmental toll of modern food production, many food companies are using models to track, improve, and subsequently communicate the environmental performance of their supply chains (Nygren and Antikainen, 2010; Röös et al., 2014).

The Harvard Business Review reports that environmental, social, and governance (ESG) performance is a top priority for executives at global institutional investing firms, including the world's three largest asset managers: BlackRock, Vanguard, and State Street (Eccles & Klimenko, 2019). In ESG investing, environmental criteria help investors examine how companies steward natural resources and address issues such as climate change. Social criteria allow investors to consider how companies treat the people along its value chain, including employees and suppliers. Governance criteria account for such factors as a company's leadership structure, executive pay, and shareholder rights.

Shareholders are not the only stakeholders interested in food companies' ESG performance. In 2019 nearly a quarter of US consumers reported that they consistently consider sustainability[1] when making a purchase (Hartman Group, 2019). Companies have responded by

---

[1] In this study, Hartman Group defined sustainability as "shorthand for a complete moral system of cultural values, beliefs, and attitudes related to a sense of responsibility for the greater good." Hartman Group, 2019, pp 8.

providing transparent communications around environmental impact. In 2019, 90 percent of Standard & Poor's 500 Index companies published a sustainability report, up from just 20 percent in 2011 (Governance and Accountability Institute, Inc. 2020). Much of the content in these reports relies on environmental models to estimate outcomes in the absence of on-the-ground measurements across complex supply chains. In the case of General Mills, a $17.6 billion dollar food company with brands such as Cheerios, Nature Valley, and Betty Crocker, an estimated 54 percent of the company's greenhouse gas footprint stems from agriculture (General Mills, 2021a), on operations that are not owned by the company (General Mills, 2020b). Given the scale and multi-nodal nature of supply chains like General Mills', models can provide a scalable approach to estimating impact upstream and downstream of companies' owned assets.

Greenhouse gas emissions, water consumption, land use, and biodiversity represent only a few of the impact areas companies track across all stages of the supply chain, from the extraction of resources to grow ingredients, to a product's consumption and the disposal of any waste generated. In this chapter, we focus primarily on the use of Life Cycle Assessment (LCA) by food companies, as LCA modeling frameworks serve as the basis of many businesses' sustainability efforts. As discussed in Chapter 3, LCA is a framework for quantifying the impacts of complex systems that provide goods and services, such as food products. It considers multiple impact areas such as climate change and water quality, and provides an opportunity for food companies to evaluate the environmental performance of their supply chains. We discuss some of the limitations that companies face in using LCA, and emerging data collection and modeling technologies that address those limitations.

Food companies use environmental models for a variety of reasons. A desire to balance profit with social and environmental metrics motivates some. Others want to capture brand value by communicating impact to attract sustainability-focused consumers and investors. Here we review two key motivations for food companies to implement environmental models: elevating environmental impact tracking on par with financial tracking, and substantiating public communications about environmental impact.

### 14.1.1 Pursuing a triple bottom line business model

In 1994, author and corporate responsibility expert John Elkington coined the term 'triple bottom line' to refer to a business's social, environmental, and economic impact. (Elkington, 1994). The framework was intended to help companies measure and address the full cost of doing business, encouraging a focus beyond traditional financial metrics to include environmental metrics such as greenhouse gas emissions and water consumption. Many companies seeking to pursue a triple bottom line framework leverage modeling tools such as LCA to understand and communicate progress against the environmental pillar of the framework.

As one example, the popular outdoor clothing company Patagonia, which launched their food line Patagonia Provisions in 2012, has used environmental models in alignment with their stated mission "to save our home planet" (Patagonia, 2021). In 2007 the company launched The Footprint Chronicles, a website that uses LCA to disclose information about Patagonia's greenhouse gas emissions and other environmental impacts across the product lifecycle. (Chouinard and Stanley, 2012). In this case, the use of LCA has helped Patagonia track progress against the environmental ambitions at the heart of their company mission.

Anglo-Dutch Unilever also uses models to estimate the environmental impact of their supply chain (Unilever, 2021). Unilever is a multinational consumer packaged goods company with food brands such as Hellman's, Lipton, and Ben & Jerry's. Recognized as a leading triple bottom line business (Elkington, 2018), Unilever uses LCA to estimate the impact of existing products in their portfolio (Unilever, 2021). The company also goes a step further to use LCA to proactively optimize for the environmental impact of ingredients and packaging at the design stage of new product development (Unilever, 2021)–a process known as ecodesign (Karlsson and Luttropp, 2006).

Annie's, a consumer packaged goods company owned by General Mills, uses models to evaluate environmental Key Performance Indicators (KPIs) that are publicly reported in an annual sustainability report (Annie's, 2019). Tracking KPIs such as pounds of synthetic pesticides replaced with USDA certified organic pest control methods, Annie's has used models to track progress across the triple bottom line framework, moving beyond traditional financial reporting metrics to also measure and disclose environmental metrics. For companies with complex supply chains and missions rooted in a triple bottom line approach, models can be useful tools to estimate progress on environmental targets.

## 14.1.2 Communicating environmental impact to diverse stakeholders

Another key motivation for food businesses to use models is to communicate environmental impact to their stakeholders, often in ways that capture economic value. This value can stem from increasing brand loyalty, outcompeting industry peers by setting ambitious environmental targets, meeting or surpassing the standards of scrutinizing and vocal NGOs, and attracting investors, among other metrics of business success (Braam and Peeters, 2017; Cone, 2017; Governance and Accountability Institute, Inc, 2020; Whelan and Fink, 2016).

Cone, a US-based agency focused on social and environmental issues, reports that purpose-driven[2] companies may have a competitive advantage in attracting consumers, employees, and investors (Porter Novelli/Cone, 2019). In a 2019 study conducted by Porter Novelli and Cone, 86 percent of respondents reported that they would purchase products from purpose-driven companies, 70 percent reported that they want to work for purpose-driven companies, and 64 percent reported that they would invest in purpose-driven companies (Porter Novelli/Cone, 2019). Notably, in Cone's 2015 Global Corporate Social Responsibility Study, over half of respondents indicated that they require tangible proof of a company's actions to provide assurance the company is acting responsibly, highlighting the importance of compelling and substantive sustainability communications (Cone, 2015).

While many companies recognize the value that sustainability marketing can generate for the business, some communications campaigns may also provide value to the public in the form of education. In recent years, companies such as Unilever, Oatly, Chipotle, and Quorn have launched campaigns to label the carbon footprint of their food products, using LCA frameworks to generate estimates (Chipotle, 2020; Oatly, 2021; Quorn, 2021; Unilever, 2020). An estimated 63 percent of Americans believe climate change affects their local community (Tyson & Kennedy, 2020) and nearly 40 percent believe that their individual actions make

---

[2] Cone defines purpose as a company's "authentic role and value in society, which allows it to simultaneously grow its business and positively impact the world" (Porter Novelli/Cone, 2019).

a real difference in addressing climate change (Cone/Porter Novelli, 2018), suggesting that carbon labels may have potential to resonate with certain consumers. While more in-market research is needed, product labels that rely on models to estimate greenhouse gas footprints could represent an emerging opportunity to engage consumers in efforts to address climate change.

The growth in carbon footprint product labeling is consistent with momentum around company-wide greenhouse gas reduction targets and the use of greenhouse gas modeling tools to track progress against these public commitments. In 2018, the Intergovernmental Panel on Climate Change (IPCC) warned of the adverse and inequitable impacts that are likely to worsen should the global average temperature rise 1.5 °C or higher above pre-industrial levels (IPCC, 2018). With the release of the special report came a flurry of commitments from companies, investors, and other stakeholders to limit global warming to under 1.5 °C compared with pre-industrial levels. The IPCC indicates that limiting global temperature increase to 1.5 °C or under requires reaching net zero carbon dioxide emissions by around 2050, with concurrent deep reductions in non-$CO_2$ emissions such as methane (IPCC, 2018). In 2020, the Science Based Targets Initiative (SBTi) reported that over 360 companies had committed to the Business Ambition for 1.5 °C, a global initiative engaging companies on their commitments to achieve net zero emissions (SBTi, 2020). The list includes top food and beverage companies such as Nestle, Unilever, General Mills, Tesco, and Danone (SBTi, 2020). Further indicating a growing commitment by food companies to publicly disclose environmental impact data, a 2019 report by the Boston-based non-profit Ceres found that 34 out of the top 50 food and beverage companies in the U.S. and Canada publicly disclose data on the greenhouse gas emissions generated by their business activities (Ceres, 2019). These disclosures rely on widely used modeling-based greenhouse gas accounting standards such as the Greenhouse Gas Protocol, demonstrating the value of models for companies' public communications around environmental performance. While environmental impact communications–sometimes described as green marketing–can generate significant value for food companies, prudent marketing teams are wise to consult the Federal Trade Commission's Green Guides to ensure that claims are compliant with stipulations around environmental claims (Federal Trade Commission, 2012).

Mounting pressure on food companies to address and disclose their environmental impacts, as well as the opportunity to appeal to stakeholders seeking to support purpose-driven companies, are two common motivations for businesses to communicate environmental impact transparently and credibly. With heightened public awareness around greenwashing–the use of false or misleading information about environmental impact–and stringent green marketing regulations from the Federal Trade Commission, models can also help food companies avoid accusations of misleading claims.

## 14.2 Applications of environmental models within food companies

Models are a familiar tool for many food companies that engage in activities such as volume demand forecasting and quarterly earnings predictions. The recorded surge in corporate sustainability reports from 2011 to 2020 could suggest that the widespread use of models to communicate environmental impact is a more recent development for many food companies (Governance and Accountability Institute, 2020).

LCA represents the most widely used tool to assess the environmental impacts generated by agricultural products (Van der Werf, Knudsen, and Cederberg, 2020). LCA frameworks are used by food companies to track environmental impact across the supply chain from ingredient and material cultivation to a product's consumption or disposal. While food supply chains have environmental impacts across all stages of production beyond agriculture, for many ingredients the majority of greenhouse gas emissions is estimated to occur at the farm level of the supply chain. In fact, about 70 percent of greenhouse gas emissions from the broader global food system is estimated to stem from agricultural production and associated land use changes (Crippa et al., 2021). The extent to which agriculture contributes to total food supply chain emissions varies across regions. Agriculture and associated land use changes represent a higher share in developing countries than industrialized countries due to factors including a higher current rate of active deforestation to establish agriculture in developing countries (Crippa et al., 2021). Greenhouse gas emissions, water consumption, land use, and biodiversity represent a few of the common impact areas that food companies include in LCAs. Applications of LCA by food companies include measuring progress against greenhouse gas reduction targets, product-level green marketing, and designing products to minimize negative environmental impacts, among other activities.

Most LCAs consider the relevant type of ingredient and the ingredient country of origin to estimate farm-level impacts. This can help companies identify ingredient hotspots to focus their sustainability efforts. However, farming practices within an ingredient category and country of origin can vary significantly, ultimately leading to different environmental outcomes. Because of this, some companies use farm-level surveys to better account for the activities and local contexts at the foundation of their supply chain.

## 14.2.1 Using LCA to track corporate greenhouse gas reduction targets

Multiple groups interact to enable the calculation, target setting, and reporting of corporate level greenhouse gas emissions. In 2001 the World Resources Institute (WRI) and the World Business Council for Sustainable Development (WBCSD) partnered to develop the Greenhouse Gas Protocol (WBCSD/WRI, 2004), which serves as the most widely used corporate-level greenhouse gas calculation standard in the world. The Science Based Target Initiative (SBTi) provides the methodology and guidance by which companies set reduction targets in line with the latest climate science, and companies, typically working with consultancy partners, report their estimated emissions through the global disclosure system Carbon Disclosure Project (CDP).

LCA models and life cycle databases serve as the basis for companies to develop corporate level greenhouse gas inventories. Recall from Chapter 3 that LCA typically refers to a study that considers multiple impact categories. In contrast, a greenhouse gas footprint uses an LCA framework that only focuses on a single impact area, in this case climate change. The Greenhouse Gas Protocol categorizes emissions as scope 1, 2 or 3. Scope 1 consists of direct emissions within the owned operations of the company, commonly stemming from fuel combustion in any company owned assets such as vehicles and manufacturing facilities. Scope 2 consists of indirect emissions stemming from the purchase of electricity for the organization; for example, the electricity used in office buildings or owned manufacturing sites. Scope 3 consists of all emissions that occur beyond a company's owned operations, such as emissions from upstream purchased ingredients and transport, or downstream consumer use and disposal of products. (WBCSD/WRI, 2004). Generally, between 75 percent and 90 percent of food companies' greenhouse gas impact occurs in the supply chain, upstream of the consumer point of sale (Tidy, Wang, and Hall, 2016). Downstream impacts of food products can vary significantly depending on whether refrigeration or cooking is required. However, the Greenhouse Gas Protocol does not require that companies include these emissions within their inventory boundary (WBCSD/WRI, 2011). The Greenhouse Gas protocol divides emissions into 12 standardized categories that allow companies to report on areas of impact such as business travel, logistics upstream and downstream of a company's owned assets in the supply chain, and purchased goods and services (WBCSD/WRI, 2011). With nearly 80 percent of consumers expecting companies to address climate change, food companies are increasingly setting and reporting against targets for greenhouse gas emissions reductions (Cone, 2017). Here we include two examples of how prominent food companies leverage LCA-based models to track corporate greenhouse gas reduction targets.

General Mills and Mars are two companies on a growing list of global food companies that publicly disclose information about their corporate greenhouse gas footprint. Both companies report progress through public channels such as websites and annual sustainability reports. In 2015, General Mills was the first company to publish a greenhouse gas target approved by the Science Based Targets initiative. The company committed to a 28 percent reduction across its full value chain by 2025, over a 2010 baseline (General Mills, 2020a). Mars set a similar goal, working in 2016 with the Science Based Targets Initiative, World Resources Institute, and other partners to set a target to reduce greenhouse gas emissions by 27 percent by 2025, and 67 percent by 2050 against a 2015 baseline (Mars, 2021). Both General Mills' and Mars' targets, set in 2015 and 2016, respectively, represented the best available science at the time, following the IPCCs recommended emissions reductions to align with the global goal of limiting warming to well below 2 °C above pre-industrial levels (Putt del Pino, Cummis, Lake, Rabinovitch, and Reig, 2016).

Because climate science and climate modeling are always evolving, setting corporate targets on greenhouse gas emissions is an iterative process. With the release of the IPCC 2018 report warning of the impacts associated with warming above 1.5 °C, General Mills set an updated target to reduce greenhouse gas emissions by 30 percent between 2020 and 2030, and to achieve net zero emissions by 2050 (General Mills, 2020b). As part of a similar iterative process of aligning with the latest climate science, Mars announced in 2019 that it would work to further cut emissions from within its own operations in line with limiting global warming to 1.5 °C (Mars, 2019a). Fig. 14.1 and Fig. 14.2 show each company's respective climate ambition pathway.

It is important to note that achieving net zero emissions will require reaching a global balance between carbon dioxide emissions and removals. Carbon removals are typically subdivided into natural carbon removals, such as carbon stored in trees or soils, and technological removals, such as direct air capture and underground storage. However, as of the time of this writing, the Greenhouse Gas Protocol focuses on emissions reductions and does not allow for companies to account for any form of greenhouse gas removals in their greenhouse gas inventory, due to a lack of consensus methods for carbon sequestration

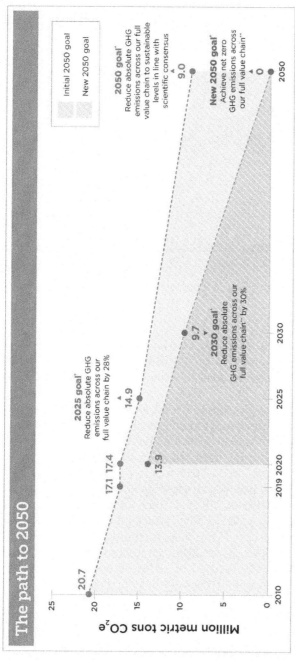

*Baseline for 2025 goal and initial 2050 goal is 2010. Baseline for 2030 goal and new 2050 goal is 2020.

**This goal focuses on the categories of GHG emissions that are the most impactful and actionable for General Mills, representing 13.9 million metric tons CO₂e in 2020 (81% of our total value chain GHG emissions footprint). The following GHG emissions are excluded from this goal, consistent with SBTi guidelines: some low volume ingredients, capital goods, employee commuting, franchises, downstream warehouse and storage at retail, consumer trips to store and end of life (consumer food waste).

**FIGURE 14.1** General Mills updated 2050 climate ambition. Figs. are subject to change with shifts in sales, acquisitions and divestitures, and other business variables. (General Mills, 2021a).

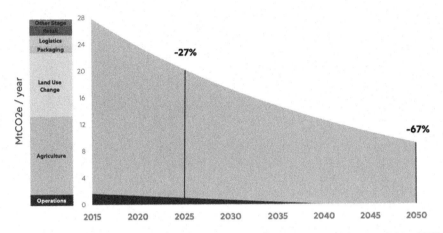

# MARS

## GLIDEPATH FOR GREENHOUSE GAS (GHG) EMISSION REDUCTION TARGETS

FIGURE 14.2    Mars' corporate glidepath for reaching their greenhouse gas reduction target (Mars, 2019b).

accounting (WBCSD/WRI, 2004). A working group has been established to develop a protocol for incorporating greenhouse gas removals into the inventory, and new guidance is underway to define accounting guidelines for companies to achieve net zero emissions (Greenhouse Gas Protocol, 2020).

Many other large food companies, including Coca Cola Enterprises, Kellogg Company, and Nestle, have also used modeling frameworks to set greenhouse gas targets and track progress. Despite an encouraging growth of science-based climate commitments by food companies since 2015, critics note that many companies fall short of comprehensively accounting for emissions across the full supply chain. A 2019 report found that while 34 of the top 50 food companies in North America publicly disclose greenhouse gas data, only 16 report on full scope 1, 2, and 3 emissions (Ceres, 2019). In other words, many companies may not be accounting for portions of their value chains associated with significant greenhouse gas emissions. Additionally, despite scope 3 emissions accounting for an average of 89 percent of a food company's total emissions, as of 2018, only nine of the top 50 North American food companies had targets explicitly focused on reducing scope 3 emissions (Ceres, 2019). While food companies have made notable headway in setting climate ambitions and holding themselves accountable through public reporting, limiting global warming to below 1.5 °C will require that all food companies–and all emitting entities for that matter–collectively commit to reducing emissions. Actions by a majority of major food companies will not be sufficient to meet the IPCCs recommendations. Using LCA-based greenhouse gas models can be an important step for companies to understand and address their greenhouse gas emissions.

## 14.2.2 Using LCA to substantiate product-level green marketing

As LCA-based models continue to support a growing number of corporate-level climate ambitions within the food industry, food companies are also increasingly using models to estimate and subsequently communicate the environmental impact of individual food products (Röös et al., 2014). In contrast to comprehensive sustainability reports or portfolio-wide greenhouse gas inventories, these product-level claims may provide a more tangible opportunity for consumers to address climate change and other environmental issues (Röös et al., 2014). In a 2019 sustainability assessment by the Hartman Group, surveys found that while nearly 80 percent of respondents were familiar with the term sustainability, only 22 percent felt they could identify a sustainable product and only 15 percent felt they could identify a sustainable company (Hartman Group, 2019). These findings, paired with the Hartman Group's estimate that 24 percent of consumers consistently make purchases based on sustainability criteria (Hartman Group, 2019), suggest that the use of environmental models and surveys as the basis to communicate product-level impact may represent an opportunity to both inform and attract consumers.

Oatly, a Swedish brand that produces oat-based dairy alternatives, works with the third-party life cycle assessment firm CarbonCloud to publish greenhouse gas emission footprint estimates for many of their products (Oatly, 2021). For example, as show in Fig. 14.3, one of the brand's one-liter oat milk products is packaged in a carton that states "climate footprint 0.38 KG CO2e/KG". Public communications from Oatly have noted that without the inclusion of analogies or comparisons to help consumers anchor the significance of the emissions Figs., these estimates may carry little meaning for a consumer beyond a gesture of transparency (Oatly, 2021). For this reason, Oatly has called on other food companies to follow suit with voluntary emissions labeling that will enable consumers to make informed choices by comparing individual products' impacts (Oatly, 2021). Of course, if companies use different assumptions in their LCAs, the results may not be comparable. Even without a strong consumer understanding of the relative significance of emissions estimates, it is possible that the mere inclusion of impact estimates can help build consumer trust in brands (Kim and Lee, 2018).

Some companies go beyond communicating greenhouse gas footprints to report environmental performance across a suite of impact areas. In 2020, the fast casual chain Chipotle Mexican Grill launched a menu sustainability impact tracker called the Real Foodprint (Chipotle, 2020). Using Latis, a modeling-based sustainability impact platform designed by the US-based research company HowGood, Chipotle sought to empower consumers to make informed food choices based on environmental impact (Chipotle, 2020). Using metrics such as gallons of water saved and square feet of organic land supported, the model compares Chipotle's ingredient sourcing impact to that of industry averages for conventionally sourced counterparts, highlighting differential procurement practices with claims of superior environmental performance. For example, Chipotle notes that some of the ranches from which the company sources beef practice rotational grazing (Chipotle, 2021), a practice that has shown promise to improve soil health and help sequester carbon (Rowntree et al., 2020). Ordering a steak burrito within the mobile ordering application generates a series of impact statements including the claim "182.9 gs less carbon in the atmosphere". With additional click-throughs, a viewer can read about the model's methodology and the comparison Chipotle is making against industry averages. Chipotle's Real Foodprint represents an emerging trend in the food industry to use

**FIGURE 14.3**  Oatly product with a greenhouse gas emission footprint estimate. (Oatly, 2021).

models to help consumers understand how their food choices impact multiple environmental outcomes.

Research suggests that communicating the results of LCAs in a way that resonates with consumers remains a challenge for companies (Molina-Murillo and Smith, 2009). Accordingly, some companies survey the farmers in their supply chain to develop qualitative communications about activities with potential environmental benefits. In 2020, organic baby food company Happy Family Organics, launched Happy Baby Organics Regenerative[3] + Organic, a line of baby food pouches made with ingredients that are traceable back to the farm sources for the ingredients. (Happy Family Organics, 2021). Using an internal farm-level survey, Happy Family Organics collected information on farmers' management practices to inform the storytelling on the product line (Crawford, 2020). The side of the package, pictured in Fig. 14.4,

---

[3] While a standardized definition of regenerative agriculture does not exist (Newton et al., 2020), Happy Family Organics defines regenerative agriculture as "a holistic approach that empowers farmers, promotes animal welfare, and builds healthy soil which can capture carbon from the atmosphere and help to reverse climate change" (Happy Family Organics, 2021).

**FIGURE 14.4** Side panel packaging from Happy Family Organics Happy Baby® Organics Regenerative + Organic line, which relies on a farm-level survey to support messaging about environmental impact.

includes an infographic of practices the company describes as regenerative, such as planting cover crops, implementing conservation tillage, and applying compost). Where LCA relies on industry average values for ingredient categories and would therefore not inform a company of the farm-level practices taking place in their supply chain, surveys such as Happy Family Organic's provide information about documented farming practices that are associated with beneficial environmental outcomes in research. It is important to note that using assessments to substantiate product-level communications can require resource-intensive processes such as segregating and identity-preserving ingredients all the way back to the farm level of the supply chain.

As food companies continue to use models and survey tools to support product-level environmental claims, we may gain a better understanding of the type of information that resonates with consumers and the tools needed to substantiate these communications. As noted earlier, most US consumers report that they would purchase products from a purpose-driven company (Porter Novelli/Cone, 2019), but less than a quarter report that they are able to identify a sustainable product (Hartman Group, 2019). Using models and surveys to substantiate compelling environmental performance claims on products may provide tangible opportunities for consumers to engage with companies' sustainability initiatives directly through their food purchases.

### 14.2.3  Using farm-level surveys to account for agricultural activities in a company's supply chain

While LCA can help identify supply chain hotspots of environmental impact, the widely used databases that feed LCA models do not capture the unique suite of agricultural practices used at the farm level of a particular supply chain. For companies seeking to support farm management decisions that can lead to beneficial environmental outcomes, it is helpful to first understand the activities taking place at the farm source. Some farm-level surveys collect data on farming practices to use as inputs for environmental outcome models, while others capture management decisions with the purpose of creating farmer engagement and education opportunities without quantifying outcomes.

One example of a tool that focuses on farmer engagement and education without modeling outcomes is the General Mills Regenerative Agriculture Self-Assessment. The open-source tool was launched by the company in 2018 with the goal to empower farmers and ranchers to understand alignment between their practices and the principles of regenerative agriculture. While a standardized definition of regenerative agriculture does not exist, General Mills defines it as a holistic, principles-based approach to farming and ranching that seeks to strengthen ecosystems and community resilience (General Mills, 2021a). Rather than use practice data to estimate outcomes such as greenhouse gas emissions, the General Mills self-assessment uses practice data to extrapolate to five broad regenerative agriculture principles: minimize soil disturbance, maximize crop diversity, keep the soil covered, maintain a living root in the ground year-round, and integrate livestock (General Mills, 2021b). These principles are consistent with many frameworks for managing soil health, and General Mills leveraged agriculture consulting firm Understanding Ag's 6 Principles of Soil Health to design the assessment. (Understanding Ag, 2021). As an example of how the tool uses practice survey data to understand the adoption of broader principles, indicating the use of low or no tillage would enhance the survey score for the minimize soil disturbance principle. Similarly, indicating the avoidance of highly hazardous pesticides, as defined by Pesticide Action Network International,[4] would also enhance the farm's score for the minimize soil disturbance principle (Pesticide Action Network, 2021).

Emphasizing the importance of general principles, rather than specific practices, gives farmers multiple pathways to improve environmental impact. This is notable because farming practices are associated with varying levels of feasibility and suitability across context-dependent variables such as climate, soil type, and certification schemes. With links to educational resources throughout the assessment, the tool encourages respondents to learn more about each principle and identify opportunities to further advance regenerative agriculture. Beyond the educational element of the tool, the survey responses help General Mills identify

---

[4] Pesticide Action Network International categorizes hazard criteria in four groups: acute toxicity, long term health effects, environmental hazard criteria, and international pesticide regulations (Pesticide Action Network, 2021).

opportunities to resource farmers with one-on-one coaching, educational field days, and other engagement opportunities.

As food companies work to understand and influence management decisions in the agricultural portion of their supply chain, many are also eager to quantify farm-level outcomes such as reduced greenhouse gas emissions. Quantifying farm-level outcomes enables companies to communicate and track progress on scope 3 emissions reductions targets. Recall that scope 3 emissions occur outside of a company's owned operations, including the cultivation of any ingredients upstream of a company's assets. To connect farm practice survey data to outcomes, companies are turning to farm-level footprinting tools such as the Cool Farm Alliance's Cool Farm Tool, Field to Market's Fieldprint Calculator, and Colorado State University's COMET-farm Tool. These tools collect information directly from farmers on parameters including farming practices, purchased inputs, and yields to model on-farm greenhouse gas emissions. The inclusion of farm-level data in these models helps ensure that the results of the model reflect the unique activities within a particular supply chain. In contrast, traditional LCA studies have used industry averages for a broad ingredient category or country of origin and therefore do not yield results that are customized to a particular supply chain. Farm-level footprint tools seek to encourage farmers to make informed management decisions that can improve environmental impact, while providing companies with data that help track and communicate progress on corporate commitments.

Engaging farmers in surveys can provide a more accurate representation of farm-level activities as compared with traditional LCA databases, but data requests from companies can be burdensome for farmers and may not scale efficiently across large supply chains. With these challenges in mind, many food companies are seeking new ways to streamline farm-level data collection from the regions where key ingredients are cultivated. Satellite remote sensing is one emerging approach food companies are pursuing to quantify farm-level environmental impact without having to engage farmers in survey responses. Satellite remote sensing measures reflected and emitted radiation to monitor physical characteristics in a geographical area, and it is becoming more cost effective and sophisticated in the type of data it can track. For example, some remote sensing technology can determine whether a cover crop has been planted and the presence or absence of living or senescent cover throughout the year, both of which impact environmental outcomes such as greenhouse gas emissions and water pollution (Hagen et al., 2020). Pairing farm management data and geographically-specific physical data from satellite remote sensing, biogeochemical models could simulate the cycling of nutrients such as carbon and nitrogen. Consider the fact that carbon dioxide emissions are influenced by factors such as local climate and weather, soil type, tillage intensity, crop rotations, net primary productivity, and other factors affected by both inherent site characteristics and management decisions (Hagen et al., 2020). A traditional LCA might assume no change in soil carbon, and therefore no soil-related carbon emissions or sequestration, regardless of the context of a specific operation. In this case, an LCA would not enable a company to account for changes in soil carbon emissions or sequestration that could stem from tillage or crop management practices in their sourcing regions. Coupled with assumptions about fertilizer use based on county or region-specific datasets like those collected by the USDA National Agricultural Statistics Service, satellite remote sensing and biogeochemical models can obviate the need for farmers to input survey data while helping companies quantify how changes in farm management are impacting environmental outcomes over time.

The Ecosystem Services Market Consortium (ESMC), for example, combines satellite remote sensing and biogeochemical models to create a cost-effective and scalable way to quantify environmental impact at the farm level. ESMC represents a multi-stakeholder collaboration among corporations, NGOs, and agricultural experts. Satellite remote sensing and biogeochemical models serve as the verification backbone of an ecosystem services market that ESMC is working to launch by 2022. The consortium is seeking to provide a pathway for companies to address their climate ambitions while simultaneously financially rewarding farmers and ranchers for achieving ecosystem services such as greenhouse gas reductions and improved water quality (ESMC Corporate Social Responsibility, 2021). While the emergence of increasingly sophisticated models to estimate environmental impact at the farm level is hopeful, the proliferation of different modeling methodologies may create challenges for companies to combine environmental impact information from different sources into one cohesive enterprise-level footprint.

## 14.3 Limitations in the application of environmental models in the food industry

Models play an important role in positioning food companies to track, communicate, and improve their environmental performance. However, limitations in the way these tools are used by companies and their LCA consultancy partners can hinder progress in achieving established targets. Here we discuss two key limitations of LCA applications within the food industry: limited characterization of geospatial and agricultural management factors at the farm level of specific supply chains, and limited ability to track changes in environmental performance over time. It is important to note that these limitations are rooted in the way companies and LCA consulting firms conduct LCA and apply results; they are not necessarily inherent shortcomings of the tool itself. Understanding these limitations may help companies identify appropriate applications of models to track and improve environmental performance.

First, many LCA studies conducted by food companies do not take into account inherent site characteristics or farming practices associated with specific farm operations in a company's supply chain. Therefore, these studies typically reflect generalized agricultural outcomes rather than supply chain-specific agricultural outcomes. This is a notable challenge for an industry where the biggest supply chain environmental impacts–and therefore the biggest opportunities for environmental performance improvement– often stem from farming and ranching. Consider two wheat operations that abut the Mississippi River where all management parameters are equal but one has taken several acres out of production to plant a riparian buffer along the water that helps filter chemical runoff and prevent water pollution. If the presence or absence of riparian buffers is not considered in the LCA, the modeled chemical runoff into the river will appear equivalent between the two operations when in reality the operation with riparian buffers may be contributing less pollution. Additionally, the operation that took acres out of production and reduced total wheat yield to plant the buffers will generate higher estimated chemical runoff impacts per unit of wheat production, even though it likely has a lower absolute nutrient load compared to the operation without a buffer. If wheat farms with riparian buffers are common in a company's supply chain but their LCA uses an assumption that wheat farms in the region exclude riparian buffers, the environmental benefit from the buffers would be obscured. This example highlights how the omission of context-specific data in LCA can overlook benefits that may stem from updated management practices at the farm level. As discussed in the previous section, the advent of farm-level surveys and biogeochemical models that leverage satellite remote sensing data represents an opportunity for companies to apply LCA in a way that better captures their supply chains' unique agricultural impacts.

Second, the LCA databases on which food companies typically rely for measuring and reporting greenhouse gas emissions are not designed to track longitudinal changes in emissions. Rather, generic LCA databases such as Ecoinvent and the World Food LCA Database were designed to provide a high level and comparative snapshot in time estimate of impact across a wide range of activities. To track greenhouse gas reductions over time within a value chain, companies must be able to capture baseline data on greenhouse gas emissions, the interventions that were implemented to reduce emissions, and the estimated reduction in greenhouse gas emissions resulting from those interventions. Because widely used LCA databases typically have a single set of estimated impacts for a given ingredient grown in a given region, introducing a change in farming practices to reduce emissions for that ingredient would yield no change in the LCA's estimated emissions. For example, relying on a generic impact entry for soy grown in the US would not allow a company to track greenhouse gas reductions that might stem from updated practices within their suppliers' soy operations, such as reducing fertilizer use or transitioning from conventional tillage to a no-till system. However, companies may choose to partner with their suppliers and LCA consultancies to supplement or replace generic database inputs with supply chain-specific information. The inclusion of this supply chain-specific data can better account for changes in management practices and therefore changes in impact over time. Because collecting data across individual operations in a multi-farm supply chain can be resource-intensive, many companies and LCA firms currently rely on generic database inputs to their models and face challenges in reliably tracking greenhouse gas reductions over time. Because farm-level surveys and biogeochemical models that use satellite remote sensing data can capture changes in supply chain management over time, these tools may help companies apply LCA in a way that dynamically tracks longitudinal progress.

While LCA databases are not currently designed to account for supply-chain specific agricultural impacts or to track greenhouse gas reductions over time, it is worth noting that scientists do have a high level of certainty around some of the interventions that will be most impactful in addressing climate change moving forward. The IPCC outlines four key actions needed to limit global warming to 1.5 °Celsius: 1) halt deforestation by 2030, 2) reduce fossil fuel[5] emissions to as close to zero as possible by 2050, 3) reduce other non-fossil fuel agricultural greenhouse gases such as nitrous oxide and methane by about 50 percent by 2050, and 4) ensure that land can serve as a net carbon sink to help draw down remaining carbon dioxide from the atmosphere (IPCC, 2018). Given these recommended actions, limiting warming to below 1.5 °Celsius would require that food companies address each greenhouse gas differently, prioritize deforestation-free supply chains, phase out fossil fuel emissions

---

[5] The IPCC AR5 report defines fossil fuels as "Carbon-based fuels from fossil hydrocarbon deposits, including coal, peat, oil, and natural gas." (IPCC DDC Glossary, 2014)

within their supply chains, and support a transition to land management practices that avoid excess non-carbon dioxide emissions while optimizing the drawdown and storage of carbon in the soil.

We have discussed two key limitations in the ways that food companies apply LCA-based models: limited characterization of the unique geospatial context and management practices on farm operations within a supply chain, and challenges in reliably tracking greenhouse gas reductions over time. Food companies need not wait for advancements in model databases, farm-level surveys, or satellite remote sensing to make urgent progress on our most pressing environmental challenges. The IPCC pathways to limit global warming have provided clear directives around the most important actions to limit warming while creating ancillary ecosystem benefits. Addressing challenges as complex as climate change will require collective action beyond the reach of any individual food company. How might an industry-level commitment to phase out fossil fuels or eliminate deforestation move the needle faster than individual companies' greenhouse gas reduction or deforestation commitments? Are there opportunities to reframe accountability in a collective way? These are just a couple of the questions that the food industry, scientists, and other stakeholders are considering as we rapidly reach environmental tipping points with irreversible climate changes. While models and farm-level surveys provide a framework for food businesses to understand and communicate about the environmental impact of their activities, it is important to avoid a narrow reliance on these tools to identify impactful paths forward.

## 14.4 Conclusion

In this chapter, we discussed common motivations for food companies to pursue environmental models, typical applications of these tools within the food industry, and key limitations in the way food companies are using LCA to track and improve environmental impact. Among other motivations, food companies use environmental models to pursue a triple bottom line approach to business that accounts for social, environmental, and financial metrics. Companies may also leverage these tools to estimate and subsequently communicate environmental impact to appeal to consumers, investors, NGOs, and other stakeholders. With awareness growing around the environmental impacts associated with food production, food companies are under increasing pressure to address and transparently communicate the environmental performance of their business activities.

LCA is the most widely used tool to model the environmental impacts of agriculture, which is the largest source of greenhouse gas emissions across the global food system. Many prominent food companies leverage LCA frameworks to track corporate greenhouse gas reduction targets and to substantiate product-level messaging about environmental impact. As of 2019, nearly 70 percent of North America's top food companies publicly disclosed their greenhouse gas footprint (Ceres, 2019). In addition to using LCA-based models to help track corporate-level climate ambitions, some food companies conduct LCAs to communicate environmental outcomes associated with the production of individual food products. These consumer-facing communications about environmental impact on product packaging may provide an accessible way for consumers to engage with companies' environmental commitments.

LCA can help companies identify hotspots of environmental impact across ingredients and supply chain nodes. However, using industry averages for farming practices can create blind spots at the agricultural level of companies' supply chains. With agriculture typically representing the largest contribution of greenhouse gas emissions across supply chain nodes, visibility to supply chain-specific farming practices and geospatial factors such as soil type can help identify meaningful opportunities to improve management decisions. Companies interested in tracking and influencing farm management decisions are turning to farm-level surveys to account for impacts in the agricultural portion of the supply chain. Some farm-level surveys focus on farmer engagement and education, while others model environmental outcomes to track and communicate progress against corporate commitments. The emerging use of satellite remote sensing and biogeochemical models to quantify environmental outcomes in agriculture shows promise to scale impact tracking over time and space while reducing the onus on farmers to complete surveys.

In their current applications within the food industry, LCA-based models have important limitations that food companies should consider. LCA studies do not typically characterize geospatial or farming practice data that are specific to a company's supply chain. Additionally, LCA does not reliably track companies' environmental progress over time. Despite these challenges, there is strong scientific consensus around the most meaningful actions companies can take to address environmental issues such as climate change. The IPCC has outlined four key actions needed to reach the global goal to limit warming to under 1.5 °Celsius: 1) halt deforestation by 2030, 2) reduce fossil fuel emissions to as close to zero as possible by 2050, 3) reduce other non-fossil fuel agricultural greenhouse gases such as nitrous oxide and methane by about 50 percent by 2050, and 4) ensure that land can serve as a net carbon sink to help drawdown remaining carbon dioxide from the atmosphere (IPCC, 2018). With this clear directive from the IPCC, food companies need not wait for advancements in models or the data that feed them to take urgently needed action to address our most pressing environmental challenges.

Models can provide important opportunities for food companies to estimate, communicate, and improve the environmental impacts of their value chains. However, a narrow reliance on models can obscure immediate opportunities for food businesses to advance global goals to reduce greenhouse gas emissions and enhance ecosystem resilience. As food companies increasingly look to models to characterize the impact of their business activities, at the same time it will be critical for the food industry to collectively and hastily commit to actions that we already know can yield vital environmental outcomes.

# References

Annie's, 2019. Sustainability Highlights Fiscal Year 2019. Annie's. https://www.annies.com/our-mission/ (accessed June 16, 2021).

Braam, G., Peeters, R., 2017. Corporate Sustainability Performance and Assurance on Sustainability Reports: diffusion of Accounting Practices in the Realm of Sustainable Development. Corporate Social Responsibility and Environmental Management 25 (2), 164–181. https://doi.org/10.1002/csr.1447.

Ceres, 2019. An Investor Guide on Agricultural supply Chain Risk, Top US food and beverage companies scope 3 emissions disclosure and reductions. https://engagethechain.org/top-us-food-and-beverage-companies-scope-3-emissions-disclosure-and-reductions (accessed June 15, 2021).

Chipotle, 2021. Chipotle Realfood Footprint: reduced Carbon Emissions. https://realfoodprint.chipotle.com/reduced-carbon-emissions.html (accessed June 18, 2021).

Chipotle, 2020. Chipotle Launches Real Foodprint, Introduces Sustainability Impact Trackers For Digital Orders. Press Release. https://newsroom.chipotle.com/2020-10-26-Chipotle-Launches-Real-Foodprint-Introduces-Sustainability-Impact-Trackers-For-Digital-Orders (accessed June 16, 2021).

Chouinard, Y., Stanley, V. 2012. The Responsible Company: What We've Learned from Patagonia's First 40 Years. Patagonia.

Cone/Porter Novelli, 2018. Cone/Porter Novelli Climate Change Snapshot. https://www.conecomm.com/research-blog/climate-change-snapshot (accessed June 16, 2021).

Cone, 2017. ConE Communications CSR Study. https://www.conecomm.com/2017-cone-communications-csr-study-pdf (accessed June 15, 2021).

Cone, 2015. Cone Communications/Ebiquity Global CSR Study. https://www.conecomm.com/2015-cone-communications-ebiquity-global-csr-study-pdf (accessed June 16, 2021).

Crawford, E., 2020. Happy Family Organics' Regenerative Organic Baby Food Addresses Concerns about Climate Change. Food Navigator. https://www.foodnavigator-usa.com/Article/2020/02/26/Happy-Family-Organics-farmed-for-our-future-baby-food-addresses-concerns-about-climate-change# (accessed July 30, 2021).

Crippa, A., Solazzo, E., Guizzardi, D., Monforti-Ferrario, F., Tubiello, F.N., Leip, A., 2021. Food Systems Are Responsible for a Third of Global Anthropogenic GHG Emissions. Nature Food 2, 198–209. https://doi.org/10.1038/s43016-021-00225-9 (accessed June 14, 2021).

Eccles, R.G., Klimenko, S., 2019. The Investor Revolution: shareholders are getting serious about sustainability. Harvard Business Review May-June Issue, 106–116. May-June 2019 Magazine https://hbr.org/2019/05/the-investor-revolution (accessed June 15, 2021).

Elkington, J., 2018. 25 Years Ago I Coined the Phrase "Triple Bottom Line." Here's Why It's Time to Rethink It. Harvard Business Review May-June Issue, 106–116. https://hbr.org/2018/06/25-years-ago-i-coined-the-phrase-triple-bottom-line-heres-why-im-giving-up-on-it (accessed July 16, 2021).

Elkington, J., 1994. Towards the Sustainable Corporation: win-Win-Win Business Strategies for Sustainable Development. California Management Review 36, 90–100. https://journals.sagepub.com/doi/10.2307/41165746. http://dx.doi.org/10.2307/41165746 (accessed July 16, 2021).

ESMC Corporate Social Responsibility, 2021. https://ecosystemservicesmarket.org/corporate-social-responsibility/ (accessed June 15, 2021).

Federal Trade Commission, 2012. Guides for the Use of Environmental Marketing Claims; Final Rule. Federal Register 77 (197), 106–116.

General Mills, 2021a. Global Responsibility Report 2021. https://globalresponsibility.generalmills.com/HTML1/tiles.htm (accessed 21.06.15).

General Mills, 2021b. General Mills Regenerative Agriculture Self-Assessment V2.0. https://www.generalmills.com/en/Responsibility/Sustainability/Regenerative-agriculture/For%20Farmers (accessed June 19, 2021).

General Mills, 2020a. General Mills to Reduce Absolute Greenhouse Gas Emissions by 30% Across its Full Value Chain Over Next Decade. Press Release. https://investors.generalmills.com/press-releases/press-release-details/2020/General-Mills-to-Reduce-Absolute-Greenhouse-Gas-Emissions-by-30-Across-its-Full-Value-Chain-Over-Next-Decade/default.aspx (accessed 21.06.15).

General Mills, 2020b. 2020 Annual Report to Shareholders. https://investors.generalmills.com/financial-information/annual-reports/default.aspx (accessed July 16, 2021).

Governance & Accountability Institute, Inc., 2020. Flash Report Standard & Poor's 500: trends on the sustainability reporting practices of S&P 500 Index companies. http://www.ga-institute.com/fileadmin/ga_institute/images/FlashReports/2020/G_A-Flash-Report-2020.pdf?vgo_ee=foVLW1LTt4o%2BO3CEnDjFfa1Fl4vtXwIDdK2ElycJgU4%3D (accessed July 16, 2021).

Greenhouse Gas Protocol, 2020. Update on Greenhouse Gas Protocol Carbon Removals and Land Sector Initiative. https://ghgprotocol.org/blog/update-greenhouse-gas-protocol-carbon-removals-and-land-sector-initiative (accessed 21.06.15).

Hagen, S.C., Delgado, G., Ingraham, P., Cooke, I., Emery, R., Fisk, J.P., et al., 2020. Mapping Conservation Management Practices and Outcomes in the Corn Belt Using the Operational Tillage Information System (OpTIS) and the Denitrification–Decomposition (DNDC) Model. Land (Basel) 9 (11), 408–441. https://doi.org/10.3390/land9110408.

Happy Family Organics, 2021. Farmed for our Future. https://www.happyfamilyorganics.com/farmed-for-our-future/ (accessed June 18, 2021).

Hartman Group, 2019. Sustainability 2019: beyond Business as Usual, 2019. http://store.hartman-group.com/sustainability-2019-beyond-business-as-usual/ (accessed June 15, 2021).

Hosonuma, N., Herold, M., De Sy, V., De Fries, R.S., Brockhaus, M., Verchot, L., et al., 2012. An assessment of deforestation and forest degradation drivers in developing countries. Environmental Research Letters 7 (4), 4009–4020.

Huang, A., Hejazi, M., Tang, Q., Vernon, C.R., Liu, Y., Chen, M., et al., 2019. Global agricultural green and blue water consumption under future climate and land use change. J Hydrol (Amst) 574, 242–256. ISSN 0022-1694 https://doi.org/10.1016/j.jhydrol.2019.04.046 .

IPCC, 2018. Global Warming of 1.5 °C. An IPCC Special Report on the impacts of global warming of 1.5 °C above pre-industrial levels and related global greenhouse gas emission pathways, in the context of strengthening the global response to the threat of climate change, sustainable development, and efforts to eradicate poverty. https://www.ipcc.ch/site/assets/uploads/sites/2/2019/06/SR15_Full_Report_Low_Res.pdf IPCC DDC Glossary, 2014. https://www.ipcc-data.org/guidelines/pages/glossary/glossary_fg.html (accessed 21.06.15).

Karlsson, R., Luttropp, C., 2006. EcoDesign: what's happening? An overview of the subject area of EcoDesign and of the papers in this special issue. J Clean Prod 14 (15–16), 1291–1298. doi:10.1016/j.jclepro.2005.11.010.

Kim, H., Lee, T.H., 2018. Strategic CSR Communication: a Moderating Role of Transparency in Trust Building. International Journal of Strategic Communication 12 (2), 107–124. https://doi.org/10.1080/1553118X.2018.1425692.

Mars, 2019a. Mars Accelerates Action to Tackle Climate Change With New #PledgeForPlanet Initiative. Press Release. https://www.mars.com/news-and-stories/press-releases/mars-accelerates-action-tackle-climate-change-new-pledgeforplanet (accessed 21.06.15).

Mars, 2019b. Climate Action Position Statement. https://gateway.mars.com/m/134f423a25bfa645/original/Climate-Position-Paper.pdf (accessed 21.06.15).

Mars, 2021. Mars Sustainability in a Generation Plan. https://www.mars.com/sustainability-plan (accessed 21.06.15).

Molina-Murillo, S.A., Smith, T.M., 2009. Exploring the use and impact of LCA-based information in corporate communications. International Journal of Life Cycle Assessment 14 (2), 184–194. https://doi.org/10.1007/s11367-008-0042-8.

Newton, P., Civita, N., Frankel-Goldwater, L., Bartel, K., Johns, C., 2020. What Is Regenerative Agriculture? A Review of Scholar and Practitioner Definitions Based on Processes and Outcomes Front. Sustain. Food Syst. 4, 577723–577743.

Nygren, J., Antikainen, R., 2010. Use of life cycle assessment (LCA) in global companies, 16. Reports of the Finnish Environment Institute, Helsinki, https://helda.helsinki.fi/bitstream/handle/10138/39723/SYKEre_16_2010.pdf?seque.

Oatly, 2021. Oat Drink with Carbon Dioxide Equivalents. https://www.oatly.com/uk/climate-footprint (accessed June 16, 2021).

Patagonia, 2021. Environmental Activism https://www.patagonia.com/activism/ (accessed June 16, 2021).

Pesticide Action Network. PAN International List of Highly Hazardous Pesticides. 2021 (accessed June 15, 2021).

Porter Novelli/Cone, 2019. Feeling Purpose: 2019 Porter Novelli/Cone Purpose Biometrics Study. https://static1.squarespace.com/static/56b4a7472b8dde3df5b7013f/t/5ce6eb8c15fcc0076a874b15/1558637485726/Biometrics+Research+FINAL+Single+Pages.pdf (accessed June 16, 2021).

Putt del Pino, S., Cummis, C., Lake, S., Rabinovitch, K., Reig, P. 2016. "From Doing Better to Doing Enough: Anchoring Corporate Sustainability Targets in Science." Working Paper. Washington, DC World Resources Institute and Mars Incorporated. Available online at http://www.wri.org/publications/doing-enough-corporate-targets

Quorn, 2021. Carbon Footprint: carbon Emissions: it's Time to Take Action. https://www.quorn.co.uk/carbon-footprint (accessed June 16, 2021).

Röös, E., Sundberg, C., PHansson, P.A. 2014. Carbon Footprint of Food Products. Assessment of Carbon Footprint in Different Industrial Sectors, 1. https://link.springer.com/book/10.1007/978-981-4560-41-2 (accessed June 16, 2021).

Rowntree, J., Stanley, P., Maciel, I., Thorbecke, M., Rosenzweig, S., Hancock, D., et al., 2020. Ecosystem Impacts and Productive Capacity of a Multi-Species Pastured Livestock System. Frontiers in Sustainable Food Systems 4. https://doi.org/10.3389/fsufs.2020.544984.

SBTi, 2020. Business ambition for 1.5 °C. https://sciencebasedtargets.org/business-ambition-for-1-5c/ (accessed 21.06.15)

Tidy, M., Wang, X., Hall, M., 2016. The Role of Supplier Relationship Management in Reducing Greenhouse Gas Emissions From Food Supply Chains: supplier Engagement in the UK Supermarket Sector. J Clean Prod *112* (4), 3294–3305. https://doi.org/10.1016/j.jclepro.2015.10.065.

Tyson, A., Kennedy, B., 2020. Two-thirds of Americans think government should do more on climate. Pew Research Center 1–39. https://www.pewresearch.org/science/2020/06/23/two-thirds-of-americans-think-government-should-do-more-on-climate/ (accessed June 16, 2021).

Unilever, 2021. Reducing our environmental impact: how we harness the latest science to manage environmental impacts and improve the health of the planet. Unilever Website. https://www.unilever.com/brands/innovation/safety-and-environment/reducing-our-environmental-impact/ (accessed June 16, 2021).

Unilever, 2020. Unilever sets out new actions to fight climate change, and protect and regenerate nature, to preserve resources for future generations. Press Release. https://www.unilever.com/news/press-releases/2020/unilever-sets-out-new-actions-to-fight-climate-change-and-protect-and-regenerate-nature-to-preserve-resources-for-future-generations.html (accessed June 16, 2021).

Van der Werf, H.M.G., Knudsen, M.T., Cederberg, C., 2020. Towards better representation of organic agriculture in life cycle assessment. Nature Sustainability 3, 419–425. https://doi.org/10.1038/s41893-020-0489-6.

Vermeulen, S.J., Campbell, B.M., 2012. Climate Change and Food Systems. The Annual Review of Environment and Resources. Ingram, J.S.I., *37*, 195–222. doi:10.1146/annurev-environ-020411-130608.

Whelan, T., Fink, C., 2016. The Comprehensive Business Case for Sustainability. Harvard Business Review. https://everestenergy.nl/new/wp-content/uploads/HBR-Article-The-comprehensive-business-case-for-sustainability.pdf (accessed June 16, 2021).

WBCSD/WRI, 2011. Corporate Value Chain (Scope 3) Accounting and Reporting Standard. https://ghgprotocol.org/sites/default/files/standards/Corporate-Value-Chain-Accounting-Reporing-Standard_041613_2.pdf

WBCSD/WRI, 2004. Greenhouse Gas Protocol: a Corporate Accounting and Reporting Standard. https://ghgprotocol.org/sites/default/files/standards/ghg-protocol-revised.pdf

# Inquiry within, between, and beyond disciplines

*Jennifer E Cross[a], Becca Jablonski[b] and Meagan Schipanski[c]*

[a]Colorado State University, Sociology [b]Colorado State University, Agricultural and Resource Economics [c]Colorado State University, Soil and Crop Sciences

## 15.1  Introduction

The burgeoning food system community of practice has many dedicated researchers, practitioners and communities who are working to better understand and address sustainability and resilience issues across the US, and in a global context. Much scientific research on food systems has goals of serving diverse communities, improving local economies, and addressing inequities in food access and quality. These efforts, however, are often disparate, sectoral, and fragmented, which limits how actionable food system research outcomes are within communities. As you have read in the previous sections, researchers and community partners face a variety of challenges in connecting research across disciplines and additional barriers for connecting research with communities and decision-makers in meaningful ways. Striving to make change in complex systems, such as food systems, requires that researchers engage in larger and more diverse teams including community stakeholders and researchers from multiple disciplines.

Food Systems research needs to be transdisciplinary rather than interdisciplinary because it reaches beyond discipline specific approaches to create new frameworks to address complex problems (Aboelela et al., 2007). Additionally, food systems research requires a paradigm shift in how knowledge is created by valuing interdisciplinarity as well as community-based knowledge production. It includes new habits of mind characterized by inclusivity versus exclusivity, collaborative versus independent exploration, and being reflective and adaptable rather than directive and inflexible. For researchers, it includes several new competencies that include systems thinking, interpersonal communication skills, collaboration skills, and adaptability(Cheruvelil et al., 2014). Achieving transdisciplinarity means that food systems research teams can co-create new questions, new data, new methods and new understandings.

In this chapter, we share key principles of team science, best practices to strengthen the evaluation, assessment and research efforts that are undertaken in food systems and use a couple of case studies to illustrate how these principles guide effective team projects.

## 15.2 Food systems research and participatory team science

Food systems research is ripe for an approach that Tebes and Thai (2018) call *participatory team science,* which is defined as public engagement in interdisciplinary team science. Over the past several decades, we have seen the expansion of participatory action research, community engaged research, and community-based participatory research, which engage community members as more than research participants but as co-creators of knowledge (Jason and Glenwick 2016). Despite the variety of names for these participatory approaches to research, they share in common several key features: 1) a focus on application of knowledge rather than only discovery, experimentation, and theory creation; 2) require interdisciplinary knowledge and methods; 3) value co-production between inter- or transdisciplinary teams and a variety of stakeholders; and 4) seek to create knowledge that is embedded in local context and cultures (Nowotny, Scott, & Gibbons, 2003).

We argue that food systems research needs a participatory team science approach. Team science is the broad term which describes interdisciplinary or transdisciplinary research in which individuals from various disciplines work together to solve major challenges, generate deeper understanding, and create scientific discoveries not possible within a single discipline (Disis and Slattery, 2010). Participatory team science, goes a step further engaging communities and decision-makers at the regional scale in order to embed community priorities in the modelling approaches undertaken by food systems research teams. A participatory team science approach should be framed and translated in a way that allows all communities to have better and more equitable access to knowledge, expertise, and tools that can help them find and connect strategies across scales, sectors, and issues. While much of the literature on team science and food systems does not use the term participatory team science, we will use it throughout the chapter in discussing team science competencies, our illustrative case, and assessment of transdisciplinary teams.

### 15.2.1 Individual competencies for participatory team science

As team science grows in necessity and popularity, a few frameworks have emerged which describe essential competencies and orientations for successful team science (Gilliland et al., 2019; Lotrecchiano et al., 2021; Nurius and Kemp, 2019). Overcoming the barriers to transdisciplinary science require the development of new competencies at three levels: individual, team, and organization (Lotrecchiano et al., 2016, 2021). Individuals need to bring a collaborative mind-set, openness to learning from others, capacity to consider divergent perspectives, and interpersonal relationship and communication skills (Lotrecchiano et al., 2021; Nurius and Kemp, 2019). These values, mindset, and interpersonal skills are the foundation needed for the more complex tasks of participatory team science.

In the health sciences, the term translational science was coined to describe research which aims to turn observations and lab experiments into actionable knowledge that is adopted

**TABLE 15.1** Characteristics of a participatory team scientist.

| | |
|---|---|
| Domain Expert | Possesses deep disciplinary knowledge and expertise within one or more of the fields relevant to the aspects of the food system under study, ranging across the basic agricultural and food sciences, applied public health and environmental sciences to the broader social sciences, arts, and humanities affecting food environments and choices. |
| Rigorous Researcher | Conducts research at the highest levels of rigor and transparency, possesses strong analytical skills, and designs research projects to maximize impact to core fields and reproducibility. |
| Skilled Communicator | Communicates, compiles feedback and reinforces shared understanding across all stakeholders in the transdisciplinary project recognizing diverse social, cultural, economic and scientific backgrounds, including community members and stakeholders team intends to impact. |
| Team Player | Practices a team science approach by leveraging the strengths and expertise and valuing the contributions of all players on the team and across engaged communities. |
| Boundary Crosser | Breaks down disciplinary silos and collaborates with others across areas and professions to collectively advance the development of integrated models that link key metrics and points where food system managers, eaters and policymakers make decisions that influence outcomes |
| Systems Thinker | Evaluates the complex external forces, interactions and relationships impacting the food system, including the regulatory and policy environment, consumer market and competitive landscape, public attitudes toward food, ag and environmental issues, and key social drivers influencing the system (equity, justice, diversity). |
| Process Innovator | Seeks to better understand the scientific and operational principles underlying transdisciplinary and community-based participatory research, as well as between rigorous researchers and community stakeholders, and innovates to overcome incongruencies and bottlenecks in the processes of engaged and applied research. |

Adapted from Gilliland et al. 2019. Fundamental Characteristics of a Translational Scientist. *ACS Pharmacol. Transl. Sci.* 2(3): 213–216. https://doi.org/10.1021/acsptsci.9b00022.

in clinical practice and developed into interventions that can be widely disseminated to improve patient and public health. Gilliland et al. (2019) argued that training within scientific fields is atomistic and maintains the silos and fragmentation so prevalent in science today. Translational scientists need to possess several skills beyond domain expertise and depth. They call these the characteristics of translational scientists (Gilliland et al., 2019). Drawing on this framework, we adapt the seven characteristics for scientists working on participatory team science projects, thinking specifically about characteristics, orientations, and competencies needed in food systems research (Table 15.1).

Where most scientists are trained to be domain experts and rigorous researchers (Gilliland et al., 2019), these characteristics are inadequate for participatory team science. A collaborative mindset and value for interdisciplinary perspectives are prerequisites for participation in any kind of transdisciplinary team science, allowing individuals to adopt a team orientation rather than a hierarchical or independent approach to science (Lotrecchiano et al., 2016; Nurius and Kemp, 2019). In addition to practicing a team approach, food systems research benefits when individual team members also build the skills to be boundary crossers, systems thinkers, and process innovators (Table 15.1).

Transdisciplinary research should include a deliberate cross-disciplinary scientific process, engagement of key stakeholder groups (a theme revisited by several chapters in this book), and authentic integration of science and practice. Thus, a strategic blend of scientists with different skills and inclinations should be assembled to create an effective team. What may be most important to highlight about these characteristics for food systems work is that both content expertise (scientific fields) and context expertise (understanding of place, community, and issue) are essential, and that various combinations of team members may share roles, depending on the size of the team. Boundary crossers may have one of the most challenging tasks in moderating the tensions that may emerge between rigor and relevancy in the team's approach and work efforts.

Food systems research includes a diverse array of scientific fields—soil and crop science, animal and meat science, economics, public health, nutrition, engineering, ecology, and rural sociology—and stakeholders who will require the team to fully understand and appreciate the context of their situation. This diversity in scientists and stakeholders requires that team members cultivate the ability to be boundary crossers, helping to translate science to community members and scientists from other fields as well as understand the diverse perspectives of others. Systems thinkers can evaluate complex forces, interactions, and relationships that need attention in food systems modelling and in applying new knowledge gained from that modeling. Being a process innovator is perhaps the most distinguishing characteristic because participatory team science faces a variety of institutional barriers, which a process innovator seeks to overcome. Later in this chapter, we will discuss tensions faced by participatory teams in search of balance between rigorous research, direct application, and policy relevance of food systems models and research.

## 15.2.2 Team capacities for participatory team science

In addition to the capabilities and characteristics of individuals, participatory team science also depends on building the capabilities of teams. Team science competencies include several domains, building trusting and genuine relationships, communication, collaborative knowledge generation, collective problem-solving, team management, and team leadership (Lotrecchiano et al., 2021). Like all community-based research methods, participatory team science begins with building trusting relationships which evolve from respect and valuing local knowledge, questions, and priorities (Cross, Pickering, and Hickey, 2015). Specifically in food systems research, respect for local community partners means seeking out current processes, blueprints or plans created by various partners in the more broadly defined team. It is essential to value work that was already done, even when the goal is to expand on that. For instance, if an agency initiative just identified six focus areas through a participatory process involving their network, that data must be used as a starting point. Discounting or ignoring past community efforts degrades trust between researchers and communities and misses an opportunity for mutual learning. Even if only focused on one aspect of the broader system you intend to map out, the priorities (and the process used to arrive at them) are part of the team building process and finding authentic ways to incorporate them builds necessary trust for long-term collaboration.

Most food system research typically emerges from a narrow group of scholars, teams, or organizations, yet it may require linkages to a broader set of sciences, partners, and initiatives.

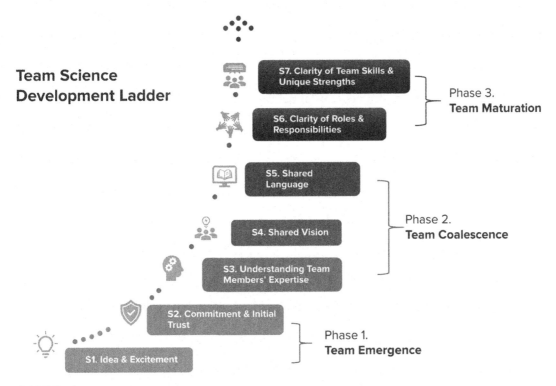

**FIGURE 15.1** Team science development ladder.

Building the network of researchers and industry, policy and community partners into a participatory team requires simultaneously strengthening existing partnerships, rebuilding trust if it has been lost between any partners, and inviting new members to address the complexity of the food system. Including existing partnerships may be as simple as developing a survey instrument that asks respondents to identify their organization, unit, collaborative and non-collaborative initiatives, and communities or areas where projects take place. Or it may require multiple steps to develop a shared vision, identification of issues that are most pressing, examination of what barriers are inhibiting development of research, and creation of strategies to build collaborative opportunities and overcome barriers.

Doing participatory team science requires commitment and patience because it takes time to build relationships and trust, create a shared vision, and overcome the many barriers to participatory research. The Participatory Team Science Development Ladder was created by the first author and CSU collaborators based on observations of 21 emerging transdisciplinary teams and literature on team science and group formation (Fig. 15.1). The ladder represents the steps and phases that complex teams progress through as they form, develop and grow.

The ladder is visualized as an ascending arrow of seven steps and three phases showing how teams progress as they expand their capacity and identity as a team. The first phase, Team Formation is the earliest stage when a group comes together around an exciting idea that inspires commitment and is based on an initial willingness to trust and form a team.

Exciting ideas may bring a group together, but unless members are able to commit time and resources and begin to form a team identity, they may disband without moving to phase 2 activities. Success in phase 1 results in team members being able to articulate a preliminary team vision, enthusiasm for the idea, and a sense of belonging and willingness to continue working together.

Once members have committed, substantial work is required to coalesce a team around a shared vision. The three steps in this phase are distinct, each must be accomplished, but teams may shift back and forth between refining a vision, developing shared language, and gaining clarity around the expertise of all members and how that contributes to the vision. We found that teams must develop a certain level of clarity around the vision before they are able to satisfactorily build shared language around their theories, methods, and unique expertise. In phase 2, new members may join and those who don't feel their expertise contributes to the vision are likely to drop out. This shifting membership is to be expected in the coalescence stage. Teams may progress through this stage in a few months, or it may take them several months to a year to feel settled around a shared vision. Success at this stage results in clear statements of team vision and direction and increased trust and sense of belonging and team identity.

Movement through each stage may involve conflict that when addressed can produce deeper levels of commitment, trust, mutual understanding, and clarity of purpose. As conflict is resolved, the team expands its capacity to engage in the next level of work. As teams begin working in earnest in phase 3, the team must develop clear roles and responsibilities that allow the team to accomplish its goals and objectives. As people work together and clarify roles and responsibilities, they deepen their knowledge of each other's individual talents as well as how to leverage the disciplinary expertise of other team members. Success at this stage is evident in team accomplishments and expressions of appreciation for each other's strengths.

Research on teams and collective capacity show that trust and collaborative work are mutually reinforcing. Trust is required to work together, and as people work together, they increase their trust and, thus, their capacity for more complex collaboration and coordination (National Research Council, 2015). Each of the phases should be seen as not as a single task or a clearly bounded step, with a distinct end, but rather as a set of work that when engaged provides the foundation for the next level of work. While some team members may be excited about an idea or immediately willing to commit to the project and trust other partners, others will take longer to trust and will need to work on small things together before they are ready for higher levels of commitment and deeper forms of collaboration (Fig. 15.1).

## Catalyst for Innovative Partnerships

The mission of the Catalyst for Innovative Partnerships (CIP) program is to facilitate and position Colorado State University, through the creation and support of interdisciplinary research teams tackling grand societal and scientific challenges, to achieve significant global impact, in accordance with our land grant heritage.

Each team is awarded $200,000 for two years with the expectation that they are using these funds to support team development and seek new and substantial external funding.

**Catalyst for Innovative Partnerships —*cont'd***

The program evaluation of the first cohort revealed that emerging teams were spending a year or more in the early stages of team development and were not ready for a $200,000 investment until they had begun to coalesce a shared vision and shared language. Therefore, a new program the Pre-Catalyst for Innovative Partnerships (PreCIP) was created.

The PreCIP program funds teams with $5,000 over nine months and provides teams with several workshops designed to support the earliest stages of team development. All PreCIP teams attend seven educational workshops designed to advance them along the Participatory Team Science Development Ladder (Fig. 15.1):

- Intellectual Property
- Scientific Basis of Team Science Lecture
- Developing Shared Language Workshop
- Value Proposition Creation and Validation
- Science Communication and Elevator Pitch
- Communication to Stakeholders
- Corporate & Foundation Grantmaking

## 15.2.3 Developmental evaluation

In 2015, the Office of the Vice President for Research (OVPR) at Colorado State University (CSU) launched the Catalyst for Innovative Partnership (CIP) awards to support emerging transdisciplinary teams (see sidebar). Dr Cross initiated an evaluation of the CIP program and the teams participating to help the OVPR refine and improve the program as well as to develop near-term markers of success for teams. This model was created based on observations and analysis how 21 emerging transdisciplinary projects progressed or disbanded during their participation in the CIP or PreCIP program in 2015 and 2016.

Large transdisciplinary teams benefit from participating in evaluation and training activities designed specifically to enhance their capacity to address challenges and foster self-awareness about best practices for working in teams (Looney et al., 2013; Wooten et al., 2014). The type of evaluation used to assess and support emerging transdisciplinary teams in the CIP program is a model called *developmental evaluation*. Developmental evaluation is an approach to team development and assessment focused on continuous improvement, supporting team adaptation under dynamic and evolving conditions in real time (Patton, 2010).

Led by Dr Cross, the evaluation team used social network analysis (SNA), team observations, informal interviews with team leaders, and social surveys to assess individual and team collaboration readiness, team management practices, and team interactions. SNA is the measuring and mapping of a variety of relationships between people, groups, and organizations. In team science, SNA examines the structure and patterns of knowledge sharing, trust, and collaboration between teams, mapping these in sociograms. The combination of team assessments, feedback, team building workshops, and management advice is designed to assist teams in progressing along the Participatory Team Science Development Ladder (Fig. 15.1).

## 15.3  Case study team: meeting the challenges of team science

Here we present a case study of an interdisciplinary food systems team (FST) from CSU. The FST provides an interesting example because they were awarded two team development awards, PreCIP and then CIP, which were specifically designed to invest in the formation and growth of new transdisciplinary teams on campus. Between 2016 and 2019, the FST was assessed and evaluated by external team science evaluators and participated in a variety of team development activities.

The first iteration of the FST started in 2015 with two seed grants: one was a Global Challenges Research Grant from CSUs School of Global Environmental Sustainability; and a Pilot Project Grant from CSUs One Health Institute. Though each of these seed funded projects had different faculty involved and different research questions, objectives and outputs, there was a core team of faculty that coalesced through these efforts. Through collaboration on these two grants, this team of had accomplished Phase 1 of the team development ladder and was eager to advance as a team.

This new supergroup subsequently applied for and received a PreCIP grant. During this award period, the team increased team membership from 10 to 21 researchers across six academic colleges at CSU. As a PreCIP team, the FST received a small financial award and participated in a variety of professional development workshops. In addition, this was the first time that team members participated in team science workshops and developmental evaluation, including team readiness surveys and social network analysis of team integration.

### 15.3.1  Phase 1. building a transdisciplinary team

The early stages of building a transdisciplinary team take time, which is why the PreCIP program was initiated. For the FST, Phase 1 (Fig. 15.1) had been completed in 2015 and the team was ready to progress to a larger more diverse team. The PreCIP award encouraged and supported the team to utilize their funds and time on activities specifically focused on Phase 2 of the Participatory Team Science Development Ladder (Fig. 15.1).

### 15.3.2  Phase 2. team coalescence

The funding provided by the CIP and PreCIP programs were instrumental in supporting team development in the early stages. Fig. 15.2 provides an overview of various team development and evaluation activities the FST engaged in during the PreCIP and the subsequent CIP award, aligned with the sidebar presented earlier.

#### 15.3.2.1  Step 3. understanding member's expertise

To build cohesion, the FST took time to have meetings that were focused not on a specific research agenda but on helping the team get to know each other's expertise and interests (Fig. 15.2). Team meetings centered around presentations of each member's disciplinary knowledge. The team also organized social events and tours with Extension Specialists to see various activities around the state. The PreCIP award provided funds and support to spend the year with no objective beyond building team cohesion and getting to know each other's expertise.

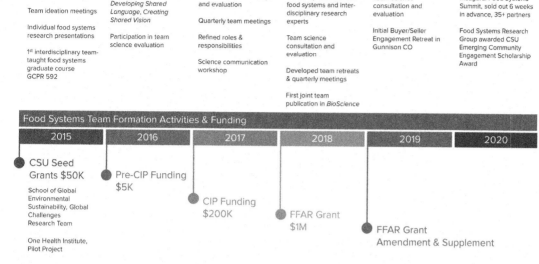

FIGURE 15.2   FST team development activities and assessment activities 2016–19.

While scientists are sometimes frustrated that no action or research is taking place, spending time on Steps 3–5 is essential for building the capacity to quickly coalesce a team vision around emerging funding opportunities or response to emergent issues and opportunities. The investment in thoroughly sharing and exploring the knowledge of diverse team members, Step 3, builds trust and mutual understanding (Read et al., 2016).

### 15.3.2.2  *Step 4 & 5. building a shared vision and language*

Similar to many investments, considering team members as a portfolio of talents and skills may provide the best context to recruit the right team. Team diversity allows each member to focus on one aspect of the project's needs (rigor, communications, modelling, or community engagement), knowing that others in the team will complement those skills with their contributions. Teams must also discuss how each of the skills are needed and valued for what they contribute to the portfolio. As the team builds a vision that includes diverse disciplinary expertise, consideration should also be made for assuring that a full spectrum of the Characteristics of Team Scientists (Table 15.1) is also present on the team. Most team members will contribute both scientific expertise as well as one or more characteristics of team scientists.

Fig. 15.3 uses social network analysis to illustrate how the FST built a team with diverse disciplinary knowledge and characteristics of team science. This network illustrates collaboration ties between team members at the beginning of the CIP grant (2016), and then at the end (2020). From 2016, when the team applied for CIP funding, to 2020 the team grew from 14 members to 28 members. These sociograms illustrate both that new members were integrated into the team (as evidenced by multiple collaboration ties to each node) and the increasing number of collaboration ties between members. The nodes have been colored based on their strongest characteristic (Table 15.1), the outer ring indicates their strongest contribution to the

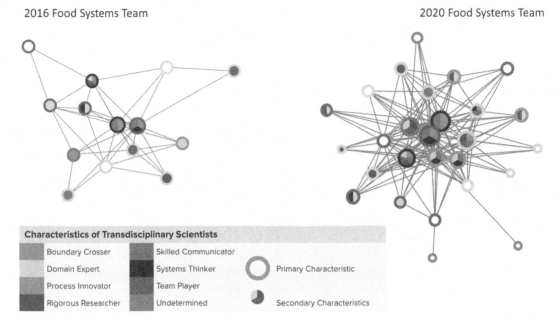

**FIGURE 15.3  Characteristics of team members and growth of collaboration ties from 2016 to 2020.** Note: Larger nodes have more connections than smaller nodes.

team, and the inner colors indicate additional characteristics enacted on this team. Members in the center were more commonly skilled communicators, systems thinkers, and boundary crossers, while those on the outside were more commonly rigorous researchers, team players, and domain experts.

In addition to the collaboration ties depicted in Fig. 15.3, members of the FST also reported growth in their personal ties and learning ties. The average degree, number of ties between members of the FST, grew from 2 to 6 when asked, "who do you learn from?". In teams with two to four dozen members, we have found that an average degree of 2 or less for research collaboration or learning ties is indicative of a new team with some pre-existing ties, and 6 or higher is a common marker of integrative learning between team members that only appears after a year or two of collaborative work. The FST network grew in size, strength, density and types of ties from 2016 to 2020.

The PreCIP provided time, space and permission to develop a shared vision and language. First, having funds to support facilitated discussions without a particular research objective was invaluable. Second, finding time for monthly discussions, including presentations by team members on their own research helped the team to understand where there was confusion around language. For example, one team member was defining "food systems" when one of the external facilitators was with the team. The facilitator turned to the team member and said, "You just defined a food supply chain, not the food system." This comment spurred an important discussion that was useful to the team in developing shared language about this core concept. It was clear that prior to this point the team, unknowingly, had not been on the same page.

The variety of workshops and assessments illustrated in Fig. 15.2 were essential to the team coalescing a shared vision and building shared language. Following a successful PreCIP experience, the team was awarded the Team Research Award from the CSU College of Agricultural Sciences in 2017, as well as a full-scale CIP grant, which provided $200,000 for the team for two years with the expectation that the team would use those funds to collect preliminary data, intentionally build the team, and prepare a research proposal $2 million or larger. Working on Steps 3, 4, and 5 during the PreCIP meant that upon award of the CIP, the FST had achieved the level of coalescence required to put together a successful proposal for external funding.

Very quickly after receiving the CIP grant, a call for proposals was released for the inaugural Tipping Points grant from the Foundation for Food and Agriculture Research (FFAR). The Tipping Points grant supports projects that identify leverage points or tipping points in food systems where specific changes can improve overall community health and the economy. Though the team was perhaps a bit early to have fully coalesced a transdisciplinary research agenda, the seed grants received between 2015 and 2017 had enabled a significant amount of trust to be built across disciplines. Thus, when the team made the decision to apply for the Tipping Points grant, they were better able to very quickly put together a credible research proposal that demonstrated strong collaboration and understanding across disciplinary lines. Organizing for a grant of this scale was only possible for such a diverse team because they had already dedicated time to Steps 1–5. Given the short timeline and busy schedules, preparing the proposal necessitated 7am convenings at coffee shops. Team members were willing to commit to this arduous schedule because they had already begun advancing up the Team Science Developmental Ladder.

## 15.4 Team management and leadership

Advancing from Coalescence to Team Maturation requires that teams dedicate time and effort to a variety of management and leadership activities. Creating a team charter is one team activity that can be especially helpful in the early stages of team development, setting the stage for later success. A team charter helps improve team management by explicitly (1) defining team purpose-mission, scope, and goal; (2) identifying skills, responsibilities, and metrics for success; (3) establishing group values, norms and how you will connect and celebrate team success; and (4) selecting communication and coordination tools that will help enable collaboration. A team charter is both road map for a team and a process of cultivating a group identity, clarity of purpose, and building trust. It should be considered a living document, not something to be created and never revisited (Hall et al., 2019) (Fig. 15.4).

### 15.4.1 Shared authentic leadership

Large, complex teams benefit from a clearly articulated leadership approach. Guenter, Gardner, Davis McCauley, Randolph-Seng, and Prabhu (2017) proposed a model, called Shared Authentic Leadership, as the ideal leadership model for transdisciplinary scientific teams. Shared Authentic Leadership occurs when leaders share leadership influence across several members and team members see it as their joint responsibility to lead the team

## Team Charter Template

| Mission | Scope | Goals |
|---|---|---|
| **Why:** reason for forming the team | **Boundaries:** within and outside team goals | **What:** the world will BE different because the team achieved its highest aspirations |

| Strengths & Skills | Roles & Responsibilities | Metrics for Success |
|---|---|---|
| **Diversity:** the variety of strengths and skills brought by all team members—disciplinary knowledge, interpersonal skills, lived experience, methodological tools | **Who:** specific roles and duties required to meet the team goals, which team members are best suited to take on each role | **What:** measurable team outcome/performance |

| Values | Norms | FUN! |
|---|---|---|
| **How:** team conducts business, treats each other, | **How:** team routines, expectations, interaction rituals, decision-making processes, and resolving conflict | **How:** events and group rituals to bond, create team identify, and celebrate success |

| Communications Plan and Coordination Tools |
|---|
| **What & How:** tools the team will use to stay connected, interact formally and informally, share data internally and externally, store data, and foster regular interaction |

**FIGURE 15.4**  Team charter template.

(Guenter et al., 2017). Authentic leadership describes a form of leadership where team leaders are "true to the self" and act according to their personal values and convictions and are thus able to garner high levels of trust with teammates (Guenter et al., 2017). This approach can be effective for large teams, but it requires clear role definition. In a large team, a small group of core-decision makers can help teams maintain momentum as the capacity of individuals ebbs and flows with other responsibilities, personal or professional. When one member of a leadership team needs to take leave or faces competing priorities, a team of leaders can more easily adapt than more hierarchical structures. Guenter et al. (2017) found that Shared Authentic Leadership and team trust were positively associated with perceived team performance and team satisfaction. Additionally, they found that team coordination was improved by Shared Authentic Leadership (Guenter et al., 2017). Specifically, team's self-awareness and relational transparency enhances team coordination.

Shared leadership can enhance transdisciplinary work. It helps leverage all the different members of the team, their disciplinary expertise and contributions to the team dynamic outlined in Table 15.1, and authentically involve the communities where food system work is focused. Shared leadership supports collaboration that includes joint problem definition and representation from community partners. This approach encourages teams to be self-reflective, and to make space for many types of experts to be leaders on teams.

### 15.4.2 Roles, responsibilities and network structures

As transdisciplinary teams grow in size, so does the diversity of roles and responsibilities and the complexity of coordination. All teams perform better when members have clear roles and responsibilities, and large teams require it. While there are a variety of different structures

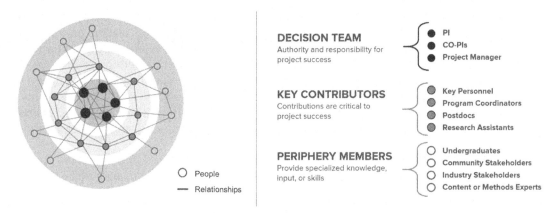

**FIGURE 15.5**  Large teams have layers with unique roles & responsibilities.

teams might choose, we propose thinking of teams as networks rather than as a group. Once teams group beyond 6–8 people, different network structures begin to appear (Read, Cross, Herman-Mercer, Oliver, and O'Reilly, 2022).

Some teams are organized around a central hub with a lot of interaction, some split into small teams with few connections between the teams, and yet others build a team that is more like a web than distinct sub-teams. Thinking of large teams as an integrative network, will likely mean that some members have weak ties or infrequent interaction with the large team, while others will have more frequent communication with more people (Cross et al., 2015). Effective teams do not need everyone to participate with the same level of depth or frequency, this is where social network analysis can help inform effective team structures.

One structure that can be highly effective for teams larger than three dozen is what is called a core-periphery network (Fig. 15.5). In the center is a small core team, which interacts frequently with each other and engages in a variety of collaborative activities. We call this the decision-making core. It handles the logistics and administration of team activities, guides the scientific direction, and has final decision-making authority. This small team has the most frequent meetings and is responsible for setting the agenda and designing larger team meetings or sub-group meetings. Around that small decision team is a densely connected core of contributors (Cross et al., 2015, Cross and Plaut 2019). This group is highly connected, actively engaged with the team, attends meetings frequently, makes substantial contributions to the team, and communicates frequently across the network. This group tends to be where the systems thinkers, boundary crossers, and process innovators are clustered in large teams.

The outer ring, or periphery, includes a variety of team members that are essential to the project but who may engage less frequently in team events. This group includes those people whose roles and interactions are more limited. It also includes community and industry stakeholders, graduate and undergraduate researchers, as well as domain experts and rigorous researchers whose roles are smaller. It is important to realize that the contributions of members in the periphery are essential to team success. Being in the periphery means that they attend fewer meetings and communicate and interact less frequently than those in the core. It does not mean their contributions are less important.

Managing from a network perspective can help team members identify what type of role is best suited to their skills and available time for the team. It can also help leaders understand that managing a large team does not require every member to be present for all discussions of team direction. One helpful team activity is to create a conceptual network map of how a team could be structured. The team discussion of this network map helps create a focused conversation about what roles need to be filled and what collaboration is needed across teams. Online systems mapping or network mapping tools and old-fashioned marker on flip charts can both be engaging and fun ways for a team to explore team roles and responsibilities (Holley, 2019).

A typical networked management strategy includes meetings of different subgroups at different intervals (Robertson, 2014). The Core Decision Team typically meets or talks on the phone weekly or bi-weekly. The Key Contributors typically meet with Decision Team on a bi-monthly basis in small sub-teams. The full network, including the Periphery members typically occur 1–3 times a year, depending on the size and geographic dispersion of the team. The FST held three full team meetings a year, with the Decision Team doing substantial preparation for these meetings. The agenda typically included: 1) revisiting team accomplishments since the last meeting, 2) celebration and recognition of team members, and 3) discussion of upcoming efforts to maintain shared vision and plans. In the earlier years, these meetings lasted for one to two days, and shortened to half-day retreats as the team progressed through the stages of team development. When working to craft shared vision and develop strategic plans, the team typically brought in a third party facilitator, which is demonstrated to improve team success (Thompson, 2003). These events were typically held off-site to help foster team rapport and focus on team development. Spending focused and social time together as a team is important for building the trust and relationships necessary for integrative teams (Cross et al., 2015).

### 15.4.3 Project management

During the first cohort of CIP teams, the OVPR recognized that teams that hired project managers were more successful that those that did not. For Cohort 2, of which the FST was part, teams were required to dedicate a portion of their budget to a project manager. While clinical sciences commonly use study coordinators and industry research and development teams widely use project managers, the hiring of project managers is less common in transdisciplinary science teams.

The size of this team and complexity of coordination activities required a dedicated person to take on the role of project manager. Once the FFAR grant was awarded, the FST conducted a national search, recognizing the important role that this person plays in coalescing the team. Unfortunately, the team had a failed search and ended up using a series of two postdocs, and an administrative assistant to formally fill this role. None of these situations were ideal. Neither postdoc was in the role for a long enough period to make a positive impact, and the administrative assistant did not have enough context to do more than ensure meetings were being scheduled. As a result, much of the project management responsibility fell on the PI, a faculty member with limited time to dedicate to this role.

Fortunately, towards the end of the project, a graduate student who had been involved with several aspects of the project, working directly with stakeholders, collecting data, conducting data analysis, and building the model, took on the role once she graduated. She had the necessary scientific understanding and the skills as a boundary crosser and skilled communicator to fill this role. She also had the benefit of familiarity with the project and history with the team that created trusting relationships, which has been tied to team performance.

Project management is a role that can be well-filled by a scientist with a master's degree, who understands the scientific program and has the skills to speak to both scientists and community stakeholders. One of the challenges for transdisciplinary teams is defining the job description for these positions and being able to hire for them. These kinds of complex projects need formal project managers, but this career track is often not well defined within universities or in graduate training programs like it is in the translational and clinical sciences. As transdisciplinary science is growing, scholars have been articulating best practices and calling for specific career paths around interdisciplinary research expertise (Bammer, 2013; Moore & Khan 2020). Hendren and Ku (2019) have named this professional role the interdisciplinary executive scientist.

There is no doubt that the overall project could have been more successful with a strong and consistent project manager or an interdisciplinary executive scientist, an even higher level of responsibility for team coordination and integration. The lack of a consistent project manager meant that much of the team's efforts were not properly utilized. For example, documents created early in the project, though they were stored in shared team cloud folders, were forgotten and not leveraged to the extent that they could be. An interdisciplinary executive scientist could have elevated the team's integration processes to an even higher level.

## 15.5 Navigating collaborative science tensions

Participatory team science includes a variety of tensions, including but not limited to:

1. Diverse knowledge and context depth of those involved
2. Varying planning and action timelines across stakeholders
3. Tensions between feasibility of community engaged research and rigorous science and
4. Divergent goals for outcomes and dissemination of research products.

Food systems modelling teams will likely face all these tensions and can only overcome them by developing trusting relationships, developing individual and team awareness, and adopting open communication habits. One of the key capacities that teams need to develop is the ability to have productive conversations and constructive conflict, where differences and issues can be discussed and explored without team members feeling that these discussions are personal attacks or don't result in resolution and common agreement (Bennett, Gadlin, and Marchand, 2018).

Participatory team science is a new mode of research, where team members need a broader set of skills and the goals and standards of research are determined by a diverse group of stakeholders. One of the primary tensions to be resolved is the tension between scientists who want to hold the project to the gold standard of scientific research versus team members who see the value of conducting research that is feasible and can meet stakeholder expectations and timelines. Researchers who strive to use perfect data and frontier scientific methods may not be a good fit for team science, unless they are seeking to branch out into more engaged and relevant work. Instead, they are often best engaged as periphery members, brought into specific meetings to determine appropriate methods for more applied studies. Using the characteristics from Table 15.1 and the model of network team management, teams can help identify members that identify themselves as systems thinking and boundary crossing, and in large teams, there can be room for varying levels of integration and participation of more narrowly focused experts to guide methods.

The usual respect given to rigor and complexity in each team member's discipline can inadvertently be a challenge to relevance and usability of community-driven research. Most understand there is a tradeoff between how the frontier of disciplinary methods can be employed and how transdisciplinary a food systems model must be to allow for a breadth of important biological, social, market and policy dimensions. With a vision to improve relevancy, some modelers (particularly systems thinkers and process innovators) will seek to incorporate more phenomena, more parameters and the context of a larger set of team members and stakeholders. Yet, there are other team members, particularly the rigorous researchers, who will raise concerns about how each variable, metric or interaction in such a multi-faceted model will be specified in an unbiased, representative and/or transparent manner. These two approaches must be reconciled for a team to move forward.

For example, in this project we used an agent-based model, which is ideal for simulating complex systems, including emergent behavior that may result from autonomous actions of food system actors. The interdisciplinary team codeveloped the agent-based model that integrates economic data, nutritional indices, social decision-making factors, biophysical crop data, and life cycle assessment. The purpose of the model was to improve our understanding of complex rural-to-urban food chains among four major Colorado agricultural products. This model simulates potential changes to the Denver food policy environment and observes resulting effects throughout the supply chain, from school purchasing decisions to potential changes in producer cropping regimes. The model enables the team to examine the effects of policy changes on everything from household health to soil health.

As one might guess, the multi-faceted nature of this model creates an important, necessary, and complex conversation of how to balance the number of parameters needed for such a model with the time needed to estimate them using cutting edge methods and data. Discussions on tradeoffs such as these need to be managed by skilled facilitators and perhaps process innovators or external facilitators who can help the team navigate and resolve these tensions with respect and care. Ideally, this responsibility falls to an integration executive scientist, who does not have a stake in one of the disciplines and has the skills to help team members integrate diverse considerations. Models such as these need to balance scientific rigor with feasibility and timelines that make the output useful for industry and policy decision-makers.

One of the outcomes of participatory team science is documenting the steps and challenges faced to resolve these tensions. In food systems research, teams should be evaluating the choices of model parameters and choices in the context of the policy or program solutions the target community of concern identified. The content expertise of the domain experts and rigorous researchers is not enough to guide the research direction. It must also be informed by the context depth that process innovators and system thinkers have developed. Only when stakeholder concerns, domain expertise, and systems thinkers' views are integrated will such projects offer up relevant outcomes. The tension between rigor, feasibility, and contextual relevance will never disappear. It is up to teams to create a culture and set of processes that allow for productive debates and then to find an acceptable path forward. The end result is high quality convergent research.

Timing is another source of tension for transdisciplinary teams. When is the right time to bring in new team members from other disciplines? Upon receiving the Future of Food and Ag Research (FFAR) award, a model needed to be built that included quantitative representations of the food supply chain in economic, environmental, and health dimensions. Because economic sciences allow for the standardized valuation of products and supply chains, initial modelling of parameters to include in the agent-based model focused on estimating costs of production, consumer values, and supply chain costs and performance. Unintentionally, this may have signaled less attention to other socioeconomic forces at play in food supply chain decision making.

At that point, the team included some strong sociologists and political scientists, whose disciplinary perspectives were pushed aside: as a result, the team lost one researcher, with others perhaps feeling less valued. Perhaps this was because the FST received the FFAR grant before the newly formed team had reached the point where they had a shared vision and language (even though there was a fairly high level of trust among most members). However, due to the trust among most, the sociologists remained part of the team and actively engaged in meetings. Over the course of the FFAR work, the socio-cultural elements were increasingly seen as more integral to the work than originally conceived. Interestingly, as the modeling work of the FFAR grant comes to a close, the socio-cultural elements that have been integrated into the model are perhaps some of the most innovative elements of the team's collective work. This experience illustrates that those disciplines that are a challenge to initially integrate can also provide the greatest opportunity for innovation. One solution for resolving this tension is prioritize team integration and iteration of visions and specific research plans, especially when researchers have expressed interest and commitment to the team.

One defining characteristic of most interdisciplinary food systems projects is their engaged nature. These projects have additional complexity related to stakeholders. Not only do we need to find ways to coalesce the team around shared research questions, methods, data, and language, we also want to integrate the context-rich knowledge and values of stakeholders. One innovative element of the FFAR call for proposals was a requirement to document and leverage existing local investments in the food system (>$1M in food system investments), and to build bridges across often-disconnected community, government, private-sector and research partners. In Colorado, this meant trying to bring urban and rural stakeholders, diverse scales of agricultural operations, policymakers and government agency leads, supply chain stakeholders, and nonprofit organizations to the Table. Most of these partners usually focus on only one aspect of the health, environment, or economic components of the food system. Integration of stakeholders with diverse knowledge and interests poses another challenge to transdisciplinary teams.

Early on in the research, partners in the City and County of Denver determined that they wanted to build the case for the Mayor to adopt the Good Food Purchasing Program (GFPP) for its procurement policy. GFPP is a food system rating metric that integrates a set of well-known, third-party sustainability, food justice, economic, and labor standards. The FST research quickly revealed that there are tradeoffs associated with adopting GFPP. In particular, the GFPP did not necessarily capture the diversity and importance of place as written. So, for example, the FST found that there are tradeoffs associated with organic certifications (one of the priority areas under the "environmental" value of the GFPP) in dryland farming systems as producers cannot practice no till and thus more erosion occurs (Jablonski, Dillon, Hale, Jablonski, and Carolan, 2020). Sharing this point, however, led to some tensions across researchers and stakeholders.

One way the team worked to navigate tensions was by providing project funds to the City and County of Denver to hire an individual to build a broader advisory team. Through working to incorporate a broader diversity of geographies, scales of producers, sectors of the food systems, etc. the "Good Food Purchasing Program Coalition" gradually fostered a culture of inclusion. For example, the Coalition was able to have conversations about issues related to agricultural labor that did not occur in any other venue in the state due to the contentious nature of the topic (National Research Council, 2015; Jablonski et al., 2019). Addressing these kinds of emerging issues and committing resources to community partners helps scientific teams realize their goals for being both transdisciplinary and inclusive.

## 15.5.1 Discussion

Participatory team science is a new paradigm of knowledge creation and thus deserves a roadmap for new ways of conducting research in complex teams. We have offered three contributions towards this roadmap: 1) description of several unique characteristics of team members, 2) exploration of the team development process, and 3) a framework for team management drawing on network characteristics. The study of two cohorts of PreCIP and CIP teams revealed that transdisciplinary teams need new roles—project managers and decisions teams—to support team success. This case, and other CIP teams, also illustrate that development of transdisciplinary teams takes time to build trust and collaborative capacity. Teams benefit when they receive capacity building grants that allow them to spend time on Phases 1 and 2 of the team development ladder, without the pressure to produce specific research products. Universities as well as the National Science Foundation offer planning grants that are designed to support teams in accomplishing these team development tasks. Teams can benefit from thinking of them as team development support rather than seed funding to start pilot research projects.

As mentioned above, the time invested to participate in successful interdisciplinary or transdisciplinary food systems teams is significant. The rewards generally do not begin to be felt for years down the road. There are a number of things that make participating in these teams rewarding:

1. Anyone who spends significant time in the "field" working with stakeholders or trying to affect policy knows that there are no silver bullets. No individual discipline has *the* answer. If the focus is on understanding a policy that is going to reduce GHG emissions, for example, it may be at the expense of the long-term viability of agricultural businesses. Thus, it is only through working together, acknowledging, and discussing tradeoffs that we can begin to address the many wicked problems across the food system. Ultimately, decisions will not be made just based on the science but based on values. Hopefully, however, understanding tradeoffs can help community members and policymakers better understand how to craft policies that are best in line with their values, vision, and goals.

2. If you are fortunate enough to participate on a strong food system team, the process of better understanding work of your colleagues from different disciplines should help to improve your disciplinary work through better recognition of the complex ways stakeholders make decisions, how other disciplines perceive issues, and the tools and data that they use for analysis. In other words, in the best situations, participation in interdisciplinary teams can help to inform and improve your disciplinary research.

3. Working in teams can be personally fulfilling as teams allow us to tackle complex problems and make substantial impact in the world. Academia can be a frustrating or lonely profession, and team science offers the opportunity to forge positive relationships with colleagues outside your department to be a valued expert and to contribute to impactful work.

Finally, we want to offer a few specific recommendations to anyone planning food systems research. It is important to get to know people and their working styles. Good teams will work together for years, so you need to know who plays well in the sandbox; being brilliant is not as important being a good teammate. Know that it is OK for the team roster to change and shift each year. If you adopt a Shared Authentic Leadership model, you'll be able to attract and keep good teammates. As you invite new team members and work to integrate them, remember that you are looking for more characteristics than domain expertise, you are also looking for hidden skills (e.g. the ability to make brilliant Figs. or maintain meticulous data records) and characteristics of transdisciplinary scientists (e.g. systems thinkers and process innovators). You can maintain high performance with regular assessments of team roles and responsibilities, making room for team members to shift roles and responsibilities and looking for new members to fill gaps in the team. Finally, good teams need time to socialize and share stories which builds the trust required to work on challenging problems, so take time to celebrate success or just hang out.

## Acknowledgements

The research team described in this chapter was supported in part by funding from the Office of the Vice President for Research at Colorado State University, the Foundation for Food and Agriculture Research, the Colorado Potato Administrative Committee, the Colorado Wheat Research Foundation, and the Colorado Agricultural Experiment Station. The evaluation research reported in this chapter was supported in part by funding from the Office of the Vice President for Research at Colorado State University. The authors also wish to thank members of the Team Science Group, Dinaida Egan, Ellen Fisher, Bailey Fosdick, Hannah Love, and Meghan Suter, for their partnership in the evaluation for the Catalyst for Innovative Partnerships in the Office of the Vice President for Research, Colorado State University. The lead author also thanks Dr. Jesse Fagan for his assistance with visualizations.

## References

Aboelela, S.W., Larson, E., Bakken, S., Carrasquillo, O., Formicola, A., Glied, S.A., et al., 2007. Defining Interdisciplinary Research: conclusions from a Critical Review of the Literature. Health Services Research 42, 329–346. doi:10.1111/j.1475-6773.2006.00621.x.

Bammer, G., 2013. Disciplining interdisciplinarity: Integration and implementation sciences for researching complex real-world problems. ANU Press, Canberra, ACT, Australia.

Bennett, L.M., Gadlin, H., Marchand, C., 2018. Collaboration Team Science: Field Guide. US Department of Health & Human Services, National Institutes of Health, National Cancer Institute, Washington DC.

Cheruvelil, K.S., Soranno, P.A., Weathers, K.C., Hanson, P.C., Goring, S.J., Filstrup, C.T., et al., 2014. Creating and maintaining high-performing collaborative research teams: the importance of diversity and interpersonal skills. Front Ecol Environ 12 (1), 31–38.

Cross, J.E., Pickering, K., Hickey, M., 2015. Community-based participatory research, ethics, and institutional review boards: untying a Gordian knot. Crit Sociol (Eugene) 41 (7–8), 1007–1026.

Cross, J.E., Plaut, J.M., 2019. Integrating social science and positive psychology into regenerative development and design processes. In: Caniglia, B., Frank, B., Knott, J., Sagendorf, K.S., Wilkerson, E.A. (Eds.), Regenerative Urban Development, Climate Change and the Common Good. Routledge, New York, NY, pp. 204–225.

Disis, M.L., Slattery, J.T., 2010. The road we must take: multidisciplinary team science. Science Translational Medicine 2 (22cm9). https://www.science.org/doi/full/10.1126/scitranslmed.3000421?casa_token=aoB0u FmPxGAAAAAA:2PSyEE8wcjyScxrj8fVFA88IwAy3N6FQbqDy4tB36qVR0SpY8lyodG-HSV7sA12j9gYJMl02PibD.

Gilliland, C.T., White, J., Gee, B., Kreeftmeijer-Vegter, R., Bietrix, F., Ussi, A.E. et al., 2019. The fundamental characteristics of a translational scientist. ACS Pharmacology & Translational Science 2 (3) 213–216. DOI:10.1021/acsptsci.9b00022.

Guenter, H., Gardner, W.L., Davis McCauley, K., Randolph-Seng, B., Prabhu, V.P., 2017. Shared authentic leadership in research teams: Testing a multiple mediation model, 48. Small group research, pp. 719–765.

Hall, K.L., Vogel, A.L., Huang, G.C., Serrano, K.J., Rice, E.L., Tsakraklides, S.P., Fiore, S.M., 2018. The science of team science: A review of the empirical evidence and research gaps on collaboration in science. American Psychologist 73 (4), 532.

Hall, K.L., Vogel, A.L., Crowston, K., 2019. Comprehensive Collaboration Plans: Practical Considerations Spanning Across Individual Collaborators to Institutional Supports. In: Hall, K., Vogel, A., Croyle, R. (Eds.), Strategies for Team Science Success. Springer, Cham. https://doi.org/10.1007/978-3-030-20992-6_45.

Hendren, C.O., Ku, S.T., 2019. The Interdisciplinary Executive Scientist: Connecting Scientific Ideas, Resources and People. In: Hall, K., Vogel, A., Croyle, R. (Eds.), Strategies for Team Science Success. Springer, Cham https://doi.org/10.1007/978-3-030-20992-6_27.

Holley, J. (2019). Synergy Commons. https://synergycommons.net/resources/tool-mapping-your-network/

Jablonski, B.B.R., Carolan, M., Hale, J., Thilmany McFadden, D., Love, E., Christensen, L.O., et al., 2019. Connecting Urban Food Plans to the Countryside: leveraging Denver's Food Vision to Explore Meaningful Rural-Urban Linkages. Sustainability 11 (7), 10 pp. https://doi.org/10.3390/su11072022.

Jablonski, K.E., Dillon, J.A., Hale, J., Jablonski, B.B.R., Carolan, M.S., 2020. One place doesn't fit all: improving the effectiveness of sustainability standards by accounting for place. Frontiers in Sustainable Food Systems 4, 145. https://doi.org/10.3389/fsufs.2020.557754.

Jason, L., & Glenwick, D. (Eds.). (2016). Handbook of methodological approaches to community-based research: Qualitative, quantitative, and mixed methods. Oxford university press. pp. 220–243.

Looney, C., Donovan, S., O'Rourke, M., Crowley, S., Eigenbrode, S.D., Rotschy, L., et al., 2013. Seeing through the eyes of collaborators: using Toolbox workshops to enhance cross-disciplinary communication. In: O'Rourke, M., Crowley, S., Eigenbrode, S.D., Wulfhorst, J.D. (Eds.), Enhancing Communication and Collaboration in Interdisciplinary Research. Sage Publications, Thousand Oaks, CA.

Lotrecchiano, G.R., Mallinson, T.R., Leblanc-Beaudoin, T., Schwartz, L.S., Lazar, D., Falk- Krzesinski, H.J., 2016. Individual motivation and threat indicators of collaboration readiness in scientific knowledge producing teams: a scoping review and domain analysis. Heliyon, 2(5), e00105.

Lotrecchiano, G.R., DiazGranados, D., Sprecher, J., McCormack, W.T., Ranwala, D., Wooten, K., ..., Brasier, A.R., 2021. Individual and team competencies in translational teams. Journal of Clinical and Translational Science 5 (1), E72. doi:10.1017/cts.2020.551.

Moore, J.E. and Khan, S., 2020. Core Competencies for Implementation Practice. The Center for Implementation and Health Canada: Toronto, Canada. The report and other implementation support resources are freely available from The Center for Implementation. (Online): https://thecenterforimplementation.com/core-competencies.

National Research Council, 2015. Enhancing the effectiveness of team science. National Academies Press, Washington DC.

Nowotny, H., Scott, P., Gibbons, M., 2003. Introduction: 'Mode 2' revisited: The new production of knowledge. Minerva 41 (3), 179–194.

Nurius, P.S., Kemp, S.P., 2019. Individual-Level Competencies for Team Collaboration with Cross-Disciplinary Researchers and Stakeholders. Strategies for Team Science Success. Springer, Cham, pp. 171–187.

Patton, M.Q., 2010. Developmental evaluation: Applying complexity concepts to enhance innovation and use. Guilford Press, New York, NY.

Read, E.K., O'Rourke, M., Hong, G.S., Hanson, P.C., Winslow, L.A., Crowley, S., et al., 2016. Building the team for team science. Ecosphere 7 (3), 1–9 e01291.

Read, E., Cross, J.E., Herman-Mercer, N., Oliver, S., O'Reilly, C., 2022. Teams, networks, and networks of networks: advancing our understanding and conservation of inland waters. LIMNO Encyclopedia, Elsevier Inc.

Robertson, B.J., 2015. Holacracy: The new management system for a rapidly changing world. Henry Holt and Company, New York, NY.

Taylor Gilliland, C., White, J., Gee, B., Kreeftmeijer-Vegter, R., Bietrix, F., Ussi, A.E., Hajduch, M., Kocis, P., Chiba, N., Hirasawa, R., Suematsu, M., Bryans, J., Newman, S., Hall, M.D., Austin, C.P., 2019. ACS Pharmacology & Translational Science 2 (3), 213–216. doi:10.1021/acsptsci.9b00022.

Tebes, J.K., Thai, N.D., 2018. Interdisciplinary team science and the public: Steps toward a participatory team science. American Psychologist 73 (4), 549.

Thompson, L., 2003. Improving the creativity of organizational work groups. Academy of Management Perspectives 17 (1), 96–109.

Wooten, K.C., Rose, R.M., Ostir, G.V., Calhoun, W.J., Ameredes, B.T., Brasier, A.R., 2014. Assessing and evaluating multidisciplinary translational teams: a mixed methods approach. Evaluation & the Health Professions 37 (1), 33–49.

# Towards a holistic understanding of food systems

*Christian J. Peters[a] and Dawn D. Thilmany[b]*

[a]USDA, Agricultural Research Service, Food Systems Research Unit, Burlington, VT, United States. [b]Department of Agricultural and Resource Economics and Regional Economic Development Institute, Colorado State University, B310 Clark, Fort Collins CO, United States

## 16.1 Introduction

Our main goal in this book has been to help readers, both novice and expert, to comprehend modeling in order to understand food systems. When we started this project, we knew that our own knowledge of food systems would grow through the process of writing and editing. Accordingly, we saw the project as a collaborative learning opportunity for the contributors, and we recruited authors who shared this vision. As a result, themes and ideas shared by the team of contributing authors influenced the organization of this book, especially its opening and closing chapters.

Thanks to support from a USDA National Institute of Food and Agriculture grant, we were able to gather the authors via videoconference several times during the writing process. Authors had a chance to peer review other chapters within the volume and to discuss their perceptions of key themes for the book as well as issues worth exploring further in future food systems research. We believe this helped us create a more coordinated and cohesive overview of the current state of food system modeling than had we worked in isolation. We hope you agree.

To conclude this volume, we offer suggestions on how to approach models and modeling literature to attain a holistic understanding of food systems. We begin with an overview of the epistemology of food system modeling, then discuss the complexity of making inferences in a field with diverse modeling approaches that sometimes offer up competing conclusions. After providing examples of how to put modeling into action, we conclude with some thoughts on future directions for the food systems modeling field framed by the entire writing team for this book.

*Food Systems Modelling.*
DOI: https://doi.org/10.1016/B978-0-12-822112-9.00016-3

## 16.2 Epistemology and modeling

Epistemology is the study of how we know what we know. Each methodology described in this book takes a distinct approach to studying food systems. Yet while the methods may differ, modelers follow some common principles to judge when you really *know* something and how to practice the art of modeling. These principles matter both when working with models and when simply reading about them.

First, modeling is an iterative process. This holds true throughout a model's lifespan, both when creating it and with each revision. At first, one might be trying to determine how to conceive of a system, working through prototypes until the elements of the system are distinct and the relationships clear and measurable. Chapter 06, for example, walks through the thinking behind the construction of the U.S. Foodprint model. Once a model, or modeling methodology, is complete, it tends to get refined to improve its accuracy or extend the questions it can answer. Consider a couple examples. While a standard methodology exists for calculating water footprints, Chapter 04 explains how one can use crop models to estimate the consumptive use of water by crops, enabling more precise partitioning of water requirements (irrigation water vs. rainwater). The IMPACT model (Chapter 10) was developed to understand the food security implications of major economic drivers of the food system, like investment in yield improving technologies. It has evolved and been integrated with biophysical models to estimate the impact of drivers like climate change. In short, models are never done, but they can be useful and sufficient for certain purposes.

Second, a model should always be used within its intended limits. This applies when adapting a model for a new situation and for interpreting its results. A transportation model (Chapter 07) developed for broccoli might serve as a good starting point for another researcher modeling a supply chain for another perishable produce item, like lettuce, but less suitable for cabbage, which can be stored, and completely unsuitable for milk, which gets processed into an array of different products. Likewise, life cycle assessments (Chapter 03) are often conducted to compare products created using different processes. Unless a deliberate attempt was made to make the analysis representative of a population, these results may not be generalizable. How do you avoid inappropriate use of a model? Heed the limitations the authors have mentioned in their description of the model and be sure to understand the sector-specific context and other assumptions at the model's foundation, so you don't misinterpret it or misrepresent its findings.

Third, the sufficiency of a model depends on the context. What question are you trying to answer? How precise does your answer need to be? How large is the difference you are trying to detect? What will you do with the information? How much is riding on the decision? These questions are easier to answer for an established method. Many people have used it, and undoubtedly, they have already discovered the common pitfalls of the method and set standards to help avoid these problems. For example, Chapter 08 discusses the types of data needed to appropriately estimate the economic impacts of local and regional food programs using IMPLAN, a widely used modeling tool. For a new method, however, these questions are harder to answer, and it may be hard for a non-expert to judge the quality of the work. Try to balance enthusiasm for new results and approaches to modeling with a healthy skepticism. Read critically. Think of what may have been missed. Resist the urge to jump to conclusions.

These three principles apply not only to modeling, but to science in general. While simulation modeling has been called a "third way" of practicing science (Harrison, Lin, Carroll, and Carley, 2007), in addition to observation and experimentation, modeling can also be seen simply as another tool in the toolbox. To make sense of the data we collect through observation, experimentation, or modeling, we must also analyze and put together a story that explains what we found. Scientists regularly share these stories in oral, visual, and written forms in lectures, conference presentations, charts, tables, posters, papers, and books. They debate and critique this work in perpetual pursuit of a clearer and truer understanding of the world. Thus, models really are no more esoteric than the rest of science, they just enable us to calculate much faster and therefore to create virtual worlds that let us test ideas 'in silica.' As a farmer once said, "I never try anything on the farm until I can make it work on paper."

## 16.3 Considering results from multiple models

Once we take away the mystique of modeling, we are left with a more familiar problem. How do we make sense of the seemingly contradictory stories that different disciplines tell? To attain a holistic understanding of food systems, one must learn how to consider and compare results from multiple models. Doing so is challenging because models vary in scope and scale, but not impossible if you consider the underlying factors contributing to differences.

### 16.3.1 Scope and disciplinary perspective

The scope, or disciplinary background, of the models in this book vary widely. For example, life cycle assessment (LCA) attempts to holistically evaluate environmental impacts in terms of ecosystem health, natural resource availability, and human health (see Chapter 03). This breadth gives the LCA modeler a very broad perspective, but other narrower methods may enable one to understand the relationship between a food value chain and one particular natural resource more deeply, such as water (Chapter 04) or land (Chapter 05). For all this breadth and depth, none of these approaches directly address the economics of a system. What is a modeler, or a reader of modeling literature, to do?

One approach is to build ever more comprehensive models, such as combining life cycle assessment and nutritional data and to simultaneously evaluate the environmental impact and healthfulness of diets (Heller, Keolian, and Willett, 2013) or integrating economic and biophysical models (Kling, Arritt, Calhoun, and Keiser, 2017) to understand the interactions between food, water, and energy. Several chapters in this book (especially Chapters 03, 09, and 10) present established models that have evolved to cover a broad range of impacts and, by doing so, to make important contributions to our knowledge of food systems. However, building better and more comprehensive models takes time and money, and depending on the quality of the model, may not always lead to greater comprehension of the food system.

A less technical approach is also available, and you do not need to be a modeler to apply it. Each person must piece together their own understanding of how food systems function. Organizing knowledge is a fundamental part of learning. As Ambrose, Bridges, DiPietro, Lovett, and Norman (2010) describe in their book on teaching, every student must organize their knowledge of a subject. Otherwise, they perceive the subject as a jumble of facts. Whereas

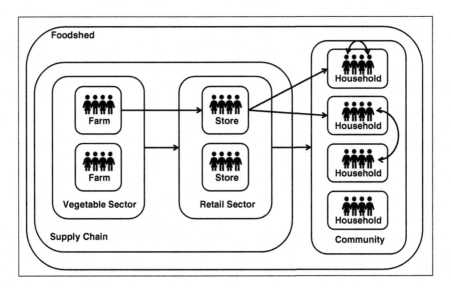

**FIGURE 16.1**  Hierarchy and spatial scales in food systems. The diagram shows people (stick figures) as actors either within businesses in a sector or households within a community. Related sectors form supply chains, and multiple supply chains feeding a community would constitute a foodshed. Arrows show that relationships, in this case material flows, occur at multiple scales: from sector to sector, business to business (or household), and person to person.

someone new to a field may see only a few, relatively simple connections between concepts, experts understand their subject in a structured way and can readily see how different concepts are connected (Ambrose et al., 2010).

Of course, one does not need to recreate the wheel. Experts have proposed frameworks for conceptualizing the food system (see Chapter 02), which can help one to fit together knowledge previously gleaned from different fields of modeling. For purposes of illustration, consider the food system framework by Ericksen (2008), which proposes that food system activities (like crop production) and outcomes (like food availability) influence social welfare and environmental security and are influenced by larger ecological and socioeconomic drivers. Using this framework, one can plausibly connect agricultural production directly to the food supply, but not necessarily to food security, which also depends on food access. Frameworks can help one to think about the connections, but each person still needs to put in the work to explore those connections in the context of the question they intend to answer.

## 16.3.2  Spatial and temporal scale

The models presented in this book cover a wide range of spatial and temporal scales. Here a picture would be helpful. Imagine a hierarchy of spatial scales (Fig. 16.1). At the smallest scale are individual people. These people work together in businesses (like farms or stores) or live together in households. Multiple farms or stores constitute a sector, like vegetable farms or food retail, and the exchange of product by many businesses across multiple sectors constitute supply chains. Likewise, communities are composed of many households. Foodsheds are composed of many supply chains directed at feeding people in a particular city, town, or community. Relevant interactions, like the physical exchange of food products, occur at multiple scales, such as actor to actor, business to business, or sector to sector. Thus, variation in scale is essential to modeling the full range of interactions that occur in food systems.

The relevant scale depends on the questions one is trying to answer. Chapter 11 shows how social network analysis can be used to understand how programs and initiatives can intentionally foster and evaluate the number and strength of connections between individuals across different sectors of the food system. Chapter 03 illustrates the estimation of environmental impact for an individual product, strawberry yogurt, across the related businesses and transactions within the supply chain. Chapter 09 estimates material flows across all major stages of the food system (e.g., farm production, processing, wholesale, retail) for all foods in the United States. Chapter 10 explains how the IMPACT model predicts food security outcomes regionally and globally for different scenarios of economic development and environmental change. There is no single correct spatial scale.

Likewise, the models in this book vary in temporal scale. Many of them present analyses relevant to a certain point in time, usually a benchmark year or series of years. This is a common unit of time to measure, in part because food systems depend on agriculture which tends to follow an annual harvest cycle. However, this is certainly not the only relevant time scale to model! Crop models, used in calculating water footprints, estimate crop water requirements in daily time steps which are totaled to estimate water use over a growing season (Chapter 04). Supply chain models, such as transport and trans-shipment models, consider intra-annual dynamics, like seasonal changes in product supply or demand (Chapter 07). The IMPACT model looks forward several decades, using estimates of future climate conditions to drive crop models which estimate variation in crop output (Chapter 10). The lesson here is that there is no one "right" temporal scale.

Looking at a point in time can reveal hot spots. Studying dynamics over short periods can reveal choke points. Studying longer time periods can highlight trends and likely outcomes of "business as usual" practices compared to alternative options perhaps being considered as a market intervention. Understanding food systems requires one to think flexibly across space and time.

### 16.3.3  Thinking across scope and scale – an example

How does one integrate across scope and scale? Consider a hypothetical example. An economic analysis, using and input-output approach, suggests that investment in regional foods would be great for regional job creation. People are already interested in buying these products, so it sounds promising, and policy makers invest more money in regional food programs. After a few years, the supply chains for certain livestock products experience some hiccups, perhaps a bottleneck in slaughter and processing capacity. At this point, a supply chain model helps illuminate the dynamics, and suggests that investment in infrastructure and coordination across the supply chain would resolve the bottleneck. But who will invest the time and money?

Given limited public funds, local and state governments want to know that investment catalyzes other public benefits, and businesses want to communicate any broader benefits to consumers. Thus, a public-private partnership funds life cycle assessment to measure the environmental impact of the regionally produced meat relative to the national average. Non-profits concerned about natural resources, like land and water, might want to ensure that local natural resources are not overburdened and are wisely and efficiently used, thereby prompting subsequent research with water footprints and various land use models.

The environmental results are promising, except for the fact that the regionally produced food is more expensive. Concerned about this trade-off, communities want to ensure that investment serves not only producers and consumers of "local food" but also improves overall food access and public health, leading to participatory research on barriers to food security and healthy eating. While this example is fictitious, it shows how a holistic understanding of the food system can emerge from inquiry across disciplines, at multiple spatial and temporal scales, and with continual effort at framing new assessments and organizing the pieces into a more complete picture of the food system.

## 16.4  Application of models – from knowing to doing

Models are most impactful when integrated into a broader context of strategic planning, decision making and program or policy refinement. As Saltelli et al. (2020) urge in their manifesto on models, "Complexity can be the enemy of relevance…there is a tradeoff between the usefulness of a model and the breadth it tries to capture." But, by being transparent with assumptions, acknowledging ignorance and uncertainty, and perhaps most importantly, recognizing the consequences and implications of reported findings, modelers can effectively work on issues of great import to society. One solution Saltelli et al. (2020) offers is including stakeholders in the research process, a theme explored in this book.

While engaging stakeholders may seem easier said than done, bear in mind that models are used to compare alternatives – something we do every day. Most models compare baseline conditions with alternative scenarios for businesses, sectors, households, or other actors, that might lead to better societal outcomes. For those in industry, for example, putting a model into action might mean mapping alternative strategies and choosing the outcome that is most profitable or most consistent with a Corporate Social Responsibility mission, weighing tradeoffs between returns and various social outcomes. Policy makers may want to identify the potential levers for attaining a policy goal, such as moving from a "low level" to a "higher level" of program participation. In this case, policy leaders may use a food systems model to understand the incentive structures program participants face, to benchmark and compare program outcomes across policy options, or to see how interactions across players in food systems can lead to unintended consequences.

Chapter 13's key focus was putting research into action in communities and suggested, "the increased use of systems mapping approaches to analyze food systems interventions may yield insights into common structures or archetypes of how these interventions are adopted, disseminated, implemented, and sustained". To illustrate this point, consider two community-based food systems initiatives, CommonWealth Kitchen (CWK) and the Greater Pittsburgh Food Action Plan (GPFAP). Food systems research has already informed their actions, but it is interesting to consider how principles highlighted in this book might address their continuing needs to refine their strategies and evaluate their impacts.

In Chapter 12, the authors explain the value of community-based systems dynamics (CBSD) to incorporate the lived experiences and wisdom of stakeholders and community members into the modeling process. Community initiatives are putting this into action. In their progress report on a standing food system assessment, the Massachusetts Food System Collaborative (2018) highlighted one example of such dynamics with CommonWealth Kitchen (CWK), whose primary mission is to "Develop financing and business support resources for food processing businesses working in incubators," one of the objectives of their standing sustainability and equity action plan. Executive Director Jen Faigel had been introduced to a human-centered design approach, focusing on the needs of the aspiring business owner, then scanning the environment to best to map the information, resources, networks, and infrastructure, eventually identifying the need to build connections all across the value chain to support their businesses and mission.

This begs the question of whether food systems approaches could leverage their community work further. Even though CWK does an impressive job of tracking basic performance metrics already, with a more structured Social Networking Analysis incorporated, CWK would be able to assess how these strong networks emerged (with lessons for others) and could make an even more compelling case of how this network building has benefited the broader community. Likewise, supply chain, partial equilibrium and economic impact models might help to evaluate how to balance "mission" with "margins," enabling the organization to explore tradeoffs between economic viability and the measurable impacts they have on job creation, market access for small and disadvantaged businesses and food security in their community.

In Chapter 12, the authors suggest that "conducting CBSD as a type of formal public participation, convened by a government agency rather than an external research institution, may help build capacity and political power around systems level change". This approach was at the heart of the process, team and engagement strategies employed to develop The Greater Pittsburgh Food Action Plan (GPFAP) in 2018 (Pittsburgh Food Policy Council, 2019). The goals, strategies and evaluation metrics that were framed by the many community stakeholders involved were perhaps guided by past research on interventions, but more importantly, set a research agenda for future work.

One goal identified in the GPFAP is to decrease the consumption of sugary beverages (Pittsburgh Food Policy Council, 2019), but due to Pennsylvania law, the Pittsburgh Food Policy Council could not recommend imposing a tax. To that point, Chapter 7 presented an example of an alternative they could consider, showing through a partial equilibrium model how Behavior Change Communication can have similar effects on decreasing the consumption of sugary drinks as taxes. Plus, an empirical and participatory research project to assess outcomes of such a program can validate their model's expected outcomes and delineate the expected implications for the community, such as consumer prices or government revenue. This is just one example to illustrate our challenge as food systems researchers to invest the time into engaging stakeholders in our modeling efforts, assuring they are communicated, translated, and put into action by industry, policymakers and communities.

## 16.5 Future directions for the field

Despite our best efforts to be as inclusive as possible when outlining this book and recruiting authors, the food systems field is vast and quickly changing. In the interest of showing how models work, this book has focused on selected modeling approaches that have been well vetted in the field, so the book is not exhaustive. Nonetheless, learning about several models in depth makes understanding new models easier. All modelers create simplified versions of real systems in hopes of better understanding them. You may encounter new methods and new data with each paper you read, but modelers constantly describe the same types of things, such as, the data they used, the structure of their model, and the limits within which the model is valid. Over time, you will become more adept at grasping the essence of a model and what lessons it has to offer.

### 16.5.1 Integrated modeling approaches

As food systems modeling evolves, expect continued innovation in the integration of modeling approaches. Like integrating scientists and disciplines within projects, integrating modeling approaches can leverage the flexibility and relevancy of multiple methods. But it leads one to consider, "When is it a benefit to mix models and have them interact with each other?" In Chapter 13, Hennessy et al. build the case for blending multiple methods from the complex systems toolkit to suit the research questions and topic domain rather than limiting our research approach within the scope of a single discipline or modeling tool. Similarly, Chapter 8 concludes with a vision of how economic impact assessments could provide richer information to interested policymakers if integrated with methodological approaches that incorporate other types of impacts, such as environmental outcomes, so that tradeoffs were easily comparable.

One strategy for "mixing models" is to integrate them. The Environmental Input-Output (EIO) model discussed in Chapter 09, for example, weaves together data on economic flows and resource use to estimate material flows. The EIO model also uses a diet model to estimate alternative scenarios of consumption that comply with nutritional guidelines and reflect the preferences shown in food intake data. The IMPACT model (Chapter 10) goes even further, linking climate models with crop models with a global model of agricultural commodities and trade. Both EIO and IMPACT show how disparate models can, painstakingly, be integrated to consider economic and environmental tradeoffs.

Another strategy for "mixing models" is to integrate the lessons or results from models into a cohesive story. For example, Chaudhary, Gustafson, and Mathys (2018) demonstrate an approach to assess food system sustainability across multiple dimensions (e.g. nutrition, environmental impact, social wellbeing) by aggregating individual indicators of each dimension and displaying the results in easy to read radar diagrams. While visuals certainly help, the mixing of models may occur by mixing of people, who can learn from one another. Chapter 15 explored the intentional planning and communication needed to navigate Team science, emphasizing the importance of taking the time to understand one another's disciplines as part of successful interdisciplinary projects.

Both hard and soft integration of models will continue to be important as the field grows. As researchers we should not shirk the effort that is needed to creatively integrate methods in our toolbox. An artful blending of methods may allow us to tell a more nuanced story about the changes, interactions, and outcomes to be expected in response to various drivers, decisions and environmental factors underlying the food system.

## 16.5.2 Interdisciplinary and transdisciplinary inquiry

Consistent with the theme of integration, expect continued calls for working in multidisciplinary, interdisciplinary, and transdisciplinary teams. Interdisciplinary and transdisciplinary work is a learning-by-doing process. Once a researcher has had the opportunity to participate in a strong food system team process, it is likely they will be more comfortable reaching out to colleagues from different disciplines to integrate their disciplinary work into more complex and nuanced models. Through a successful project, researchers recognize the scale and scope of decisions being made across food systems, see how other disciplines frame research problems, and appreciate the theories, tools and data that cross-discipline colleagues use for analysis. As Chapter 15 highlights, participation in interdisciplinary teams can elevate a modeler's disciplinary research. Moreover, the whole field of food systems benefits from the growing portfolio of projects that illustrate the structure, process, and outcomes of more complex modeling.

However, the field of food systems modeling requires both technical elegance and context-based relevance to be impactful.

For example, one of the challenges Chapter 15 highlights is the need for transdisciplinary teams to thoughtfully develop research questions, management plans, and roles that allow team members to leverage their comparative advantage within a project. So, just as there will be a need for rigorous researchers, there will also be important roles for boundary crossers, systems thinkers, and process innovators. Yet, these roles are not well defined within most graduate majors as they are more translational in nature. Figuring out how to more predictably create effective teams who can increase the sustainability of food systems remains an open subject for the field to resolve.

## 16.6 Closing points

We believe in models as learning tools. Throughout this book, authors have emphasized the value of models both to estimate impacts, to understand how systems function, and to make predictions about how systems might behave under different conditions. Both are important, and they can reinforce each other. In the process of estimating impacts, one might discover areas that are poorly understood, prompting further research, that leads to better predictions.

However, all modeling efforts are prone to the limitations discussed in Chapter 01. Results must be interpreted in context, considering the assumptions and scope, the spatial and temporal scale, and the purpose of the model. Models are limited by the accuracy and precision of their underlying data, and the uncertainty of the model estimates may be poorly understood. Do not fall prey to false precision. Just because a model gives a concrete answer, like a number, does not mean that answer is right.

Yet, uncertainty does not equal ignorance. Precise estimates can be elusive, especially in cases where important model parameters may be measured with low precision (one significant digit) or perhaps not measured at all. Consider predictions of future climate change, which depend on estimates of the drivers of change. Each of these estimates has uncertainty, and that uncertainty gets compounded with each variable. Thus, predictions get less precise the farther into the future climate is predicted. However, if multiple models show change moving in the same general direction, then the uncertainty in the estimate may be less important than the certainty about the direction. Different models developed by different people with different data yielding similar results generally increases our confidence that the results are accurate, even if imprecise.

We recognize the challenge to resolve how modelers and stakeholders communicate with each other to share a unified vision of what a food system means, how it behaves and where innovations and impacts can be made and measured. Hopefully, the themes, tools, and findings covered in this volume will enable researchers and practitioners alike to better understand models and have richer discussions of food systems as a result. We look forward to the future research that emerges from those who study from or reference this work.

## Acknowledgements

This work was supported in part by the U.S. Department of Agriculture, Agricultural Research Service.

## References

Ambrose, S.A., Bridges, M.W., DiPietro, M., Lovett, M.C., Norman, M.K., 2010. How Learning Works: Seven Research-Based Principles for Smart Teaching. John Wiley & Sons, Inc., San Francisco, California, p. 301.

Chaudhary, A., Gustafson, D., Mathys, A., 2018. Multi-indicator sustainability assessment of global food systems. Nature communications 9, 13 p. doi:10.1038/s41467-018-03308-7.

Ericksen, P.J., 2008. Conceptualizing food systems for global environmental change research. Global Environmental Change 18, 234–245.

Harrison, J.R., Lin, Z., Carroll, G.R., Carley, K.M., 2007. Simulation modeling in organizational and management research. Academy of Management Review 32 (4), 1229–1245.

Heller, M.C., Keolian, G.A., Willett, W.C., 2013. Toward a life cycle-based, diet-level framework for food environmental impact and nutritional quality assessment: a critical review. Environmental Science & Technology 47, 12632–12647.

Kling, C.L., Arritt, R.W., Calhoun, G., Keiser, D.A., 2017. Integrated assessment models of the food, energy, and water nexus: a review and an outline of research needs. Annu Rev Resour Economics 9, 143–163.

Massachusetts Food System Collaborative. (2018). Sustainability and Equity in the Massachusetts Food System: a Progress Report. Downloaded at: https://mafoodsystem.org/projects/2018progress/

Pittsburgh Food Policy Council, 2019. Greater Pittsburgh Food Action Plan. Retrieved from https://foodactionplan.org/new-home-page/

Saltelli, A., Bammer, G., Bruno, I., Charters, E., Fiore, M.D., Didier, E., et al., 2020. Five ways to ensure that models serve society: a manifesto. Nature 582, 482–484. doi:10.1038/d41586-020-01812-9.

# Glossary of terms

**Activity**: Terminology used in a social accounting matrix (SAM) to describe a grouping of establishments that produce one or more types of products using a similar production process. Agriculture is an example of an activity. (Chapter 9)

**Agent-based modeling**: A computational tool that models the actions or interaction of "agents" (i.e., people) and allows for the exploration of dynamic mechanisms that link individual behavior to overall outcomes among populations. (Chapter 13, see reference for Allender et al., 2016)

**Arable**: A term to describe land capable of being cultivated for crop production. (Chapter 6)

**Areas of protection**: In lifecycle impact assessment, these are the three fundamental areas of concern or attention: human health, ecosystem health, and resource depletion. (Chapter 3)

**Blue water footprint**: The volume of surface and groundwater (i.e., blue water resources) consumed during the production of a good or service. (Chapter 4, see reference for Hoekstra et al., 2011)

**Carrying capacity**: Maximum population size of a particular species that an area can support. Within ecology, this concept can be applied to any species. Within food systems, this term typically refers to the capacity of an area to feed people. (Chapter 6)

**Characterization**: In life cycle assessment, it is the process of calculating indicator results, which involves the conversion of life cycle inventory (LCI) results to common units and the aggregation of the converted results within the same impact category. This conversion uses characterization factors, and the outcome of the calculation is a numerical indicator result. (Chapter 3, see UNEP, 2009. Guidelines for Social Life Cycle Assessment of Products. Available at https://www.unep.org/resources/report/guidelines-social-life-cycle-assessment-products.)

**Characterization factors**: Factors for converting lifecycle inventory into lifecycle impacts. Essentially, life cycle impact assessment frameworks are tabulated values of characterization factors. For example, methane, a greenhouse gas, has a characterization factor of 34 kilograms of carbon dioxide per kilogram of methane. The characterization factor thus converts methane emissions into an equivalent carbon dioxide emission. (Chapter 3)

**Classification**: In life cycle impact assessment, it is the assignment of life cycle inventory (LCI) results to selected impact categories. (Chapter 3)

**Commodity**: Terminology used in a social accounting matrix (SAM) to describe the result from the assembly of one type of product, acquired from one or more activities that produce this product. Crop and animal products are examples of a commodity. (Chapter 9)

**Community-based system dynamics (CBSD)**: **(CBSD)** A participatory method for involving communities in the process of understanding and changing social systems. This method emphasizes the development of community member skills to create, understand, and use models to promote systems thinking and guide action. (Chapter 13, see reference for Hovmand, 2014)

**Comparative statics**: The comparison of changes in market equilibrium conditions to either a change in market conditions or a change due to an intervention, without considering dynamics or the process of adjustment. (Chapter 7)

**Complex adaptive system (CAS)**: A system composed of many heterogeneous pieces whose interactions drive system behavior in ways that cannot easily be understood from considering the components separately. (Chapter 2, see reference for Institute of Medicine, 2015)

**Consumer surplus**: The aggregated value for consumers when they are able to purchase a product for a price that is less than the highest price that they would be willing to pay. (Chapter 7)

**Consumptive water footprint:** The volume of freshwater that is used and either evaporated or incorporated into a product. It is the sum of the green and blue water footprint. (Chapter 4)

**Contribution analysis:** A method used to estimate the value of an industry sector or group of sectors in a region at their current levels of production. (Chapter 8)

**Crop model:** A model that simulates crop growth from the known biophysical relationships between the atmosphere, the crop, and the soil. These relationships often influence each other via feedback loops. (Chapter 4)

**Crop water use:** Also known as evapotranspiration (ET), it is the water used by a crop for growth and cooling as well as the water lost from soil evaporation. (Chapter 4)

**Crop yield:** Weight of harvested crop per unit of harvested area. (Chapter 4)

**Direct environmental impacts:** The environmental impacts associated with production and consumption activities that provide direct output to meet a specific final demand, or first-tier transactions to provide the output. For example, if final demand of tomatoes was increased, the direct environmental impacts linked to the farm production and retailing of tomatoes would also be increased. (Chapter 9)

**Direct requirement matrix:** In the context of an economic input-output (EIO) model, it is a rectangular array of data measuring the direct use of all commodities and activities in the model for each unit of activity and commodity output. (Chapter 9)

**Disability-adjusted life years (DALYs):** One DALY represents the loss of the equivalent of one year of full health. DALYs for a disease or health condition are the sum of the years of life lost to due to premature mortality and the years lived with a disability due to prevalent cases of the disease or health condition in a population. (Chapter 3, see World Health Organization definition at https://www.who.int/data/gho/indicator-metadata-registry/imr-details/158)

**Downstream:** In the context of lifecycle assessment, it denotes activities in the supply chain that occur following a particular activity of interest. Analogous to water flowing from upstream to downstream, materials flow-through supply chains in a similar manner. (Chapter 3)

**Elementary flows:** Material or energy entering or leaving the system that has been drawn from or released into the environment without previous human transformation. (Chapter 3, see reference for ISO, 2006a)

**Emancipatory and community-based policy change:** Policy change that is designed with input from and in partnership with community members affected by the problem the policy seeks to address and contributes to dismantling systems of oppression, such as racism and classism. (Chapter 12)

**Embodied environmental factors:** Environmental factors that may not be physically part of the final product but are used to help facilitate the production or delivery of a product acquired as final demand. For example, the water use associated with livestock feed production is embodied in a purchased hamburger. (Chapter 9)

**Emission factor:** A simplified model of emissions from an activity. For example, it is the assumption that thirty percent of applied inorganic nitrogen fertilizer will be volatilized as ammonia. (Chapter 3)

**Endogenous:** Variables from a model that are outputs are considered "endogenous." The values are determined from exogenous variables and the model's equations. (Chapter 8)

**Endpoint impact:** A category indicator near or at the end of the causal chain of environmental mechanisms leading to damage to one or more of the areas of protection. For example, the adverse effects of climate change on human health expressed in disability-adjusted life years (DALYs). (Chapter 3)

**Environmental factor:** A measurable attribute of the environment whose level, in the context of an environmental input-output (EIO) model, contributes to the economic value of the environment. (Chapter 9)

**Environmental flow requirements:** The quantity, quality, and timing of water flows required to sustain freshwater and estuarine ecosystems and the human livelihoods and well-being that depend on these ecosystems. (Chapter 4, see reference for Hoekstra et al., 2011)

**Environmental input output (EIO) model:** A top-down, economy-wide modeling approach that extends a conventional input-output model by translating (1) gross activity and commodity outputs associated with any final demand scenario and (2) direct final demand of environmental factors into measures of direct and indirect environmental flows embodied in that final demand. (Chapter 9)

**Environmental mechanism:** System of physical, chemical, and biological processes for a given impact category, which link the life cycle inventory analysis results to category indicators and to category endpoints. (Chapter 3, See European Union. 2018. Knowledge For Policy. Available at: https://knowledge4policy.ec.europa.eu/glossary-item/environmental-mechanism_en.)

**Environmental metric:** Measurable changes to the stock or flow of services brought on by any economic input-output (EIO) model transaction. (Chapter 9,)

**Environmental multiplier matrix:** A rectangular array of numbers that represents environmental impacts of an industry on a per-unit of output basis. (Chapter 9)

**Environmental services:** The qualitative functions of the natural, non-produced assets of land, water, air, minerals, related ecosystems, and the animal and plant life within. (Chapter 9)

**Environmental, social, governance (ESG) investing:** A category of financial investing wherein environmental, social, and governance criteria help investors examine how companies steward natural resources, treat the people along its value chain, including employees and suppliers, and account for such factors as a company's leadership structure, executive pay, and shareholder rights. (Chapter 14)

**Ex post:** Analysis after observing the outcome of the event under study. (Chapter 9)

**Exogenous:** Variables that are inputs into a model are considered "exogenous". The values of exogenous variables are determined outside of the model being used. (Chapter 8)

**Feed:** A generic term for material fed to animals. Sometimes used to refer specifically to grains fed to livestock. (Chapter 6)

**Feedback loop:** A closed chain of causal connections from a stock, through a set of decisions, rules, physical laws or actions that are dependent on the level of the stock and back again through a flow to change the stock. (Chapter 7, See Meadows, D. 2008. Thinking In Systems: A Primer. Chelsea Green Publishing, White River Junction, VT.)

**Feed-food competition:** Direct competition occurs when biomass suitable for direct human consumption is fed to animals instead of humans. Indirect competition occurs when feed is cultivated on areas where food for direct human consumption could be cultivated. (Chapter 5)

**Final demand:** In the context of economic input-output (EIO) models, it is the purchase or use of a commodity by an institutional actor, such as a private household, a government agency consumption or investment purchase, or an international or regional export sale. (Chapter 9)

**Final demand vector:** A mathematical representation (i.e., column of data) quantifying final demand for one or more commodities. (Chapter 9)

**Food systems:** The aggregate of food-related activities and the environments (political, social, economic, and natural) within which supply chains and activities occur. Food systems are inherently multilevel (hyper-local, local, regional, national and global) and multiscale (spatial, temporal, jurisdictional, and constitutional). (Chapter 2, see references for Cash et al., 2003 and Ericksen et al., 2010)

**Foodshed:** Geographic area from which a population supplies its food. It can be modified to refer to a subset of the food supply, such as the local foodshed. (Chapter 6)

**Forage:** A cellulose-rich plant material eaten by ruminants that is either fed to the animals in an enclosure or grazed by the animals in the field. Examples include hay, silage, and pasture. (Chapter 6)

**Forecast:** A term commonly used to describe a predictive exercise that identifies a most likely future, as in shorter-term contexts such as monthly market forecasts. In contrast, the term projections are often used to describe the results of scenarios without assigning probabilities to the different outcomes. (Chapter 10, see reference for Weibe et al., 2018)

**Foresight:** The act of thinking about the future to guide decisions in the present. It may involve relatively simple decisions or more complex challenges, where more formal and systematic foresight exercises involving scenario development and analysis can play an increasingly important role. (Chapter 10, see reference for Weibe et al., 2018)

**Functional unit:** Represents the function or purpose of a system in quantitative terms. It must quantify the properties of the product in a manner that enables comparison to other products with the same function or purpose. (Chapter 3)

**Global burden of disease (GBD)**:  A tool to quantify health loss from hundreds of diseases, injuries, and risk factors, so that health systems can be improved and disparities can be eliminated. (Chapter 3, see Institute for Health Metrics and Evaluation definition at http://www.healthdata.org/gbd/about)

**Green marketing**:  Communications that celebrate the environmental benefits of products or services. (Chapter 14)

**Green water**:  Precipitation on land that does not run off or recharge the groundwater but is either stored in the soil or temporarily stays on top of the soil or vegetation. Eventually, this precipitation evaporates or transpires through plants. (Chapter 4, see reference for Hoekstra et al., 2011)

**Green water footprint**:  The volume of rainwater consumed during the production process. It refers to the total rainwater evapotranspiration from fields and plantations plus the water incorporated into the harvested crop or wood. (Chapter 4, see reference for Hoekstra et al., 2011)

**Greenwashing**:  The use of false or misleading information about environmental impact. (Chapter 14)

**Grey water footprint**:  An indicator of freshwater pollution that can be associated with the production of a product over its full supply chain. It is the volume of freshwater that is required to assimilate a load of pollutants based on natural background concentrations and existing ambient water quality standards. It is calculated as the volume of water that is required to dilute pollutants to such an extent that the quality of the water remains above agreed water quality standards. (Chapter 4, see reference for Hoekstra et al., 2011)

**Group model building**:  A participatory modeling process, typically using system dynamics modeling, that involves iterative meetings with stakeholders to define the outcomes that interventions should improve, map system interactions, develop values of unknown parameters and undertake model evaluation. Group model building improves understanding of the problem, increases engagement in systems thinking, builds confidence in the use of systems ideas, and creates consensus for action among diverse stakeholders. (Chapter 7, Chapter 13)

**Hard systems thinking**:  An approach to problem solving that assumes the objective reality of systems, presents a well-defined problem, places technical factors in the forefront, takes a scientific approach to problem-solving, and has an ideal solution. (Chapter 2)

**Hotspots**:  An activity within a supply chain that has a large contribution to a specific impact category. These are activities may be targeted for improvement interventions. (Chapter 3)

**IMPLAN**:  A company that has developed a software and data platform that combines a set of extensive databases, economic factors, multipliers, and demographic statistics with a highly refined modeling system that is customizable. (Chapter 8)

**Indirect environmental impacts**:  These are the environmental impacts associated with interactivity transactions or second-tier activities to support first-tier activities. For example, if demand for beef was reduced, then the demand for livestock feed would be reduced and, thus, the associated environmental impacts associated with the livestock feed activity would also be reduced. (Chapter 9)

**Induced environmental impacts**:  These are the environmental impacts that result from all subsequent final demand created by the distribution of income paid to primary factors, such as labor, for their role in meeting initial final demand under study. For example, if the final demand for tomatoes was increased, all the wages paid to workers whose activity outputs are directly or indirectly impacted by the increased spending on tomatoes (and all subsequent spending) would also increase, leading to induced new spending by these workers households. (Chapter 9)

**Injection vector**:  In the context of an economic input-output (EIO) model simulation, it is the final demand vector for a scenario under study that is multiplied into the total requirement matrix to determine the level of gross activity and commodity outputs required to meet the final demand scenario. It is usually measured in monetary units such as dollars. (Chapter 9)

**Input-output model**:  A top-down, economy-wide modeling approach that measures the level of gross activity, commodity production, or gross output required throughout the economy to meet all existing final demand or changes to existing final demand. It also measures all transactions among activities and commodities plus payments to primary factors to facilitate all gross output. Transactions are usually measured in monetary units such as dollars. (Chapter 9)

**Interdisciplinary research**: Research that integrates information tools, perspectives, or theories from two or more disciplines to solve problems whose solutions are beyond the scope of a single discipline. (Chapter 2)

**Intergovernmental Panel on Climate Change (IPCC)**: The IPCC provides regular assessments of the scientific basis of climate change, its impacts and future risks, and options for adaptation and mitigation. (Chapter 3)

**Investment**: In the context of economic input-output (EIO) models, a final use category that represents the acquisition of commodities for use at a later time period. (Chapter 9)

**Key performance indicator (KPI)**: Metrics used to measure a company's performance, typically, including financial and operational performance. (Chapter 14)

**Land cover**: A description of the type of surface found at a given geographic location, such as forest, wetlands, or agricultural plants. Distinct from land use, land cover characterizes the attributes of the surface, but not necessarily the use. For example, land in herbaceous cover might be classified as pasture whether or not that location is used to graze animals. (Chapter 6)

**Land use**: The human purpose to which land is put. For example, this could be agriculture, residential development, or transportation. (Chapter 6)

**Land use ratio (LUR)**: A land use efficiency assessment method developed to account for land suitability and as such considers feed-food competition in its assessment. (Chapter 5)

**Leakage vector**: In the context of an economic input-output (EIO) model simulation, it is the total activity and commodity use of all primary factors, not broken out into categories, that result from a level of gross activity and commodity outputs required to meet the final demand scenario such that no further transaction results from this primary factor use. It is usually measured in monetary units such as dollars. (Chapter 9)

**Life cycle assessment**: A term used to describe modeling approaches that account for impacts over a product's entire lifecycle, although specific boundaries of the lifecycle may differ. (Chapter 9)

**Lifecycle impact assessment (LCIA)**: This is a phase of life cycle assessment aimed at understanding and evaluating the magnitude and significance of the potential environmental impacts for a product system throughout the life cycle of the product. (Chapter 3, see reference for ISO, 2006a)

**Lifecycle impact categories**: Class representing environmental issues of concern to which life cycle inventory analysis results may be assigned. (Chapter 3, see reference for ISO, 2006a)

**Life cycle perspective**: The adoption of a full system view of product or service under study. This implies the consideration of multiple impact categories so that the potential trade-offs between activities within the supply chain and between different impacts can be identified and understood. (Chapter 3)

**Linear homogeneity**: A mathematical property of proportional change that is an assumption made in input-output modeling about production. For example, if two hours of labor were required to produce one widget, then four hours of labor would be required to produce two widgets. (Chapter 9)

**Low-opportunity-cost biomass**: This is biomass not suitable for human consumption, which includes biomass from grassland and "leftovers": crop residues left over from harvesting food crops, coproducts left over from industrial processing of plant-source and animal-source food, and losses and waste in the food system. (Chapter 5)

**Margining**: Margins allow for consumer expenditures to be traced though retail, wholesale, and transportation industries back to the industries who manufactured the product, allowing the appropriate allocation to the producing industries. (Chapter 8, from IMPLAN. 2021. Margins & Deflators. Available at https://support.implan.com/hc/en-us/articles/115009506007-Margins-Deflators.)

**Midpoint impact**: A category indicator at an intermediate point within the causal chain. For example, global warming potential is an indicator for the endpoint impact of climate change. (Chapter 3)

**Model**: A simplified rendition of an object, a process, or a system. It can take a variety of forms such as a mental model, a physical model, or a computational model. (Chapter 1)

**Monte Carlo simulation**: A simulation technique for estimating the uncertainty in an impact category that results from the uncertainty in the input information for the activities in a product system. (Chapter 3)

**Multidisciplinary research**: Several different academic disciplines researching one theme or problem with multiple disciplinary goals. Participants exchange knowledge but do not cross subject boundaries to create new knowledge and theory. (Chapter 2, see reference to Cronin, 2008)

**Multifunctionality**: This refers to a situation in which an activity in a supply chain has more than a single function. For example, a dairy farm produces both milk and meat. (Chapter 3)

**Multiplier**: In the context of economic input-output (EIO) modeling, it is a type of analysis that uses fixed coefficients to translate gross activity or commodity output levels into measures of change to one or more environmental metrics. It is calculated as the ratio of the sum of direct, indirect, and induced effects to the direct effects. (Chapter 8 and Chapter 9)

**North American Industry Classification System**: (NAICS) Categories used by the federal government to classify different types of businesses. (Chapter 8)

**Negative feedback loop**: Feedback loop in which a change in a given direction causes change in the opposite direction. (Chapter 7)

**Net-balance**: A budget approach to estimating flows. Within food systems, it usually refers to the mass of food produced in a geographic area compared to the mass of food consumed within that area. (Chapter 6)

**Node**: A point within an interconnected network of points. (Chapter 13)

**Opportunity cost**: In economics, it is the cost of the foregone alternative. In input-output modeling, these could be: (1) the economic impact of the purchases that are displaced from the new project or (2) the economic impact of the production that will no longer occur due to a new project. (Chapter 8)

**Optimization**: In the context of modeling, optimization refers to a mathematical approach to finding the solution to a set of variables that either minimizes or maximizes a desired outcome within certain constraints, such as a least-cost diet meeting recommended daily intake of essential nutrients. (Chapter 6)

**Personal consumption expenditures (PCE)**: A final demand category that represents consumption expenditures for the acquisition of various commodities by private households. (Chapter 9)

**Positive feedback loop**: A feedback loop in which a change causes additional change in the same direction, such as growing interest in a back account. (Chapter 7)

**Primary factor**: An input into current production by an activity such that the input was not produced in the current period by another activity. An example would be a natural resource, hired labor, or refrigeration. (Chapter 9)

**Producer surplus**: Aggregated value for producers when they can sell at a market price that is higher than the lowest price at which they would be willing to sell. (Chapter 7)

**Product system**: A collection of unit processes with elementary and product flows that performs one or more defined functions and models the life cycle of a product. (Chapter 3, see reference for ISO, 2006a)

**Production function**: An equation that stipulates how inputs are combined for that sector or firm to produce output. Standard factors of production include labor, human capital, physical capital, and land. (Chapter 8)

**Projection**: A term used to describe the results of scenarios without assigning probabilities to the different outcomes. In contrast, the term forecast is commonly used to describe a predictive exercise that identifies a most likely future in shorter-term contexts such as monthly market forecasts. (Chapter 10, see reference for Weibe et al., 2018)

**Qualitative functions**: In the context of environmental services, they are disposal services of residuals by the natural environment, productive services of natural resource inputs (including space for production and consumption), and consumption services providing for the physiological and recreational needs of human beings. (Chapter 9)

**Real prices**: Inflation-adjusted prices or wages that allow for more accurate comparison over time. (Chapter 9)

**Reference flows**: The amount of material or energy at each stage in a supply chain or for each modeled activity that are necessary to provide the functional unit of the system under study. (Chapter 3)

**Reflexivity**: The act of critically examining the assumptions underlying our actions, the impact of those actions, and from a broader perspective, what passes as good. In the case of food systems research, this means individuals conducting research must examine their personal assumptions, how those assumptions shape their work, and the underlying values and norms of food systems research at large. (Chapter 12, see reference to Cunliffe, 2016)

**Regenerative agriculture**: A newly codified approach to agriculture that emphasizes reducing reliance on exogenous inputs, as well as restoring and enhancing ecosystem services such as soil carbon sequestration. (Chapter 14, see reference to Rowntree et al., 2020)

**Satellite accounts**: Supplementary statistics that allow analysis of a particular aspect of the economy, such as spending on travel and tourism. (Chapter 8)

**Scenario:** Specific representations of the future to facilitate thinking about the possible consequences of different events or courses of action within a systematic foresight exercise. (Chapter 10, see reference to Weibe et al., 2018)

**Scope 1 greenhouse gas emissions:** A company's direct greenhouse gas emissions within the owned operations of the company, commonly stemming from fuel combustion in any company-owned assets, such as vehicles and manufacturing facilities. (Chapter 14)

**Scope 2 greenhouse gas emissions:** A company's indirect greenhouse gas emissions stemming from the purchase of electricity for the organization. (Chapter 14)

**Scope 3 greenhouse gas emissions:** All company greenhouse gas emissions that occur beyond the company's owned operations, such as emissions from upstream purchased ingredients and transport, or downstream consumer use and disposal of products. (Chapter 14)

**Social accounting matrix (SAM):** An extension of the input-output accounting framework whereby payments to a subset of primary factors (e.g., hired labor and business proprietors) are recorded in their own data row, and the distribution of these payments to institutional factor owners (e.g., private households), are recorded in their own data column. In this manner, SAM multiplier models measure direct, indirect, and induced environmental impacts. (Chapter 9)

**Social network analysis:** A way to map and measure the relationships between people, groups, or organizations. This analytic tool has recently been applied to the study of interventions, including obesity prevention interventions. (Chapter 13, see reference to Scott, 1988)

**Soft systems thinking:** An approach that focuses on organizational problems that are poorly defined. It assumes that stakeholders interpret problems differently, that human factors are important, and that the outcomes are learning and better understanding rather than a solution. (Chapter 2, see reference to Cairns, 2020)

**Soil water balance model:** A model that simulates water availability in the root zone as well as water fluxes entering (e.g., rainfall and irrigation) and leaving (e.g., evapotranspiration and drainage) the root zone, using expected crop development as one of the inputs. (Chapter 4)

**Stakeholder-driven community diffusion:** A theory that posits that stakeholder (e.g., community coalition member) knowledge of and engagement with a complex issue, such as the food system, diffuses through stakeholder social networks, which facilitates evidence-based policy, system, and environmental change. (Chapter 13, see reference to Kasman et al., 2019)

**Stock management structure:** A decision making structure used in system dynamics models that is used to maintain the value of a stock, often inventories of product, in a certain range. (Chapter 7)

**Supply chain:** A set of structures and processes that a group of organizations uses to deliver an output to a customer. (Chapter 7, see reference to Sterman, 2000)

**Supply chain stages:** The organized sequencing of multiple activities that take place to assemble a final product. (Chapter 9)

**Supply line:** The amount of product that is either currently in production but not yet finished or not yet delivered to a customer. (Chapter 7)

**Sustainable agriculture:** A way of practicing agriculture that seeks to optimize skills and technology to achieve long-term stability of the agricultural enterprise, environmental protection, and consumer safety. It is achieved through a number of different management strategies to conserve resources while providing a sustained level of production and profit. (Chapter 2, see reference to Gold, 2007)

**Sustainable development:** An effort to guarantee a balance among economic growth, environmental integrity, and social well-being. (Chapter 2, see reference to Mensah, 2019)

**Sustainable food systems:** A food system that delivers food security and nutrition for all in such a way that the economic, social, and environmental bases to generate food security and nutrition for future generations are not compromised. (Chapter 2, see reference to FAO, 2018)

**System:** An interconnected set of elements that are coherently organized in a way that achieves something. (Chapter 7 See Meadows, D. 2008. Thinking In Systems: A Primer. Chelsea Green Publishing, White River Junction, VT.)

**System boundaries:** A set of criteria specifying which unit processes are part of a product system. (Chapter 3, see reference for ISO, 2006a)

**System dynamics:** An approach to understanding the nonlinear behavior of complex systems over time using informal maps and formal models with computer simulation to understand endogenous sources of system behavior. (Chapter 13, see reference to National Cancer Institute, 2007)

**System dynamics modeling:** A quantitative modeling approach that focuses on stocks, flows and feedback processes that create outcomes varying over time. (Chapter 7)

**Systems mapping:** A qualitative systems modeling approach that looks at how variables interact over time and form patterns of behaviors across the system. It is a visual tool to help share a simplified conceptual understanding of a complex system. Causal loop diagrams, a causal diagram that aids in visualizing how different variables (or elements) in a system are interested, are a type of systems mapping. (Chapter 13, see reference to Forrester, 1975)

**Systems thinking:** A way of approaching problems that asks how various parts within a system are arranged and interact to yield a specific function or outcomes over time. (Chapter 13, see reference to Meadows, 2008)

**Product flow:** Products entering from or leaving to another product system. This is distinguished from elementary flows which are generally direct exchanges with the environment rather than between activities or unit processes in a supply chain. (Chapter 3, see reference for ISO, 2006a)

**Total requirement matrix:** In the context of a Type I multiplier model, it is a rectangular array of data measuring the combined direct and indirect output effects for all commodities and activities caused by a single unit increase in the final demand for each activity and commodity in the model. (Chapter 9)

**Transdisciplinary research:** Research that integrates academic researchers from different and related disciplines and nonacademic participants to research a common goal and create new knowledge and theory. (Chapter 2, see reference to Cronin, 2008)

**Transportation problem:** Problems that involve choosing the least cost way to transport products from supplier locations to customer locations when the amounts at supply and demand locations are fixed. (Chapter 7)

**Transshipment problem:** Problems that involve choosing the least-cost way to move product through a supply chain when there are additional locations (i.e., transshipment nodes) between producer and consumer locations. (Chapter 7)

**Triple Bottom Line:** A framework that refers to a business's social, environmental, and economic impact. The term, coined in the 1990's, was intended to help companies measure and address the full cost of doing business, encouraging a focus beyond traditional financial metrics to include environmental metrics such as greenhouse gas emissions and water consumption. (Chapter 14)

**Type I multiplier model:** A modeling approach that measures the total direct and indirect (but not induced) activity and commodity requirements linked to any modeler-specified final demand. (Chapter 9)

**Unit process:** The smallest element considered in the life cycle inventory analysis for which input and output data are quantified. (Chapter 3, see reference for ISO, 2006a)

**Upstream:** In the context of lifecycle assessment, it denotes activities in the supply chain that occur prior to a particular activity of interest. Analogous to water flowing from upstream to downstream, materials flow through supply chains in a similar manner. (Chapter 3)

**Value added multiplier vector:** A mathematical representation of total payments to primary factors (e.g., labor, land, natural resources, etc.) for each unit of activity output. It is used to translate the total requirements of each activity into total income to primary factors employed by each activity. (Chapter 9)

**Virtual water flow:** The virtual water flow between two geographically delineated areas (e.g., two nations) is the volume of virtual water that is being transferred from one area to another area as a result of product trade. (Chapter 4, see reference for Hoekstra et al., 2011)

**Walras' law:** An economic principle stating that the market value of all excess supply (i.e. supply minus use) across products in a market economy equals zero. In input-output modeling, this means that the sum of all values in the injection vector must equal the sum of all values in the leakage vector. (Chapter 9)

**Water footprint:** An indicator of freshwater use that looks at both direct and indirect water use of a consumer or producer. The water footprint of an individual, community or business is defined as the total volume of freshwater used to produce the goods and services consumed by the individual or community or produced by the business (Chapter 4, see reference for Hoekstra et al., 2011)

**Water footprint accounting:** The phase in water footprint assessment that refers to collecting factual, empirical data on water footprints with a previously defined scope and depth. (Chapter 4, see reference for Hoekstra et al., 2011)

**Water footprint assessment:** The full range of activities to quantify and locate the water footprint of a process, product, producer or consumer or to quantify in space and time the water footprint in a specified geographic area; assess the environmental, social and economic sustainability of this water footprint and formulate a response strategy. (Chapter 4, see reference for Hoekstra et al., 2011)

**Water footprint response formulation:** The phase in water footprint assessment that aims to formulate response options, strategies or policies based on the outcomes of the previous water footprint accounting and sustainability assessment phases. (Chapter 4, see reference for Hoekstra et al., 2011)

**Water footprint sustainability assessment:** The phase in water footprint assessment that aims to evaluate whether a certain water footprint is environmentally sustainable, efficient and equitable. (Chapter 4, see reference for Hoekstra et al., 2011)

# Index

Page numbers followed by "*f*" and "*t*" indicate, figures and tables respectively.

Services